SUBCELLULAR COMPONENTS
Preparation and Fractionation

SUBCELLULAR COMPONENTS
Preparation and Fractionation

Edited by

G. D. BIRNIE, B.SC., PH.D.

BUTTERWORTHS LONDON
UNIVERSITY PARK PRESS BALTIMORE

THE BUTTERWORTH GROUP

ENGLAND
Butterworth & Co (Publishers) Ltd
London: 88 Kingsway, WC2B 6AB

AUSTRALIA
Butterworth & Co (Australia) Ltd
Sydney: 586 Pacific Highway Chatswood, NSW 2067
Melbourne: 343 Little Collins Street, 3000
Brisbane: 240 Queen Street, 4000

CANADA
Butterworth & Co (Canada) Ltd
Toronto: 14 Curity Avenue, 374

NEW ZEALAND
Butterworth & Co (New Zealand) Ltd
Wellington: 26–28 Waring Taylor Street, 1

SOUTH AFRICA
Butterworth & Co (South Africa) (Pty) Ltd
Durban: 152–154 Gale Street

First published 1969
Second Edition 1972

ISBN 0 408 70360 1

Published 1972 jointly by
BUTTERWORTH & CO (PUBLISHERS) LTD, LONDON
and
UNIVERSITY PARK PRESS, BALTIMORE

Library of Congress Cataloging in Publication Data
Main entry under title:

Subcellular components.

Includes bibliographies.
1. Cell franctionation—Congresses. 2. Cytology —Technique—Congresses. I. Birnie, G. D., ed.
QH585.S8 1972 591.8'7 72-565
ISBN 0-8391-0581-9

Filmset by Filmtype Services Limited, Scarborough
Printed in England by The Chapel River Press, Andover, Hants

PREFACE TO FIRST EDITION

The burgeoning interest in the application of large-scale zonal rotors to the problem of separating subcellular particles by density-gradient centrifugation is evident from the recent literature. However, there are two questions which, in this context, are of immediate interest and importance. First, how best can particular tissues be treated so that the maximum percentage of the cells are broken while, at the same time, the minimum amount of damage is caused to the subcellular components? Second, what criteria should be applied in assessing how successful any preparation of a subcellular component has been, particularly as far as the absence of contamination and structural damage is concerned? These two questions are, of course, just as pertinent when classical methods are used for isolating subcellular components. For this reason, a Symposium was held in November, 1967 so that experts in this field could discuss a number of these classical methods, and also the criteria of purity and integrity which have been applied to subcellular preparations in the past. Prof. J. Berthet was unable to prepare a manuscript of his paper on the isolation of lysosomes and peroxisomes. The Editors have added the final paper which describes a number of large-scale rate-zonal separations of subcellular components to enable some comparison of classical methods and this new technique to be made.

Acknowledgements—Many people contributed to the organization of this Symposium. In particular, we are grateful to Dr G. F. Marrian (former Director of Research, Imperial Cancer Research Fund), Dr R. J. C. Harris (Head, Division of Experimental Biology and Virology, Imperial Cancer Research Fund) and Dr T. S. Work (National Institute for Medical Research) for their help and encouragement throughout. We are indebted to Measuring and

Scientific Equipment Ltd for most generously financing the Symposium and, in particular, to Mr J. C. Z. Martin and Mr E. W. Young for their interest and support. We also thank Mrs G. M. Dunmore and Miss J. Salinger for their invaluable assistance before and during the Symposium and in the preparation of this publication.

Division of Experimental Biology and Virology,
Imperial Cancer Research Fund, London, NW7

G. D. Birnie
Sylvia M. Fox

PREFACE TO SECOND EDITION

Much of our understanding of the mechanisms underlying the functions of eukaryotic cells is based on studies of the enzymic activities of isolated subcellular fractions. It is clear that highly-purified preparations of undamaged subcellular particles are required before really meaningful results can be obtained from such studies. This immediately raises two obvious but fundamental questions. First, how best can particular tissues be treated so that the maximum proportion of the cells are broken while the minimum amount of damage is caused to the subcellular components? Second, what criteria should be applied to assess how successful any preparation of a subcellular component has been, particularly so far as the absence of contamination and structural damage is concerned? It is difficult to give definitive answers to the first of these questions since the problems encountered depend on many interacting factors which are often far from easy to control. Unhappily, methods for disrupting cells in a reproducible and controlled manner have not progressed at anything like the same pace as methods for separating their constituents. On the other hand, much of the information necessary to give answers to the second question is available in the literature but, unfortunately, it is not always easily found. Even when it is, in too many cases subcellular fractions are not examined critically to determine the degree of intactness and purity of the particles, partly as a consequence of lack of awareness of the difficulties inherent in the preparation of subcellular components in a purified and undegraded state.

The purpose of this book is not merely to present a collection of methods for the preparation of the major constituents of cells. Its aim is rather to examine and compare methods critically, indicate the problems and unsuspected difficulties, and suggest criteria which can be used to assess the quality of preparations of subcellular particles. Hopefully, it may also increase awareness of the many problems attendant upon choosing suitable methods for obtaining purified, undamaged, subcellular components and so help towards the attainment of some degree of standardisation of techniques. To these ends, the material in the first edition has now been brought up to date and in some cases greatly expanded. Also, three new contributions have been added to fill the more obvious gaps in the first edition. These new chapters not only illustrate the use of zonal centrifugation for preparing and fractionating subcellular components on a scale hitherto impossible, but also show that some subcellular fractions, which previously would have been considered quite satisfactory, in fact consist of a mixture of components with different properties.

I am grateful to the authors for their ready and willing co-operation in the preparation of this book, and I am indebted to Mrs Rae Fergusson for her skilled and invaluable assistance, particularly in the preparation of the index.

The Beatson Institute for Cancer Research,
132 Hill Street,
Glasgow, C3,
Scotland

G. D. Birnie

CONTRIBUTORS

H. R. V. ARNSTEIN, D.I.C., PH.D., D.SC., F.I. BIOL.,
Dept. of Biochemistry, King's College, London, WC2R 2LS

P. J. G. AVIS, M.I. BIOL.,
Faculty of Medicine, Memorial University of Newfoundland, St Johns, Newfoundland

G. D. BIRNIE, B.SC., PH.D.,
Beatson Institute for Cancer Research, Royal Beatson Memorial Hospital, Glasgow, C3

M. L. BIRNSTIEL, DR.SC.NAT.,
Institute of Animal Genetics, University of Edinburgh, Edinburgh 9

S. A. BONANOU-TZEDAKI, B.SC., PH.D.,
Dept. of Biochemistry, King's College, London, WC2R 2LS

J. B. CHAPPELL, B.A., PH.D. (CANTAB.),
Dept. of Biochemistry, University of Bristol, Bristol, BS8 1TD

W. G. FLAMM, M.SC., PH.D.,
Cell Biology Branch, National Environment Health Sciences Center, Research Triangle Park, N.C. 27709, USA

SYLVIA M. FOX, M.A., M.I. BIOL.,
Dept. of Cell Biology, Sloan-Kettering Institute for Cancer Research, New York, N.Y. 10021, USA

R. G. HANSFORD, B.SC., PH.D.,
Dept. of Biochemistry, University of Bristol, Bristol, BS8 1TD

D. R. HARVEY,
Dept. of Environmental Carcinogenesis, Imperial Cancer Research Fund, Mill Hill, London, NW7

R. H. HINTON, B.A. (CANTAB.), PH.D.,
Wolfson Bioanalytical Centre, University of Surrey, Guildford, Surrey

I. R. JOHNSTON, M.SC., PH.D.,
Dept. of Biochemistry, University College, London, WC1E 6BT

A. P. MATHIAS, B.SC., PH.D.,
Dept. of Biochemistry, University College, London, WC1E 6BT

J. H. PARISH, M.A., D.PH.,
Dept. of Biochemistry, University of Leeds, Leeds 2

J. W. PORTEOUS, B.SC., PH.D.,
Dept. of Biological Chemistry, Marischal College, University of Aberdeen, Aberdeen AB9 1AS

E. REID, PH.D., D.SC.,
Wolfson Bioanalytical Centre, University of Surrey, Guildford, Surrey

D. B. ROODYN, M.A., PH.D. (CANTAB.),
Dept. of Biochemistry, University College, London, WC1E 6BT

J. R. TATA, D.SC. (PARIS),
National Institute for Medical Research, Mill Hill, London, NW7

P. M. B. WALKER, B.A., PH.D.,
Dept. of Zoology, University of Edinburgh, Edinburgh 9

CONTENTS

1

PRESSURE HOMOGENISATION OF MAMMALIAN CELLS

P. J. G. Avis

Most of the procedures for isolating the subcellular components of cells make use of some form of density-gradient centrifugation. The development of the zonal ultracentrifuge (Anderson, 1966) has meant that such separations can be done with much larger amounts of material. The zonal centrifuge gives very consistent results and can handle up to 5 g of tissue or cells as an homogenate or up to 12 g of material in about 50 ml of fluid if residual whole cells, nuclei and large debris are first removed. However, one of the key problems is to find an efficient and reproducible method of homogenising suspensions of single cells which can handle at least 50 ml of suspension. It is also necessary to be able to control the extent of the disintegration.

There are at present a variety of instruments in general use for the preparation of tissue homogenates. Of the mechanically-driven types, the Potter–Elvehjem homogeniser appears to cause least damage to the cellular components. The Dounce homogeniser, a simple hand-operated device, has also been used frequently, especially for the preparation of nuclei and plasma membranes. Ideally, a homogenate prepared for the subsequent isolation of subcellular components should consist only of whole cells and the intact organelles. All procedures fall short of this ideal, and homogenates contain not only mixtures of cells and their components but also products of the disruption of cell organelles. In order to obtain optimum results for a given method, the progress of homogenisation must be closely monitored by microscopic examination.

Furthermore, the high shearing forces which develop in a viscous homogenate can produce local heating, so that it is necessary to ensure that the apparatus and material are kept as near 0°C as possible during the procedure.

Suspensions of single cells seem particularly difficult to homogenise. The efficiency of the Potter–Elvehjem and Dounce type of homogenisers in breaking a sufficient proportion of the cells within a reasonable time is quite low. It diminishes markedly with volumes greater than 10 ml. Prolongation of the homogenisation time means that those components which are liberated early in the procedure are exposed to shear forces for a longer time. This was shown clearly by an experiment in which BHK21 cells grown in tissue culture were homogenised in a volume of 10 ml in a Potter–Elvehjem glass homogeniser. After 3 min homogenisation, 33% of the cells had ruptured. At this point the cells and nuclei were removed by centrifugation and the post-nuclear supernatant fluid analysed for acid phosphatase activity in the soluble and particulate (intact lysosome) fractions. The acid phosphatase activity of the particulate fraction was 40% of the total. Doubling the homogenising time, which caused the rupture of 75% of the cells, resulted in only 10% of the total phosphatase activity being recoverable from the particulate fraction. In small-scale experiments using rat-liver tissue, a value as high as 56% of the phosphatase activity could be obtained in the bound form by homogenising for 1 min although, with this short homogenisation, only a small percentage of the cells were disrupted.

When the nature of the experiment, or the scarcity of material, requires that a high percentage of cells are disrupted, there are difficulties in deciding at what point to discontinue homogenisation in order to preserve the integrity of labile components such as lysosomes and at the same time ensure that sufficient cells have been disrupted. A method of homogenising that would expose an individual cell once only to the disrupting force would overcome some of the disadvantages. Such a method is pressure homogenisation.

Homogenisation of mammalian cells by means of applied pressure and its sudden release was mentioned some years ago by Nirenberg and Hogg (1958) as a method for Ehrlich acites-tumour cells but very few details were given. Wallach (1960) again mentioned pressure homogenisation by the application and release of a pressure of 800 lb/in^2 within a stainless-steel cylinder. The first detailed account of a suitable apparatus and method was published by Hunter and Commerford (1961), who used it successfully for the disruption of several types of mammalian tissues but found that yeast cells, which have a tough, thick wall, were resistant to this method of homogenisation. The method is based on the fact that at equilibrium the internal pressure of a gas within a cell must equal the external pressure. If a cell suspension is placed under high pressure in an inert gas such as nitrogen, relatively large amounts of the gas dissolve in the medium and will diffuse into the cells until the external and internal pressures reach equilibrium. If the external pressure is suddenly released, the internal pressure is then the higher, and if the difference is sufficiently great, cavitation ensues and rupture of the membrane leads to the release of the cellular contents. Since the nitrogen is presumably dissolved under pressure in the water space of the cells, it might be supposed that anything in the cells with a similar water content, for example, mitochondria, lysosomes and nuclei, will have the same amount of dissolved nitrogen and so be disrupted when the pressure is released. In practice, the biochemical and physical properties of mitochondria, lysosomes and nuclei measured in these studies seem little altered by the exposure to high pressure. It may be that the membranes of these organelles are impervious to nitrogen. Although very high pressures can cause protein denaturation (Curl and Jansen, 1950), pressures of the order of 1000 lb/in^2 stabilise and render proteins more difficult to denature (Johnson and Campbell, 1946). Also, adiabatic cooling results from the sudden release of pressure, and this avoids the local heating at the moment of cell rupture which may occur when disruption involves shearing.

APPARATUS AND METHOD

The pressure homogeniser illustrated in *Figure 1* was made by Baskerville and Lindsay Ltd., Chorlton-cum-Hardy, Manchester, England, from details supplied by us, based on the description given by Hunter and Commerford (1961). The lower section is a heavy-walled hollow cylinder made from stainless steel; the upper is a sealing lid held in place by eight heavy-duty bolts and fitted with two high-pressure valves, a pressure gauge and a stirrer, the shaft of which is admitted through a pressure-tight gland. The diameter of the orifice in the outlet valve is 3·0 mm and that of the delivery tube is 6·0 mm.

Tissues are first minced, preferably by some form of hand press which removes most of the connective tissue. An alternative method which has been used successfully for all soft tissues and even relatively tough organs such as rat heart is to force coarsely-chopped tissue through a screen of expanded stainless steel (specification: 978; 1·5 × 0·01 × 0·005 mm; FDP, manufactured by Expanded Metal Ltd., Hartlepool, Durham). A gloved finger has proved the most satisfactory method of forcing the tissue through the screen. The mince is then suspended in a volume of homogenising medium equal to at least 10 times the wet weight of tissue used.

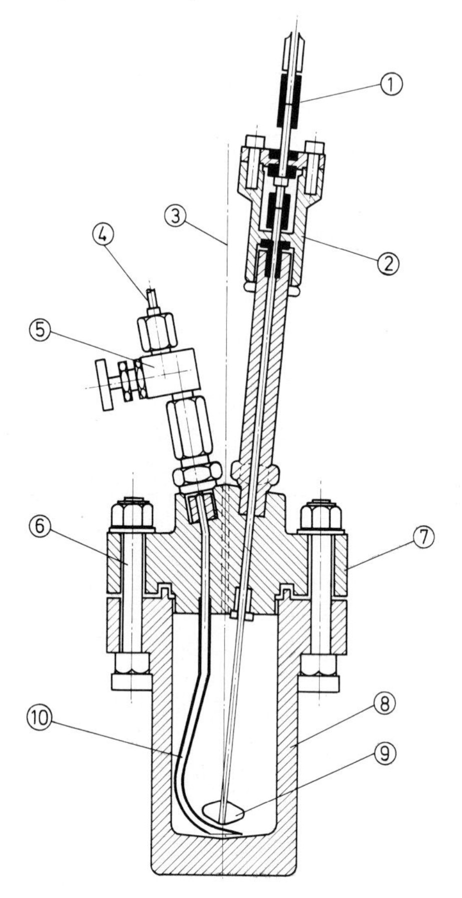

Figure 1. Cross-section diagram of pressure vessel

1 Stirrer coupling; 2 pressure-tight gland; 3 position of inlet valve; 4 delivery-tube connection; 5 pressure valve; 6 high-tensile bolts; 7 upper section; 8 lower section; 9 stirrer paddle; 10 outlet tube. All parts in contact with sample are fabricated in stainless steel

Volumes of 20 to 500 ml may be used at one time. The suspension is poured into the lower half of the apparatus and the top fitted and bolted on firmly. A cylinder of nitrogen is connected to the outlet valve and both inlet and outlet valves are opened. Nitrogen is bubbled through the cell suspension for 5 min to displace dissolved gas and air from the space above the liquid. The nitrogen is turned off at the cylinder and both valves on the homogeniser are closed. The gas line is transferred to the inlet valve which is then opened. Nitrogen is slowly admitted to the homogeniser at a rate such that it takes about 5 min to reach the required pressure inside the homogeniser. The inlet valve is closed and the cylinder disconnected. The stirrer is switched on at low speed for 15 to 20 min to ensure that equilibration of the nitrogen in the system takes place.

The sudden change in pressure which causes the homogenisation is produced by expelling the tissue suspension from the pressure vessel via the outlet valve and delivery tube (*Figure 1*). The homogenate is collected in a measuring cylinder which has a capacity at least 5 times the volume of the homogenate. (This prevents the homogenate being blown round the room by the rush of gas that follows the last of it). One drop of decanol can be used in the collecting vessel to limit foaming. The outlet valve is closed immediately the homogenate has been collected. The pressure is released by opening the gas-inlet valve. Any attempt to do this by opening the homogenate outlet valve causes any liquid in the outlet tube to freeze and can lead to a complete blockage which will persist until the ice melts. For the same reason, it is inadvisable to release the pressure while the homogenate is in the pressure vessel, as adiabatic cooling is sufficient to freeze the homogenate. This should be avoided, since freezing and thawing affect the permeability of some organelles such as mitochondria and lysosomes.

Cell suspensions can be treated in the same way but, since there are no large pieces of tissue, the suspension can be introduced into the pressure chamber through the outlet valve using a hypodermic syringe. The chamber can be washed out between samples in the same way so that, if a series of homogenisations is required, they can be done without removing the sealing lid, provided, of course, that some small amount of cross-contamination between samples can be tolerated.

Although the contents of the pressure vessel are forced through a small orifice in the outlet valve, shearing forces do not appear to contribute to the disruption of the cells since small samples of cells and tissue have been disrupted successfully *in situ* by opening the inlet valve to release the pressure. When this procedure is used, the vessel must be warmed to about 30°C to prevent freezing of the homogenate which can disrupt some cell organelles.

RESULTS WITH SOLID TISSUES

The tissue used in most of our studies to compare the effect of pressure homogenisation with other homogenisation procedures was the liver from rats of various strains. In addition, a few experiments on other tissues have been carried out. In general, tissues which are soft enough to be minced in a hand press or extruded through the stainless-steel gauze are amenable to pressure homogenisation; the exceptions are kidney, where the tubules remain largely intact although the rest of the tissue disintegrates, and heart tissue, in which most of the homogenate consists of bundles of muscle fibres. Liver, brain, AKR leukaemic lymph nodes, spleens and thymuses, and thigh tumours from AKR female mice have all been satisfactorily homogenised.

The exact conditions of pressure, time and buffer composition have been varied according to the tissue and the nature of the experiment, particularly if there was a special need to preserve one component even at the expense of others. In general, pressures less than 700 lb/in^2 did not homogenise the tissue completely, while those in excess of 1300 lb/in^2 resulted in considerable disintegration of nuclei. The composition of the suspending medium is very important but its choice presents a problem. Results, with few exceptions, parallel those obtained by other methods designed to preserve the integrity of subcellular components. The use of buffers of ionic strength less than 0·05 results in the loss of nuclei. Hunter and Commerford (1961) found that, in isotonic media, as little as 0·2 mM $CaCl_2$ was sufficient to preserve the nuclei. Certainly this is our experience, as illustrated in the next Section. We have used sucrose buffered with EDTA at pH 7·4 for the preparation of lysosomes from rat and guinea-pig liver with excellent results. With this medium, up to 60% of the lysosomal enzyme acid phosphatase was found in the sedimentable fraction and over 90% of the cells were ruptured. The use of EDTA, however, precludes the isolation of nuclei from the homogenate.

Hunter and Commerford (1961) compared the results of homogenisation of rat liver by pressure and by the Potter–Elvehjem homogeniser, using two of the more labile components of the cell, the deoxyribonucleoprotein of the nucleus and the enzyme systems of the mitochondria involved in oxidative phosphorylation, to measure the relative amounts of damage caused by the two methods. The former was extracted from isolated nuclei obtained by homogenising perfused rat livers in an isotonic buffer (0·12 M NaCl, 0·008 M K_2HPO_4, 0·002 M KH_2PO_4, 0·2 mM $CaCl_2$, pH 7·4; ionic strength, 0·15). The sedimentation coefficient of the deoxyribonucleoprotein prepared from pressure-homogenised liver was 56S, in good agreement with values obtained for that prepared from nuclei which were isolated without pressure homogenisation. Mitochondria were isolated by both methods of homogenisation from rat livers which had been perfused with isotonic saline. The homogenisation medium (0·25 M sucrose, 0·015 M potassium phosphate, pH 7·5) was one commonly used for the preparation of mitochondria. The P/O ratio of the mitochondrial suspensions prepared by both methods was 2·4. There was no measurable difference in the uptake of phosphate by these mitochondrial suspensions. In similar experiments performed in our laboratory the cytochrome *c* oxidase activity of rat and guinea-pig mitochondria prepared by the two methods have been compared. In both cases all the activity was found in the sedimentable fraction and there was no significant difference in the specific activities of the enzyme. Isopycnic banding of mitochondria prepared by both methods has also been done. The distribution of the mitochondria in the density gradient was similar for both methods of homogenisation.

RESULTS WITH SINGLE-CELL SUSPENSIONS

In dealing with single-cell suspensions, the picture is complicated by any pre-treatment the cells have received and by the tendency of the cells to stick together under certain conditions. It has already been stated that the composition of the suspending medium plays an important role in determining the character of the homogenate. With single-cell suspensions the choice of suspending medium may be influenced by considerations such as the maintenance of respiration during the period when the cells are being harvested, or the prevention of agglutination. Cells may be transferred from one medium to another, say from a maintenance medium such as medium 199 to one suitable for homogenisation, by centrifuging and re-suspending, but it is necessary to keep centrifuging and washing to a minimum. With single-cell suspensions the evidence of exposure to non-physiological conditions can readily be observed microscopically. Centrifuging in buffered sucrose solutions frequently results in agglutination, although cells vary in this respect depending on the tissue and species of origin. Attempts to homogenise agglutinated cells by pressure have inevitably resulted in a homogenate that consists of a frothy gel which separates from the suspending medium and contains most of the cellular components. It is virtually impossible to carry out any fractionation procedure on this material.

Cells such as BKH21 or Hep 2, grown in tissue culture, unless adapted for suspended culture, grow as confluent sheets on the glass walls of the containing vessels. The accepted procedure for removing the cells is to treat them with medium containing 0·05% (w/v) EDTA or, more effectively, with one containing a mixture of EDTA and crude trypsin (1·5 mg/ml). Obviously one does not want trypsin in a tissue homogenate, or even at times EDTA, so that this necessitates washing the cells. However, once removed from the trypsin-EDTA medium, the cells have a tendency to agglutinate, particularly if centrifuged. This can be overcome by washing in medium containing EDTA although it must be pointed out that over-exposure to EDTA results in what can be described as a gel-like structure of agglutinated cells. This particular type of agglutination has proved practically irreversible. Scraping the cells from the vessel walls with a rubber scraper is one solution, but clumping occurs almost immediately and is virtually complete after one washing in the centrifuge. It is possible to get a large proportion of the cells back into suspension by passage through a hypodermic needle. This does not halt the progress of clumping which will continue even when the cells are left to stand.

The clumping of cell suspensions is something that has occurred to a greater or lesser degree with all types of cells investigated. The worst type in this respect is the mouse thymocyte. In this case the procedure adopted has been to collect the cells in a modified Ringers solution containing 30 times the usual amount of Mg^{2+} and with the addition of crystalline DNAase I (10 μg/ml). After exposure to this mixture for periods of up to 2 h at 4°C the cells can be washed and suspended in the homogenising medium with virtually no clumping. Without this procedure, mere exposure to the medium, be it either unmodified Ringers or a sucrose solution buffered with a tris-K^{+}-Mg^{2+} buffer, is enough for the cells to start clumping and, once this has begun, within a very short time up to 70% of the cells have formed clumps.

The effect of the composition of the suspending medium on the appearance of R.1 lymphoma cells and pressure homogenates of these cells can be seen from *Figures 2* and *3*. The cells were harvested from the peritoneal cavity of the host mice and collected in medium 199 (*Figure 2a*). The cell suspension was centrifuged at 100*g*, washed once, and the cells were resuspended in homogenising medium (*Figure 2b*).

Figure 3a shows the appearance of a pressure homogenate prepared in 0·25 M sucrose, 0·05 M tris, 0·025 M KCl, 0·005 M $MgCl_2$, pH 7·4. Total absence of divalent cations results in complete disintegration of nuclei (*Figure 3b*), though mitochondria and lysosomes can be successfully isolated from this homogenate. The effect of added Ca^{2+} can be seen in *Figure 3c*. The plasma membrane

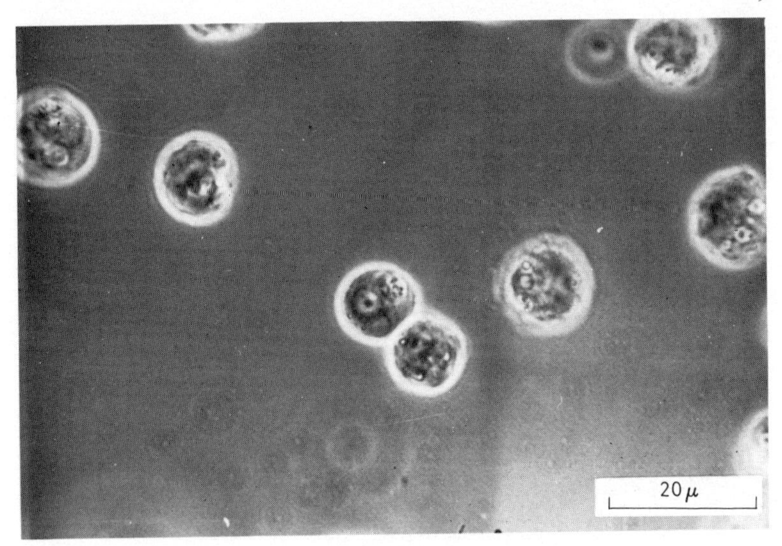

Figure 2a. *Mouse lymphoma R.1 cells suspended in tissue culture medium 199; phase contrast*

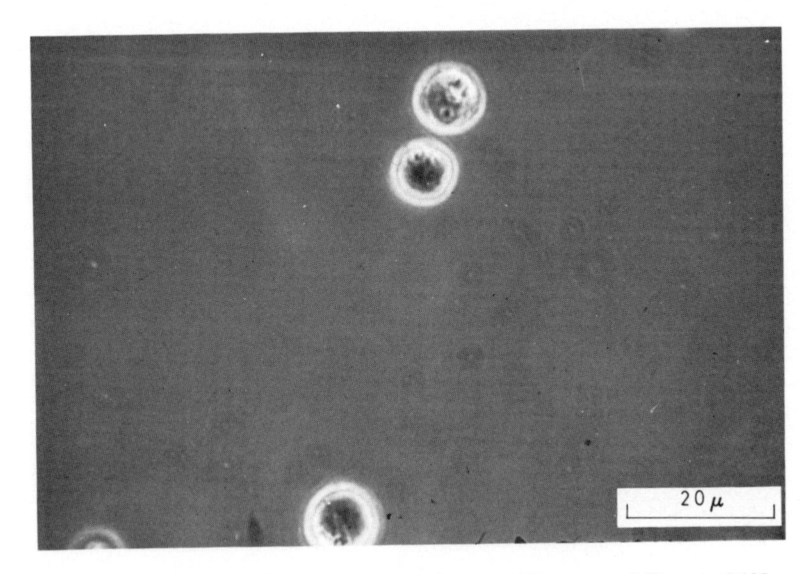

Figure 2b. *Mouse lymphoma R.1 cells after suspension in 0·25 M sucrose, 0·05 M tris, 0·025 M KCl, 0·005 M $MgCl_2$, pH 7·4; phase contrast*

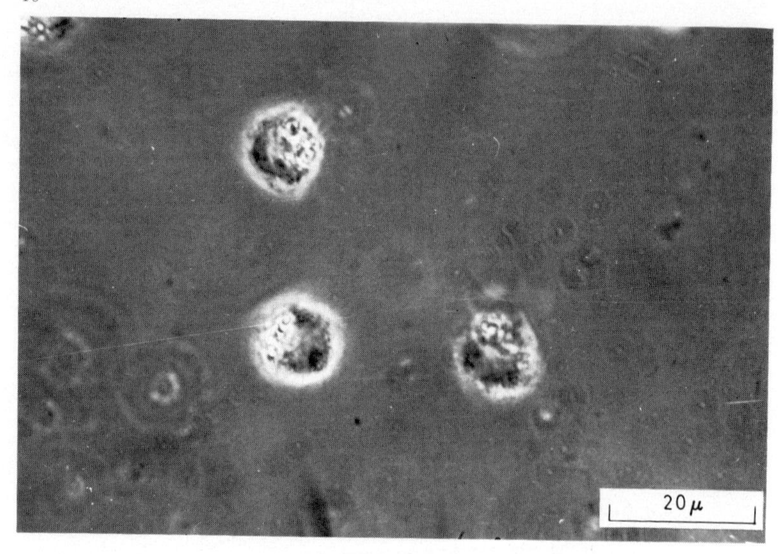

Figure 3a. Appearance of homogenate after release from pressure vessel. Suspending medium: 0·25 M sucrose, 0·05 M tris, 0·025 M KCl, 0·005 M $MgCl_2$, pH 7·4. The suspension was subjected to a pressure of 900 lb/in² for 20 min; phase contrast

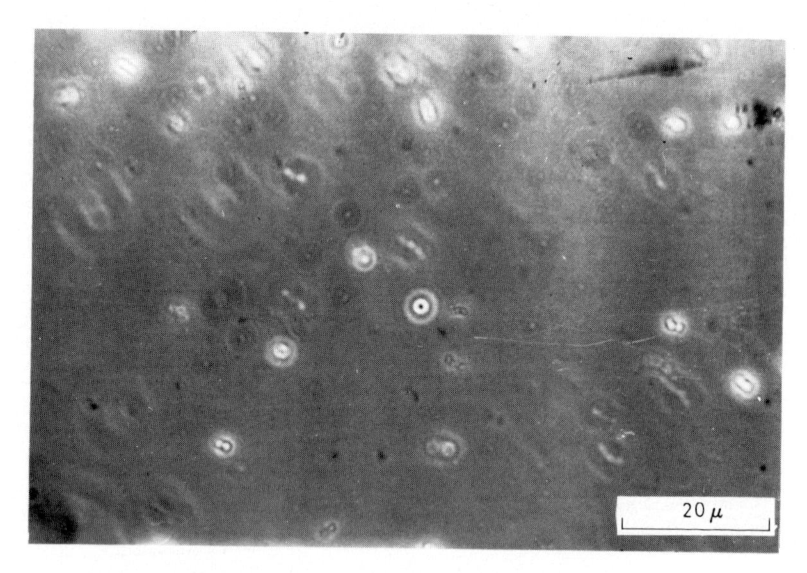

Figure 3b. Appearance of homogenate prepared in 0·25 M sucrose, 0·001 M EDTA, pH 7·4. Pressure: 900 lb/in² for 20 min. No nuclei are visible in this preparation; phase contrast

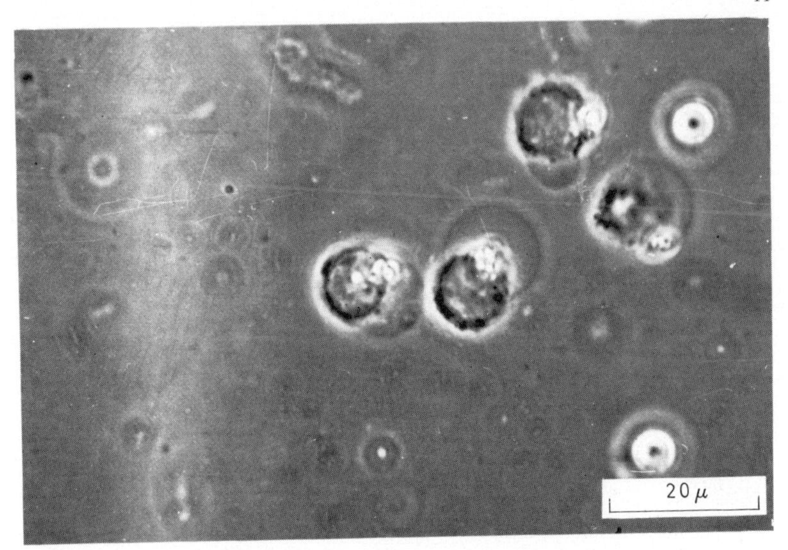

Figure 3c. Appearance of pressure homogenate prepared in 0·25 M sucrose, 0·05 M tris, 0·02 M KCl, 0·005 M $MgCl_2$, pH 7·4, with addition of 0·2 mM $CaCl_2$. Pressure: 900 lb/in² for 20 min. The plasma membranes have not completely ruptured. Shaking this preparation vigorously for 5 s yields a homogenate similar to Figure 3d *in appearance; phase contrast*

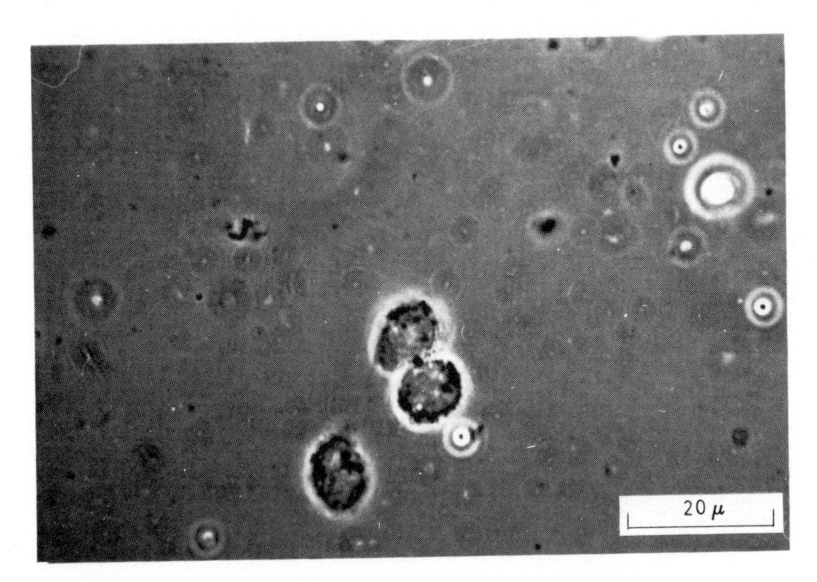

Figure 3d. Appearance of homogenate from Figure 3a *after being shaken vigorously for 5 s. Much of the cytoplasmic material originally adhering to the nuclei has been removed; phase contrast*

has not been completely ruptured.. The nuclei in all these homogenates (*Figure 3a–c*) are contaminated with cytoplasmic material, much of which can easily be removed by shaking the homogenate vigorously for 5 s (*Figure 3d*).

CONCLUSIONS

Homogenisation by the use of pressure of the order of 1000 lb/in^2 and its subsequent release in the apparatus illustrated (*Figure 1*) has proved to have considerable advantages over previous methods for dealing with large volumes of either tissue brei or suspensions of cells. Where comparisons have been made with other methods, the results obtained by pressure homogenisation have, in general, been favourable to this method. There is, however, one point that must be made. Several authors have given details of the preparation of plasma membranes in which these are identified by their morphology as open or folded sheets (Neville, 1960; Warren *et al.*, 1966; Emmelot *et al.*, 1964). These structures are not seen following pressure homogenisation. Instead, the membrane fraction appears to consist of vesicles of varying size. Kamat and Wallach (1965) have published a method for the isolation of plasma membranes based on pressure homogenisation. Wallach (1967) has also dealt very fully with the isolation of plasma membranes and some of the attendant difficulties.

Pressure homogenisation has proved efficient since, in most cases, over 90% of the cells can be ruptured given the right conditions. In addition, the cells and their liberated components are not subjected to prolonged exposure to disrupting forces. Reproducibility of the homogenates so produced is an exceptional feature of the method and exempt from individual operator variations. The parameters of the system, pressure, time and suspending medium, lend themselves readily to control by the operator.

REFERENCES

ANDERSON, N. G. (1966). Ed., *Natn Cancer Inst. Monogr.*, 21

CURL, A. L. and JANSEN, E. F. (1950). *J. biol. Chem.*, **45**, 184

EMMELOT, P., BOS, C. J., BENEDETTE, E. L. and RUMKE, P. H. (1964). *Biochim. biophys. Acta*, **90**, 126

HUNTER, M. J. and COMMERFORD, S. L. (1961). *Biochim. biophys. Acta*, **47**, 580

JOHNSON, F. H. and CAMPBELL, D. H. (1946). *J. biol. Chem.*, **163**, 689

KAMAT, V. B. and WALLACH, D. F. H. (1965). *Science, N.Y.*, **148**, 1343

NEVILLE, D. M. (1960). *J. biophys. biochem. Cytol.*, **8**, 413

NIRENBERG, M. W. and HOGG, J. F. (1958). *Cancer Res.*, **18**, 518

WALLACH, D. F. H. (1967). In *The Specificity of Cell Surfaces* (Ed. Davis, B. W. and Warren, L.), p. 129, New Jersey (Prentice Hall)

WALLACH, D. F. H., SODERBERG, J. and BRICKER, L. (1960). *Cancer Res.*, **20**, 397

WARREN, L., GLICK, M. C. and NARS, M. K. (1966). *J. cell. comp. Physiol.*, **68**, 269

2

SOME METHODS FOR ISOLATING NUCLEI

D. B. Roodyn

The study of the enzyme activities of subcellular fractions, a pursuit which I have called 'enzyme cytology' (Roodyn, 1967), has since its inception been bedevilled by poor standardisation of techniques. The most extreme case is in the study of isolated nuclei, where it is not an exaggeration to say that, until recently, almost every worker seemed to use his own method of isolation. The reasons for this state of affairs are not hard to see. Anyone who has isolated nuclei immediately becomes aware of the difficulty of breaking the cell but not the nucleus, the problems of clumping and swelling, poor recoveries of nuclei, a variety of forms of nuclear damage and last, but certainly not least, the presence of variable amounts of cytoplasmic contamination of one kind or another. The usual reaction to this is to introduce one or more 'modifications' to fractionation schemes already published. Since the final condition of the nuclear fraction depends on many interacting and often uncontrollable factors, at least with our present technology, it often happens that one or other of these modifications results in a 'better' nuclear fraction. The result is then frequently published as 'an improved method for the isolation of nuclei', usually with the claim that those so obtained are intact and essentially free of cytoplasmic contamination. While this may well be the case in some methods, it is unfortunately a depressing and, indeed, frustrating task to obtain objective evidence in support of the various claims for 'improved' methods.

CRITERIA FOR ASSESSING ISOLATION PROCEDURES

Some years ago I analysed published methods (Roodyn, 1963), comparing the morphological appearance, chemical composition, yield, extent of cytoplasmic contamination and biological activity of nuclear fractions which had been obtained by a variety of means. Although we now have a much greater wealth of information about the biochemical functions of the nucleus (see below), it is still difficult to make precise and objective comparisons between the various published isolation procedures. In many ways, the situation is not greatly different from that in 1963, when no clear picture emerged of the relative value of the various methods because of the fragmentary nature of the published evidence. It was surprising to find that only a few of the nuclear fractions studied had in fact been examined systematically under the electron microscope. Also there were very few low-power phase-contrast views of the fraction as a whole, that is, of a field containing many hundreds of nuclei rather than just 10 or 20. In some methods, no photomicrographs at all were published. Thus the morphological evidence was inadequate. What is required in each case is a low-power view under phase contrast, a high-power view to show the appearance of unfixed nuclei and, finally, electron micrographs of several fields. Although there are now several excellent studies of this type (e.g. Maggio *et al.*, 1963) it is still the exception rather than the rule to have detailed electron microscopy of the nuclear fractions in papers that are biochemically orientated. The need for electron microscopy is well illustrated by some experiments of Zentgraf *et al.* (1969). They found that the use of a Potter–Elvehjem homogeniser to isolate nuclei from bird erythrocytes gave nuclear preparations that appeared reasonably pure under the light microscope. However, examination under the electron microscope revealed that 95% of the nuclei were actually enveloped in fragments of plasma membrane. It was necessary to homogenise with high-speed blades to give nuclei free of these 'ghosts'.

The results of the chemical analyses were unsatisfactory from several points of view. The most satisfactory basis on which to express chemical analyses on nuclei is, of course, DNA. However, it is also important to know the DNA content per nucleus to allow for the effect of chromosome ploidy. It was found that some results were based on DNA, some on protein, others on total nitrogen, on dry weight or on nuclear counts. A standard procedure for expressing the results of chemical analyses on tissue fractions would be most useful, and it would be convenient if the values did not depend on ill-defined and variable standards, such as 'serum albumin' or 'thymus DNA', but on chemical parameters that can be determined with some degree of reproducibility. For example, protein analyses based on total nitrogen measurements are more reliable than those based on a colorimetric analysis using a commercial preparation of protein as a standard. Apart from these difficulties of technique, the results of the chemical analyses were not very revealing. This was because differences in chemical composition resulted from conflicting and often unrelated causes. Thus a low protein to DNA ratio could have been due to extraction of nuclear protein or to absence of cytoplasmic contamination or to both. Indeed Dounce and Ickowicz (1969) have shown that the isolation medium has a profound influence on the amounts of globulin, total histone and residual protein per unit of DNA. For example, the ratio of histone to DNA was about 1·0 in rat-liver nuclei isolated in the presence of $CaCl_2$ or $MgCl_2$ and about 2·0 in nuclei isolated in 0·44 M sucrose without added cation. The addition to the isolation medium of low concentrations of heavy metal salts (e.g. of lead, cadmium or indium) appeared to prevent this loss of histone (Dounce and Ickowicz, 1970). Similarly, Price (1970) has shown that ferric salts cause precipitation of histone in nuclei isolated from wheat embryos, giving more compact nuclei with dense chromatin. It is clear, therefore, that the chemical composition of the nucleus is readily altered by changes in the suspension medium. However, rigorous interpretation of such changes also requires precise knowledge of the purity of the nuclear fraction in each case. Unfortunately, this is often not available.

The results with yield of nuclei were perhaps the most satisfactory, in that two fairly reliable indices were available, nuclear count and recovery of DNA. The absolute amount of DNA per nucleus is a value of considerable biological interest (see, for example, Szarski, 1970). However, when one examined the actual DNA measurements, it was found that few papers gave much information about yield, and certainly very little as to whether the method used was selecting one type of nucleus rather than another. Careful correlations between yield of nuclei and that of DNA were seldom made. Indeed, one might imagine that cytoplasmic, in particular mitochondrial, DNA may well have been discovered earlier if it had been realised that in many of the simpler fractionation schemes there were no visible nuclei in any but the nuclear fraction, yet some of these fractions contained DNA.

It is fortunate that in recent years there has been a growing awareness that tissues used for cell fractionation (in particular rat liver) are far from homogeneous in cell type. Fortunately improved techniques allow for the separation of various classes of nuclei, so that much more meaningful studies may now be carried out on nuclear fractions. This problem is discussed by Johnston and Mathias in *Chapter 3* of this volume, and the use of the zonal centrifuge to separate various classes of rat-liver nuclei is described in detail by Johnston *et al.* (1968). Albrecht (1968, 1969) has separated diploid and tetraploid nuclei by sucrose-gradient centrifugation (cf. Falzone *et al.*, 1962) and McBride and Peterson (1970) describe the isolation of nuclei from mammalian tissue culture cells at various stages in the growth cycle. Since different classes of nuclei prepared by such methods appear to have different properties, it would now appear no longer possible to ignore the variation in cell type and in the stages of the growth cycle in the population of cells taken for sub-fractionation. A good example of this is the observation of Sneider *et al.* (1970) that during induction of liver tumours by azo dyes there are alterations in the relative amounts of stromal and parenchymal nuclei and in the distribution of DNA between these two classes of nuclei. In these experiments the two types of nuclei were separated by centrifuging in a discontinuous sucrose gradient.

When we consider the problem of cytoplasmic contamination, it is clear that there are no standard methods for assessing it. Some papers make rather bald statements such as 'The nuclei at this stage were essentially free of contaminant material' or 'Nuclei of a high degree of purity were obtained'. Such descriptions are unsatisfactory for several reasons. What is 'pure' to one worker may not be to another. Even 'negligible' levels of contamination may be significant if the contaminant has high activity and the nuclei none. For example, intact cells even in small numbers could greatly confuse incorporation experiments *in vitro*. It is obviously desirable to have some objective estimate of the purity of nuclear fractions. Many workers are conscious of this problem and present more or less detailed estimates of the various forms of contamination. It is quite common, for example, to use mitochondrial 'marker' enzymes, such as cytochrome oxidase, as indicators for mitochondrial contamination. Unfortunately, precise interpretation of such experiments is often very difficult. Let us say, for example, that we find that the specific activity of a typical mitochondrial enzyme in the nuclear fraction is only $\frac{1}{50}$ of its activity in mitochondria. Are we justified in concluding that the mitochondrial contamination is only 2% or less of the total protein? First we must ask whether the recovery of enzyme was satisfactory during the fractionation. Suppose that it was in fact 96%. At first sight this would appear a most gratifying answer. However, if we then discover that only 1% of the total enzyme was found in the nuclear fraction, can we be certain that the 4% loss of enzyme was in fact not due to inactivation of enzyme in the mitochondria contaminating the nuclear fraction? If this is so, we may have underestimated the level of contamination by a *factor of 5*. In order to interpret results involving low levels of enzyme activity, we need an order of precision and reproducibility in our fractionation techniques and in the enzyme-assay procedures that is probably rarely attained. Conversely, one cannot necessarily conclude that if, for example, 98% of an enzyme is found in the mitochondrial fraction, *all* the residual activity found in other fractions is due to contamination with mitochondria. Are we to be certain that the 'residual' cytochrome oxidase in nuclear fractions, for example, is not a non-mitochondrial enzyme, with its own special significance? Indeed Berezney *et al.* (1970) claim that isolated nuclear membranes contain cytochrome oxidase and that the observed activity is not due to mitochondrial contamination. There have also been many studies on nuclear oxidative phosphorylation which indicate clear differences from the mitochondrial system (see, for example, Betel and Klouwen, 1967; Betel, 1969).

To conclude, accurate assessment of cytoplasmic contamination requires great precision, great caution and a wealth of parallel and complementary chemical and morphological analyses. However, the results of experiments with isolated nuclei will never have a sound basis until these difficult problems are solved. The task is rendered more difficult by the paucity of suitable nuclear 'marker' enzymes (see Ringertz (1969), Swift (1969) and Simard (1970) for some general accounts of contemporary views on nuclear structure and function).

The most likely candidate for a routine nuclear 'marker' is the enzyme NAD pyrophosphorylase (or to use its correct Enzyme Commission designation (E.C.2.7.7.1) 'ATP:NMN adenylyltransferase') which was shown by Hogeboom and Schneider (1952) to be highly concentrated in the nuclear fraction and essentially absent from the cytoplasmic fractions. However, subsequent research has shown that the biosynthetic pathway for NAD synthesis is not as simple as originally thought when NAD pyrophosphorylase was discovered. Haines *et al.* (1969) have presented a useful survey of the enzymes involved in NAD synthesis and breakdown in nucleus and cytoplasm. DeamidoNAD is first made in the nucleus from ATP and nicotinic acid mononucleotide by the action of an enzyme other than NAD pyrophosphorylase, namely deamido NAD-pyrophosphorylase (E.C.2.7.7.18 or ATP: nicotinate mononucleotide adenylyl transferase). The deamidoNAD can migrate to the cytoplasm where it reacts with ATP and glutamine to give NAD under the action of NAD synthetase [E.C.6.3.5.1 or deamido-NAD: L-glutamine amido-ligase (AMP)]. NAD made in the nucleus may be broken down to ADP-ribose by a nuclear NAD-nucleosidase, or polymerised to form the interesting product poly-(ADP-ribose) discovered in 1967 by two Japanese groups of workers. This material is synthesised by a particulate fraction isolated from nuclei, and appears to consist of repeating adenine-ribose-pyrophosphate-ribose units, joined by ribose–ribose linkages (Nishizuka *et al.*, 1967; Reeder *et al.*, 1967; Hasegawa *et al.*, 1967; Fujimura *et al.*, 1967). No doubt because of these complexities, it has unfortunately not become the custom to estimate NAD pyrophosphorylase routinely in nuclear fractions and the enzyme has not lived up to its early expectations of becoming a standard marker enzyme.

The most striking advances in our knowledge of nuclear function have been in the elucidation of biosynthetic pathways, particularly in DNA and RNA synthesis. Unfortunately, because of their very nature, such biosynthetic reactions are difficult to use as routine enzyme markers to assess recovery, purity of nuclei, etc. Apart from the actual complexities of the assays required, the reactions are influenced by many subtle factors, such as the physiological state of the tissue and the hormone balance of the animal. Many of the assays follow the incorporation of labelled precursors into insoluble macromolecules and, apart from the notorious lack of standardisation in the method of presenting such results, it is often impossible to convert the final radioactivities observed into absolute molar concentrations because of uncertainties as to the effective pool size. Although the study of the biosynthetic reactions may not therefore be readily amenable to the 'balance-sheet' approach of cell fractionation, it is at least concerned with fundamental aspects of nuclear function and is of great use in making some assessment of the 'biological' activity of the isolated nuclei. I will therefore give a brief account of some relevant experiments in this field.

Since the early experiments of Allfrey *et al.* (1957) demonstrating protein synthesis in isolated nuclei, there has been an increasing awareness of the great technical difficulties in studying such systems. The major difficulty is that the outer layer of the nuclear membrane is in continuity with the rough endoplasmic reticulum of the cytoplasm. As a result, it is very difficult to avoid contamination of the surface of the isolated nucleus with attached fragments of membrane, with the associated cytoplasmic ribosomes. Fortunately, the nuclear protein synthesis system appears to require Na^+ rather than K^+ ions, and it is possible to remove the surface bound ribosomes with a suitable neutral detergent, such as Triton X-100. A recent example of some of the difficulties involved in such work, and also of the complications that may arise from using a heterogeneous tissue, is given by the experiments of Løvtrup-Rein (1970) on protein synthesis with rat-brain nuclei.

Studies on DNA and RNA synthesis probably have less complicating factors than experiments on protein synthesis, although even here the systems are far from simple. Considering DNA polymerase first, there are a number of practical difficulties. The heterogeneity of the cell population discussed above raises particular problems in the interpretation of measurements of DNA synthesis in bulk nuclear fractions, since Haines *et al.* (1970) have shown that the various classes of rat-liver nuclei differ considerably in their DNA polymerase activity. The precise intranuclear localisation of the enzyme is still under study, and Ballal *et al.* (1970) showed, for example, that a range of fractions isolated from tumour nuclei could synthesise DNA, but these differed in their response to ATP and added DNA template. The DNA synthesised *in vitro* appears to be added to strands that were growing *in vivo* (Lynch *et al.*, 1970), so it is not surprising that the activity observed in the isolated nuclear fraction is affected by previous treatments of the whole animal [cf. the stimulation caused by previous injection with Zn^{2+} or Hg^{2+} ions (Weser, 1970)]. Unfortunately, DNA polymerase is not uniquely located in the nucleus since there is considerable activity in the soluble fraction. Although there are certain differences in behaviour between the nuclear and cytoplasmic enzymes (see, for example, Friedman, 1970 and Lindsay *et al.*, 1970), it would certainly be dangerous to ignore the possible contribution of cytoplasmic contamination to DNA synthesis observed in isolated nuclear fractions. Finally, it appears that the activity of the system is affected by the nuclear isolation medium. For example, Wager *et al.* (1971) have shown that nuclei isolated in the presence of divalent cations have enhanced priming activity.

RNA polymerase in nuclei has been studied extensively in the last few years and it is beyond the scope of this article to give a full account of our present knowledge of this extremely important enzyme system. (The most recent findings on the different forms of this enzyme are clearly set out by Chesterton and Butterworth (1971), and these authors conclude 'It is apparent that the status of the enzyme systems involved in the overall process of RNA synthesis in mammalian cells is complex and is, as yet, not clearly defined'.) I will briefly mention some experiments relevant to the problem of methods of nuclear isolation.

In early studies on nuclear RNA synthesis, Pogo *et al.* (1967) found that the biosynthetic activity was very sensitive to the salt concentration in the medium and many subsequent studies revealed that there was a complex response to the addition of, for example, ammonium sulphate and Mg^{2+} or Mn^{2+} ions. After some confusion, the matter appeared to be clarified by the observations of Roeder and Rutter (1970) that there appear to be at least two RNA polymerases, *Form I* which is probably responsible for the synthesis of ribosomal RNA in the nucleus (see, for example Hurlbert *et al.*, 1969) and *Form II*, which is in the nucleoplasm, and makes DNA-like RNA. As discussed by Chesterton and Butterworth (1971) the two forms differ in their response to added salts. (For example, the optimum ammonium sulphate concentration for *Form I* is 0·01 to 0·05 M, and for *Form II* is 0·10 to 0·14 M). The precise assay of RNA polymerase in isolated nuclei is therefore by no means easy.

Another problem is the possible loss of RNA polymerase during isolation. Liao *et al.* (1968) describing a method for the extraction of RNA polymerase from nuclei reported that the enzyme (presumably Roeder and Rutter's *Form II*) is not firmly bound to the nuclear chromatin complex and can be released by relatively mild treatments. Read and Mauritzen (1970) have investigated various methods for isolating and preserving nuclei without affecting their RNA-polymerase activity, and recommend the disruption of the tissue with saponin, and storage of the isolated nuclei in 70% glycerol. The effect of the isolation medium is clearly shown by an interesting recent study by Moulé (1970). It appears that the activity of the RNA polymerase system *in vitro* is not only affected by the cationic composition of the incubation medium, but also by that of the *isolation* medium. Thus the addition of Mg^{2+} ions to the isolation medium alters the A/G ratio of the RNA subsequently synthesised by the nuclei *in vitro*. Apparently the synthesis of DNA-like RNA is suppressed, but optimal synthesis of ribosomal RNA still occurs. In conclusion, although RNA biosynthesis is an extremely important function of the nucleus, the rate of RNA synthesis actually observed *in vitro* by a given nuclear preparation is the result of complex and interacting factors and cannot yet be used as a simple measure of nuclear integrity.

Other activities of nuclei could well be used ultimately as criteria in nuclear isolation procedures. There have been several recent studies on nuclear polynucleotide phosphorylase (see, for example, See and Fitt, 1970), as well as increased examination of the system for acetylation of histones (Pogo *et al.*, 1968; Bondy *et al.*, 1970; Gallwitz, 1970; Wilhelm *et al.*, 1971). There have also been studies of specific hormone binding effects. For example, Zigmond and McEwen (1970) have shown that [^{3}H]oestradiol injected *in vivo* into ovariectomised rats is specifically bound by brain nuclei. The amount of binding varied with the region of the brain, and was greatest with nuclei from the pre-optic hypothalamic region. There have also been reports of specific binding of androgens in prostrate nuclei (see Bashirelahi and Villee (1970) for a recent study of this). It remains to be seen whether these observations will be employed more widely to characterise nuclear fractions.

Perhaps the future trend will be towards using 'biological' criteria to estimate the degree of damage inflicted on nuclei during isolation, rather than quantitative enzyme assays. The recent increase in nuclear transplantation and in the formation of heterokaryons has opened the way to a much more specific and meaningful analysis of nuclear function. To give a specific example, Bolund *et al.* (1969) studied the effect of transplanting chick erythrocyte nuclei into the cytoplasm of Hela cells. They found that the cell fusion reactivated the chick nuclei which proceeded to synthesise RNA and DNA again. As this kind of experiment has attracted considerable interest, techniques for formation of hybrid cells and nuclear transplants have become more sophisticated. Some experiments by Ladda and Estensen (1970) illustrate this. Mouse L cells grown in tissue culture and chick erythrocytes were fused, after treatment with inactivated Sendai virus. The fused cells were then treated with an active agent, cytochalasin B. This material causes the protrusion of nuclei from cells. As a result, in some cases, the original mouse nucleus was extruded from the fused cell, giving a true nuclear transplant of a chick nucleus into a mammalian cell, without the need for microdissection. It is clear that as these kinds of techniques develop, there will be greater opportunities for a more fundamental analysis of the true biochemical function of the nucleus. For example, by transplanting non-labelled nuclei into labelled anucleate cytoplasm of amoebae, Goldstein and Trescott (1970) were able to make a detailed study of the migration of RNA between nucleus and cytoplasm.

Ord and Bell (1970) have used the survival of amoebae after nuclear transplantation as an interesting criterion for nuclear isolation media (cf. Burnstock and Philpot, 1959). By using 35 mM KCl as an isolation medium they were able to achieve reasonable survival of transplants (e.g. 65% survival after the nuclei had been exposed to KCl for 5 to 10 min, compared to 90% survival with the direct transplants). More prolonged exposure of the nuclei to the isolated medium, however, resulted in greatly reduced survival (e.g. only 10% after 1 h). Since nuclear isolation procedures usually take far longer than this, we may perhaps have to conclude that even the 'best' bulk nuclear fractions so far obtained have lost their most sensitive function, that is, the ability to restore the viability of an enucleate cell.

DETAILS OF SOME ISOLATION PROCEDURES

Having discussed some of the difficulties in developing adequate criteria for nuclear isolation, I shall now give an account of various published methods for nuclear isolation, indicating the major types of technique that have been, and are currently being, used. I have endeavoured to make a fairly representative selection from the many methods in the literature, although space does not allow a fully comprehensive treatment of the subject. Some of the methods are really 'collector's pieces'. They had a profound influence in their day and are of significance as the parents of more contemporary ones. The methods will be presented in the form of simple flow-sheets giving only the essentials of the technique. However, for the convenience of the experimentalist, detailed descriptions of three are presented in the Appendix.

They are best described in historical sequence. Methods were available for the preparation of mammalian nuclei in the 1930s and, indeed, earlier. There are some very early references to the separation of nuclei from cytoplasm. In 1856, Francis Gurney Smith observed that nuclei of skin and tumour tissue cells were released by the action of acetic acid, and he was probably the first person to have separated the mammalian nucleus from cytoplasm. The work of Meischer in the 1870s on nucleic acids is frequently quoted. However, it is often forgotten that he was also interested in isolated nuclei and developed a method based on the use of pepsin and hydrochloric acid. The separation of the nuclei from avian erythrocytes was achieved at about the same time by Ploz (1871), and there were several chemical and metabolic studies on these nuclei in the late nineteenth and early twentieth centuries (e.g. Ackerman, 1904; Warburg, 1910). (For a modern study of nucleated erythrocytes using the zonal centrifuge, see Mathias *et al.*, 1969.) In the 1930s, Behrens published a series of papers on the isolation of nuclei with organic solvents (e.g. Behrens, 1932), and Crossman (1937) observed that nuclei could be released in a clean state from muscle tissue by the use of 5% citric acid. This observation was exploited by Stoneburg (1939) and Marshak (1941) for the bulk isolation of nuclei.

Without question the work of Alexander L. Dounce in the 1940s laid the basis and was a stimulus for a great deal of subsequent work on isolated nuclei (see Dounce, 1948). Using the citric acid method, he developed a variety of techniques for the isolation of pure nuclei in bulk from various tissues. These varied in the strength of citric acid used and in the final purity of the nuclear fraction. In general, the lower the pH, the purer the nuclei. Thus much of the exploratory work on the DNA content of nuclei that was done in the period 1945 to 1955 was carried out on nuclei isolated at pH 4 or even pH 2·5 by various adaptations of Dounce's techniques. For enzyme work, a milder pH was clearly required, and the method of Dounce (1943) using a pH of 6–6·2 was popular (*Chart 1*). Although lowering the pH is a useful and simple method for improving the final purity of the nuclear fraction, it should be used with caution, because there is little doubt extreme changes of pH have a serious effect on the chemical composition of the nucleus. For example, Dounce *et al.* (1966) found that the pH of the isolation medium strongly influenced the amount and properties of histones that could be subsequently extracted from rat-liver and calf-thymus nuclei.

Chart 1. Method of Dounce (1943)

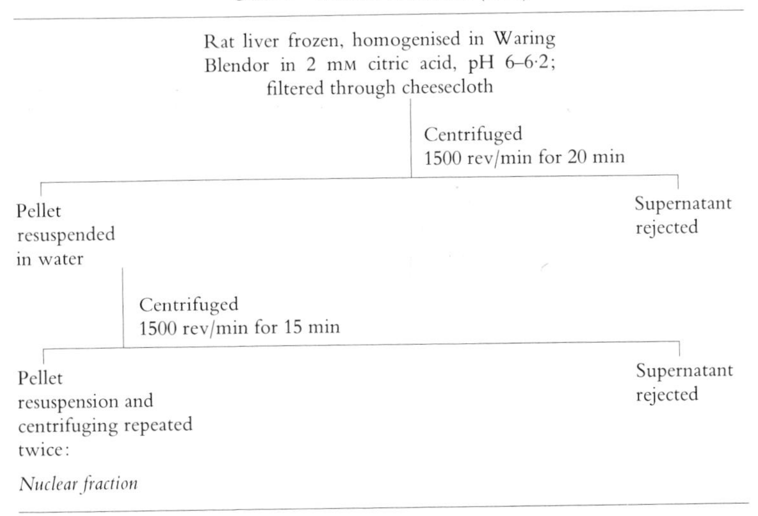

Certain points in this method are of general interest. First, the liver is frozen, which would today be regarded as a rather drastic procedure before subcellular fractionation. However, I am not certain that there has been a serious study of the effect of freezing of the tissue on the state of the nuclei finally isolated. As will be seen later, it is claimed that nuclei isolated from freeze-dried material with organic solvents are 'intact' and indeed, are the most 'intact'

of all nuclei. Second, the homogeniser used is the Waring Blendor. While mitochondria can be released from the cell simply by breaking the cell membrane, usually this is not enough in the case of nuclei. If a crude nuclear fraction from rat liver prepared with a Potter–Elvehjem homogeniser is examined, nuclei are frequently seen to be trapped in a mass of empty collapsed membranes. We know, of course, that the outer nuclear membrane is attached to, and indeed is continuous with, the cytomembrane system of the endoplasmic reticulum, so that it is easy for membranous material to remain attached to nuclei. It is the experience of many workers that 'clean' nuclei are only obtained with rather violent methods of cell breakage. The method of cell homogenisation is well known to be the least controlled and reproducible part of the cell fractionation technique. Thus it is not very accurate to talk of 'Waring Blendors' or 'Potter–Elvehjem homogenisers'. With the former the exact make should be specified, including the speed of rotation of the blades. Of course, the actual physical forces applied to a tissue suspension by a blender would depend on many things, including the volume and viscosity of fluid in it, the shape of the vessel, that of the blades, their sharpness and the speed of rotation, that is, the general hydrodynamics of the apparatus. Similarly, the breakage achieved in a conventional Potter–Elvehjem homogeniser will depend on many factors which are usually controlled only to a small degree, if at all. They include the nature of the material of the pestle and mortar and hence the frictional forces between the moving surfaces of the homogeniser and the fluid phase, the speed of rotation, the gap between pestle and mortar, the alignment of the axis of rotation of the pestle and the axis of the mortar (if the two are not perfectly parallel, direct abrasion between the two surfaces occurs), the composition and particularly the viscosity of the medium, the concentration of the tissue suspension and, of course, the nature of the tissue itself. When the simple homogenisers are used, the operator applies a force to the apparatus in the up and down strokes. The degree of homogenisation clearly depends on this force, and one has only to watch someone operating a Potter–Elvehjem homogeniser to realise how much the process depends on personal skill and how little on the design of the instrument. More controlled types of homogenisers are described in the literature, but it is a regrettable fact that the majority of experiments on nuclei have been carried out with these two types. The situation is quite anachronistic, since there is a great deal of evidence that controlled and reproducible cell breakage is essential for reproducible cell fractionation.

Third, the homogenate is filtered through cheesecloth (*Chart 1*).

In different methods a variety of filter materials are used, including cheesecloth, single- or double-napped flannel, gauze, nylon cloth, lint and, in some cases, various meshes. Again there is a problem of standardisation, as it is difficult to compare one piece of gauze with another. This is quite important, since filtration is very effective in removing pieces of fibre, unbroken cells and aggregates of clumped nuclei. Here again, a plea for standardisation of filtration materials can be made. Fourth, after filtration the homogenate is centrifuged at 1500 rev/min (corresponding to about 600*g*). It is, of course, mandatory nowadays to present centrifugal conditions in *g* or, better, g_{av} rather than rev/min. It is interesting to note that, in the various methods to be cited, the centrifugal force applied to sediment the nuclei varies considerably. In simple bulk sedimentation this is not of great importance, since all that one needs is a force sufficient to sediment all the nuclei. However, it is preferable to use a force that is sufficient just to do this, with the minimum of sedimentation of other particles. It is sometimes forgotten that the distance sedimented is as important as the centrifugal field. Thus applying 600*g* for 10 min in a 250 ml centrifuge bottle is very different from applying the same field to a 13·5 ml centrifuge tube in a fixed-angle rotor. One suspects that many methods use 1000 rev/min or 600*g* for convenience, without careful investigation of the force required. It is quite likely in many cases that the nuclei could be sedimented with less prolonged centrifuging than is actually used.

Fifth, the nuclei are then put through various cycles of washing, by repeated resuspension and resedimentation. Again there is much variation in the number of washes employed in the various methods. It is usual to wash the initial nuclear pellet at least once, and usually twice, and the efficiency of the wash will depend on that of drainage of the pellet, the volume of fluid added in relation to its size and, most important, the efficiency of resuspension of the pellet. Rehomogenisation of the pellet *in situ* is the most satisfactory method of resuspension. If a Potter–Elvehjem homogeniser is used, it is most convenient to have centrifuge tubes made of precision-bore tubing, so that well-fitting pestles can be used for the resuspension. In some of the methods described by Dounce, the final nuclear fraction is allowed to stand for about 1 h. Contaminating whole cells sediment, and the nuclei are then decanted from the upper layers of the suspension. Since it is usually necessary to obtain them as rapidly as possible, this method is rarely used today. However, it emphasises the point that separation of nuclei from intact cells is difficult and, in media of low density, requires very low centrifugal forces.

The method devised by Schneider (1948) is summarised in

Chart 2. It was not specifically designed for the preparation of pure nuclei but was part of a general scheme for subcellular fractionation of mammalian tissues. The method consists simply of three sedimentations at 600*g* for 10 min, and there is little doubt that the final nuclear fraction is extremely impure and grossly contaminated with red blood cells, fibrous material, intact cells and cytoplasmic particles. The method is of importance historically, because a very large number of fractionations have been carried out with it (Roodyn, 1965) or with similar ones. Since it is part of a larger

Chart 2. Method of Schneider (1948)

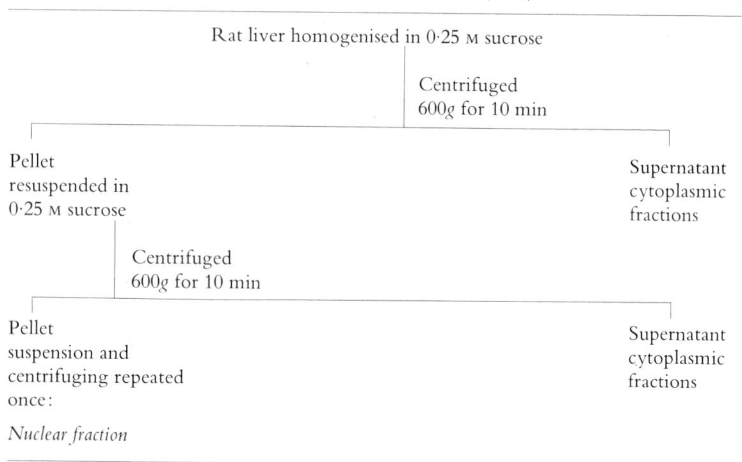

scheme of fractionation, many of those carried out with this method had the 'balance-sheet' approach, that is, it is possible to sum all the amounts in each subcellular fraction and express them as a percentage of the amount in the original homogenate. It was shown by this approach that for a large number of enzymes only a small percentage of the total activity was found in the nuclear fraction (see the detailed tabulations in Roodyn, 1959, 1965). It is most likely, therefore, that these enzymes are absent from the nuclei. Unfortunately, only a few of these fractionations have now been repeated using more refined methods for isolating nuclei. We are, therefore, in the unsatisfactory position of having a great deal of suggestive evidence based on very impure nuclear fractions, with a limited number of rigorous analyses on highly purified nuclei. The situation is clearly improving with the years. Nevertheless, there is still a great back-log of incomplete data that it would be most interesting to clarify. For these reasons, the simple fractionation schemes such as that in *Chart 2* are still of considerable importance.

The next method raises the problem of the soluble enzymes of the nucleus and the prevention of loss of soluble protein from nuclei during isolation. Georgiev (1967), in a detailed survey of the enzymology of the nucleus, discussed the advantages and disadvantages of organic solvents for isolating nuclei and, on the whole, concluded that those isolated in organic solvents retain their soluble components and may, indeed, be regarded as primary standards with which other nuclear fractions may be compared. I do not wish to discuss this matter at any great length here, but it is perhaps correct to state that there is still considerable doubt about the significance of studies on nuclei prepared in organic solvents, mainly because of the profound effect such treatment has on the morphology of the other structural components of the cell. A typical organic solvents method, due to Dounce *et al.* (1950), is summarised in *Chart 3*. The liver is freeze-dried and ground in a ball mill, and the nuclei are purified by a cycle of sedimentations and flotations in benzene–CCl_4 mixtures of various specific gravities. Filtration through appropriate cloths and gauzes is carried out at various stages. While the method is no doubt rather tedious, it has the advantage that once the tissue is freeze-dried,

Chart 3. Method of Dounce *et al.* (1950)

Rat liver freeze-dried, ground in ball mill with light petroleum (b.p. 50–60°C) and filtered through cheesecloth. Cycle of sedimentations and flotations (with filtration at various stages):

Step	*Medium*	*Fraction collected*	*No. of washings*
1	Benzene–CCl_4 (1:1)	Sediment	2
2	Benzene–CCl_4 (1:1·44)	Sediment	1
3	Benzene–CCl_4 (1:1·86)	Floating material	–
4	Benzene–CCl_4 (1:1·44)	Sediment	–
5	Benzene–CCl_4 (1:1)	Sediment (*nuclear fraction*)	2

purification can be carried out over an extended period, rather as if one was working with fixed material. However, it is essential to determine the effect of the organic solvents on the component that is being measured. Thus, although some enzymes are unaffected by them, others are either inactivated or, in a few cases, activated. There are various modifications of the organic solvents technique.

For example, Behrens and Taubert (1952) describe a method in which freeze-drying is omitted. However, most of these methods are similar in design to the method in *Chart 3*. Kirsch *et al.* (1970) have recently described an interesting revival of the non-aqueous technique. Freeze-dried cells were suspended in chilled glycerol and homogenised in suitable milling devices (e.g. the 'Polytron' or the Sorvall 'Omni-Mixer'). The homogenate was filtered through glass wool, layered over 85% glycerol, 15% 3-chloro-1,2-propanediol (β-chlorohydrin) and centrifuged at 120 000*g* for

Chart 4. Method of Maver *et al.* (1952)

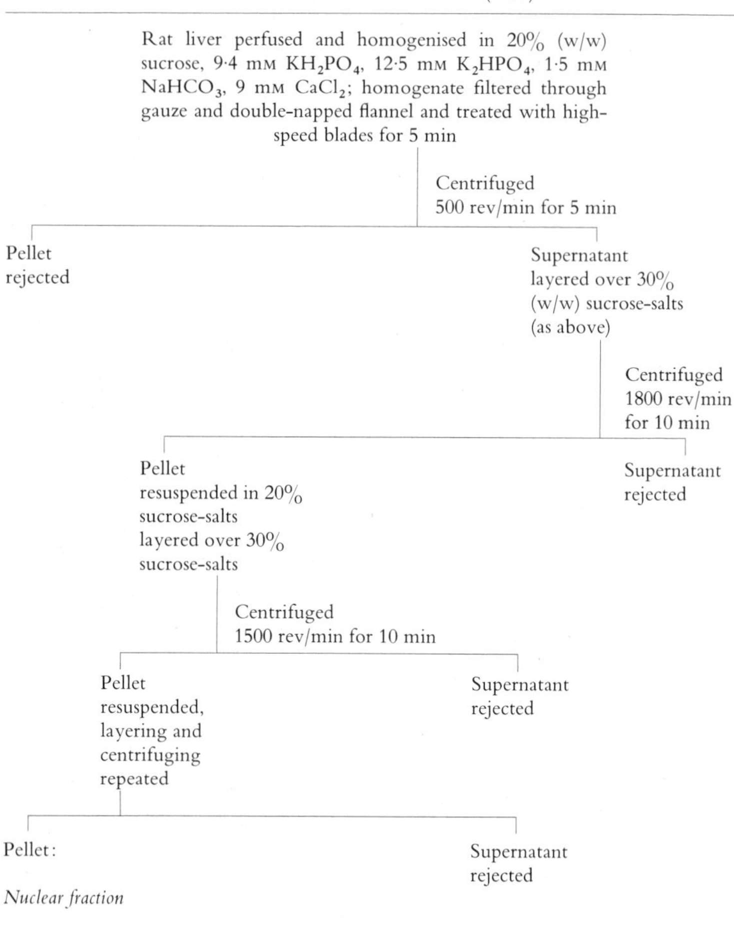

30 min. The cytoplasmic material floated, and the nuclei ($\rho = 1 \cdot 33$ g/cm^3) sedimented as a gelatinous pellet. (The purity of the nuclear fraction was checked by scanning electron microscopy.)

The next is not very well known and, to my knowledge, was only used by its originators. It is included because it employs several advances in technique for isolating nuclei that were introduced in the early 1950s and are still relevant now. The method is due to Maver *et al.* (1952) and is given in *Chart 4*. The liver is first perfused to remove erythrocytes, since contamination with them of nuclei isolated in isotonic aqueous media can be a serious problem. The citric-acid methods result in total haemolysis, and there is possibly secondary re-adsorption of haemoglobin on to nuclei. In the methods using dense sucrose, which are described later, the erythrocytes float above the nuclear pellet. However, if aqueous media of moderately low density are used, it is essential to remove as many erythrocytes as possible before starting the fractionation. The homogenisation medium used by Maver *et al.* (1952) is a rather complex salts mixture. It is based partly on the studies of Wilbur and Anderson (1951) on the effects of salts on the appearance of nuclei, and the important observation of Schneider and Petermann (1950) that addition of low concentrations of divalent cations, in particular Ca^{2+} ions, to the isolation medium results in less fragmentation of nuclei, less clumping and considerably eases isolation of nuclei. The method also employs a simple layering procedure, similar to that of Wilbur and Anderson (1951), which is very efficient in reducing contamination of nuclei with cytoplasmic particles. High-speed blades are used to break any whole cells that escape the initial homogenisation. There does not seem to be any great need when isolating nuclei to maintain conditions of isotonicity. Indeed, the appearance of nuclei depends more on the salt composition of the medium, in particular the content of divalent cations, than on the osmolarity. There is no evidence that mammalian nuclei swell in hypotonic media. On the contrary, they may swell in hypertonic media, if these are free of divalent cations.

The method of Hogeboom *et al.* (1952) (*Chart 5*) is similar in design to that of Maver *et al.* (1952) and has been used in a large number of studies of mammalian nuclei. The layering technique is employed, as well as low concentrations of $CaCl_2$. Note that that of $CaCl_2$ in the original homogenising medium is 1·8 mM but in all subsequent media only 0·18 mM. This is because the protein in the total homogenate combines with a great proportion of the added Ca^{2+} ions, thus lowering their effective concentration. The effect of Ca^{2+} ion concentration is indeed quite critical, and it is important to follow the suggested concentrations rather carefully in order to repeat methods based on the use of Ca^{2+} ions. This method gives nuclei of reasonable purity, provided that the efficiency of cell breakage is high. There is no means of separating whole cells from nuclei and, therefore, if a loosely-fitting homogeniser is used, very serious contamination with whole cells can result.

Chart 5. Method of Hogeboom *et al.* (1952)

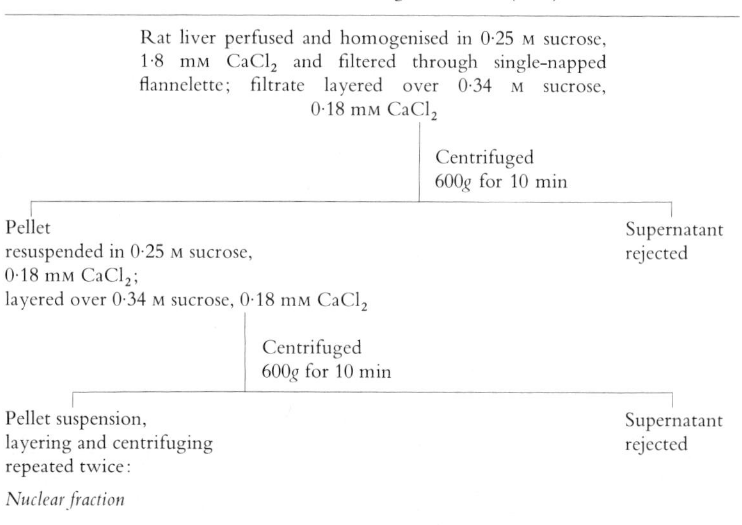

The difficulty of whole cell contamination is overcome in the next method (Roodyn, 1956, *Chart 6*). A nuclear fraction, prepared essentially by the method of Hogeboom *et al.* (1952), is layered over a gradient of 11 to 40% (w/w) sucrose containing 0·18 mM $CaCl_2$. The nuclei are centrifuged for only 4 min at low speed (300*g*), and purified nuclei are obtained as a band about 3 cm from the meniscus. The whole cells sediment to the bottom of the tube, and there is a certain degree of separation from erythrocytes. Contaminating mitochondria are found above the nuclear band. One of the difficulties in this method, however, is that, despite the presence of $CaCl_2$, nuclei tend to clump during the preliminary isolation. If the degree of clumping is small, the gradient system works well, but if large aggregates of nuclei are formed, the separation is not successful. The tendency of nuclei to aggregate is no doubt the reason for the difficulty in applying gradient differential centrifugation to their preparation. Falzone *et al.* (1962) describe a method for separating nuclei on a sucrose gradient. However, they were prepared originally in 0·04 M citric acid, and this would not be acceptable for many studies.

Chart 6. Method of Roodyn (1956)

Rat liver perfused and homogenised in 0·25 M sucrose, 1·8 mM $CaCl_2$ and filtered through lint; homogenate layered over 0·34 M sucrose, 0·18 mM $CaCl_2$

Centrifuged 600*g* for 10 min

Pellet resuspended in 0·25 M sucrose, 0·18 mM $CaCl_2$; layered over 0·34 M sucrose, 0·18 mM $CaCl_2$	Supernatant rejected

Centrifuged 600*g* for 10 min

Pellet resuspension, layering and centrifuging repeated; layered over linear gradient of 11–40% (w/w) sucrose, 0·18 mM $CaCl_2$	Supernatant rejected

Centrifuged 300*g* for 4 min

Band 3 cm from meniscus collected: *Nuclear fraction*

The high density of nuclei makes isopycnic centrifugation unnecessary, since all contaminating material is lighter. Thus, in the next method (*Chart 7*), Chauveau *et al.* (1956) made use of the important observation that if liver is simply homogenised in 2·2 M sucrose and centrifuged at 40 000*g* for 60 min, the pellet consists of highly.purified nuclei, and all the cytoplasmic components float above the sedimented nuclei. Although it was some time before the usefulness of this observation was generally realised, it is now very common practice to purify nuclei by centrifuging through very dense sucrose solutions. Currie *et al.* (1966) have made a detailed study of nuclei prepared in dense sucrose and confirm that cytoplasmic contamination is very low. However, some caution is necessary with dense sucrose solutions. First, because of the high concentrations used, any contaminant in the sucrose could reach high concentrations. For example, if this contains only 0·01% $CaCl_2$, 2·2 M sucrose will contain 0·8 mM $CaCl_2$. Such a concentration would have a definite effect on the nuclei. The properties of dense sucrose solutions also need to be considered. In *Figure 1* the viscosity of different concentrations of sucrose at 0°C and 5°C is plotted from data given by de Duve *et al.*

Chart 7. Method of Chauveau *et al.* (1956)

Rat liver homogenised in 2·2 M sucrose

Centrifuged
40 000*g* for 60 min

Pellet:	Supernatant rejected
Nuclear fraction	

(1959). There is a striking increase in viscosity at high sucrose concentrations so that, while a small error at 0·3 M has a negligible effect on the viscosity of the solution, small differences in molarity at high concentrations have a profound one. Also, the increase in

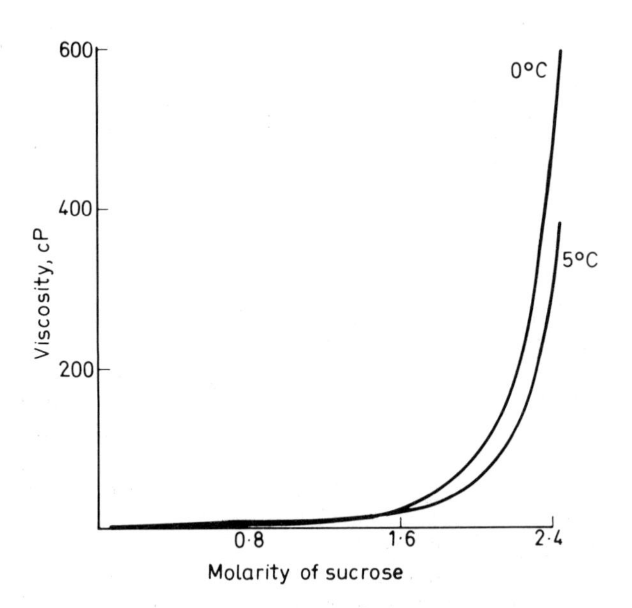

Figure 1. Viscosity of sucrose solutions at 0°C and 5°C (data from de Duve et al., *1959)*

viscosity on cooling from 5°C to 0°C is much greater at high than at low sucrose concentrations, so that very small variations in rotor temperature will have a profound effect on the rate of sedimentation of the particles. Thus, if very concentrated solutions of sucrose are used, particular care must be taken to control its purity, the precise concentration of the solution and the rotor temperature during centrifugation, otherwise results are quite likely to be variable.

The next method is included because of its importance in studies on nuclear protein synthesis. Allfrey *et al.* (1957) reported that isolated thymus nuclei, prepared by the method outlined in *Chart 8*, can incorporate amino acids into protein. This method uses

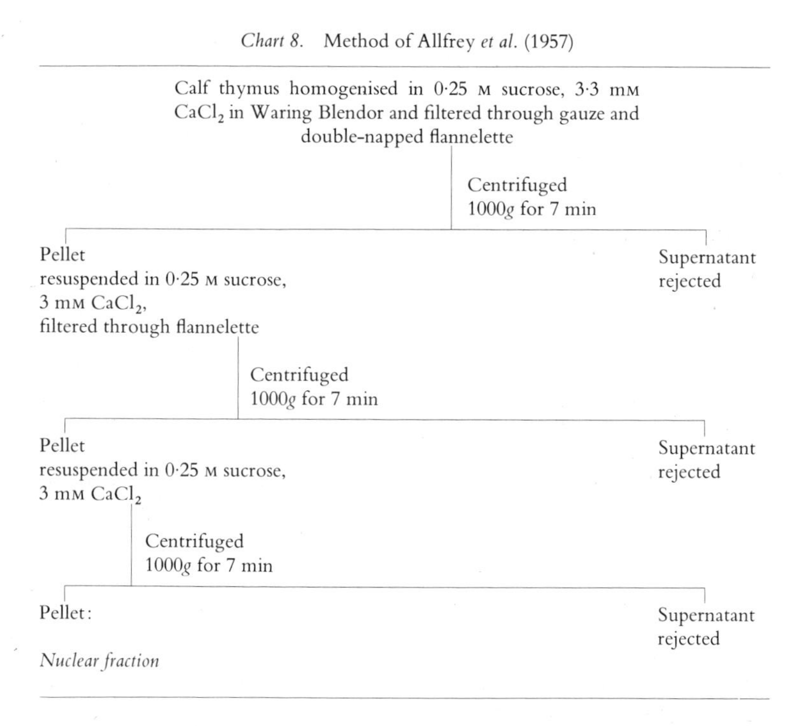

Chart 8. Method of Allfrey *et al.* (1957)

$CaCl_2$–sucrose media, the concentrations of $CaCl_2$ being higher than in the method of Hogeboom *et al.* (1952), and no layering is used. The Waring Blendor used was of special design and more efficient at breaking thymus tissue than the normal commercial blenders. The method of Allfrey *et al.* (1957) has been used for a large number of studies on the biosynthetic activity of nuclei, and there has been some discussion as to whether intact cells contribute to the observed activities. More recently, however, Allfrey *et al.* (1964) resolved this question by using autoradiography and an improved method of isolating nuclei, involving purification with dense sucrose solutions.

One of the objections to the use of $CaCl_2$ in the isolation of nuclei is that Ca^{2+} ions cause severe damage to mitochondria. Stirpe and Aldridge (1961) substituted $MgCl_2$ for $CaCl_2$ (*Chart 9*), but otherwise the method is very similar to that of Hogeboom *et al.* (1952),

Chart 9. Method of Stirpe and Aldridge (1961)

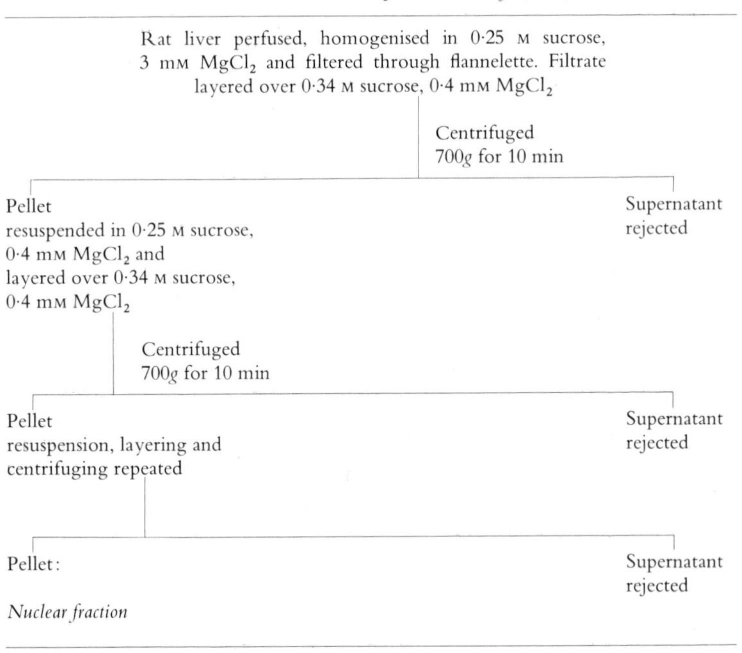

and it appears that Mg^{2+} ions are as effective as Ca^{2+} ions in preventing clumping and preserving nuclei. It is thus possible to develop fractionation schemes in which nuclei, mitochondria and other cytoplasmic particles may be isolated in a reasonable state of preservation from the same homogenate.

Because of the very high viscosity of 2·2 M sucrose used in the method of Chauveau *et al.* (1956), homogenisation of the tissue is rather difficult, and usually rather low cell breakages are obtained. Maggio *et al.* (1963), therefore, modified the method by homogenising the tissue in 0·88 M sucrose, 1·5 mM $CaCl_2$ and layering this homogenate over the dense sucrose. The scheme is shown in *Chart 10*. Their method was carefully monitored by electron microscopy and is without doubt one of the best for isolating nuclei currently available. An interesting point made by these authors is that isolated nuclear fractions are often *better* than they appear to be under the electron microscope, because much of the damage is inflicted not by the isolation procedure but by faulty fixation techniques. The isolated nuclei shown in their paper in fact retain many of the morphological features of the nucleus in the intact cell.

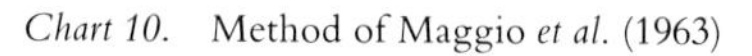

Chart 10. Method of Maggio *et al.* (1963)

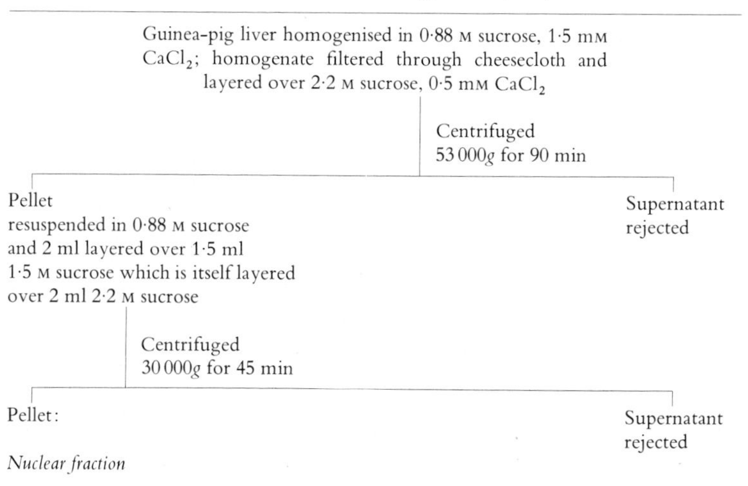

Thus it may well be that some at least of the recent improvements in the appearance of isolated nuclei may be ascribed more to improvements in electron microscopy techniques than to the isolation method.

There has been an increasing tendency to use neutral detergents during isolation of nuclei. For example, a rather complex method was described by Hubert *et al.* (1962) using neutral detergents, an Ultra–Turrax homogeniser and various mixtures of sucrose and Earle's medium for fractionation. The final nuclear fraction appeared to be very pure. Another method, of Rappoport *et al.* (1963), for the isolation of nuclei from rat brain (*Chart 11*), is simple and again gives nuclei that appear to be reasonably pure. A similar procedure for rat-liver or rat-kidney nuclei was described by Hymer and Kuff (1964). We do not yet have sufficient evidence as to their metabolic integrity, although they seem to be less damaged than one would have imagined. It appears that nuclei prepared in this way retain discrete nucleoli but have lost the outer membrane with attached ribosomes (Holtzman *et al.*, 1966). Neutral detergent destroys most, if not all, of the cytoplasmic membrane system and, in particular, probably results in the rupture of lysosomes, with concomitant release of their hydrolytic enzymes. Nevertheless, the detergent method is probably less drastic than the use of dilute acids such as citric acid, and the need for dense sucrose solutions is obviated.

Chart 11. Method of Rappoport *et al.* (1963)

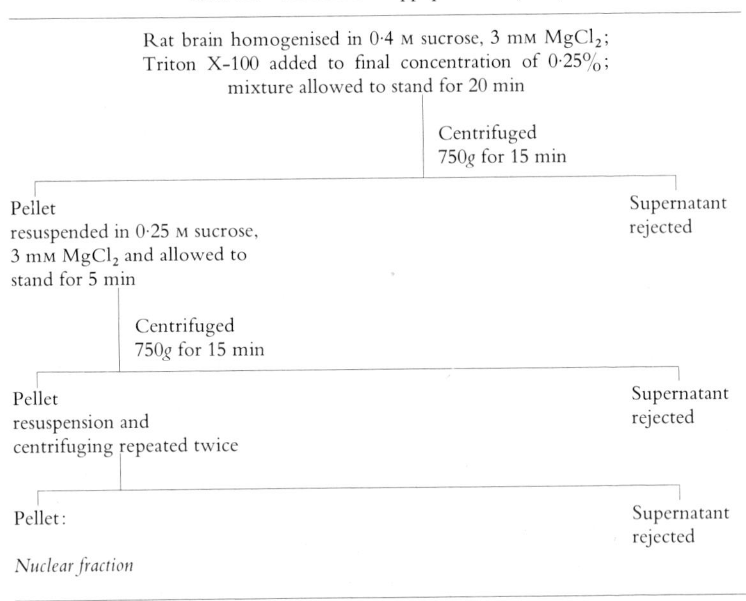

The remaining approaches are modifications and adaptations of those described so far. *Chart 12* summarises the method of Kodama and Tedeschi (1963) in which that of Allfrey *et al.* (1957) for the isolation of thymus nuclei is improved by introducing a stage in which the nuclei are centrifuged through 1·9 M sucrose. The procedure of Widnell and Tata (1964) (*Chart 13*), a hybrid of those of Chauveau *et al.* (1956) and Stirpe and Aldridge (1961), has been used by many workers recently. The crude nuclear fraction prepared in dilute sucrose–$MgCl_2$ solution is purified by layering over 2·4 M sucrose. This is probably preferable to the method of Maggio *et al.* (1963) in which the entire homogenate is layered over the dense sucrose.

Chart 12. Method of Kodama and Tedeschi (1963)

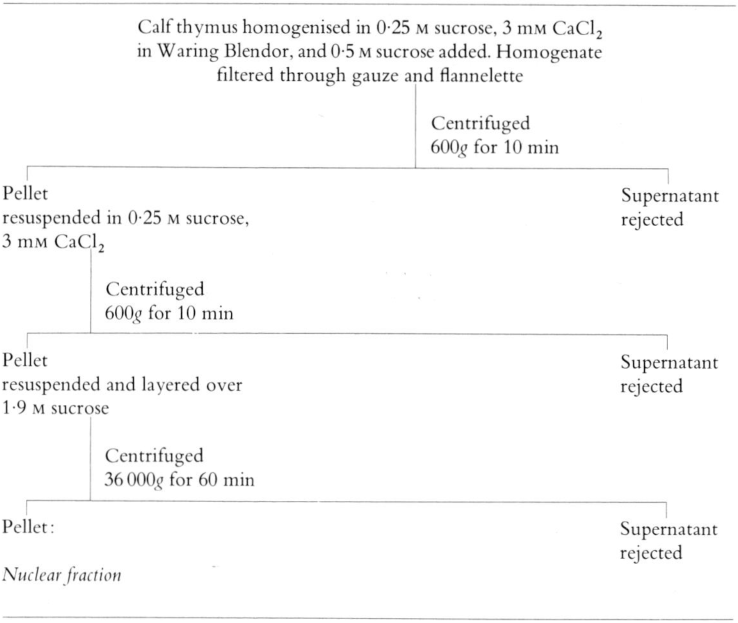

Calf thymus homogenised in 0·25 M sucrose, 3 mM $CaCl_2$ in Waring Blendor, and 0·5 M sucrose added. Homogenate filtered through gauze and flannelette

Centrifuged
600*g* for 10 min

Pellet
resuspended in 0·25 M sucrose, 3 mM $CaCl_2$

Supernatant
rejected

Centrifuged
600*g* for 10 min

Pellet
resuspended and layered over 1·9 M sucrose

Supernatant
rejected

Centrifuged
36 000*g* for 60 min

Pellet:

Nuclear fraction

Supernatant
rejected

Chart 13. Method of Widnell and Tata (1964)

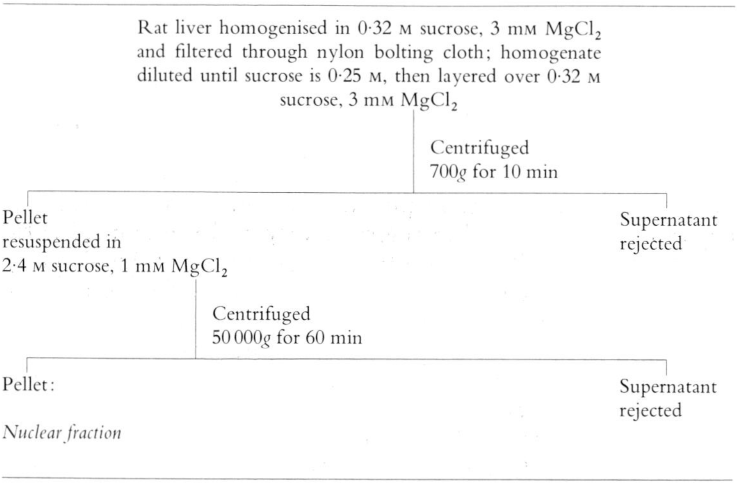

Rat liver homogenised in 0·32 M sucrose, 3 mM $MgCl_2$ and filtered through nylon bolting cloth; homogenate diluted until sucrose is 0·25 M, then layered over 0·32 M sucrose, 3 mM $MgCl_2$

Centrifuged
700*g* for 10 min

Pellet
resuspended in
2·4 M sucrose, 1 mM $MgCl_2$

Supernatant
rejected

Centrifuged
50 000*g* for 60 min

Pellet:

Nuclear fraction

Supernatant
rejected

That of Gill (1965) (*Chart 14*) is interesting because it revived the Dounce procedure of isolating nuclei at pH 6, combining this with the use of $CaCl_2$ and purification through dense sucrose solutions. It is noteworthy that very few of the methods so far described use buffered solutions for the isolation of nuclei. There is no doubt that even small variations of pH can have a profound effect on the morphology, chemical composition and metabolic

Chart 14. Method of Gill (1965)

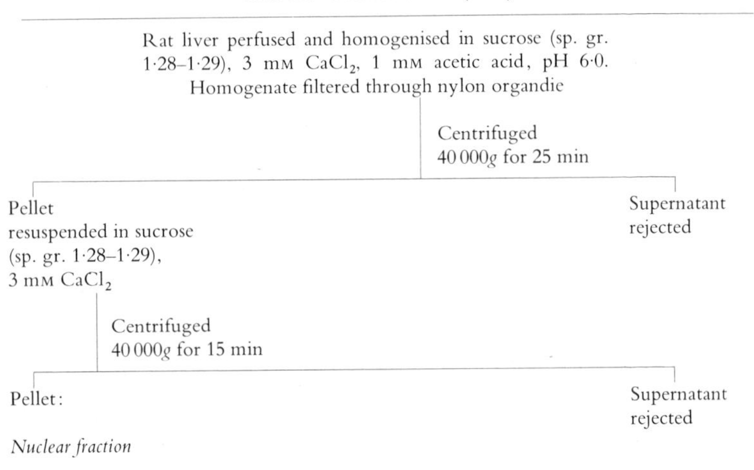

activity of the nucleus, so that it is a welcome development to see the reintroduction of a method for the isolation of nuclei at a known pH.

Blobel and Potter (1966) maintain the pH by the use of a tris-HCl buffer, in the presence of KCl and $MgCl_2$ (*Chart 15*). Purification of nuclei is effected by centrifuging through 2·3 M sucrose.

The method of *Chart 16* is included as a reminder of the difficulties that are still encountered in isolating nuclei from certain tissues. In order to obtain reasonably pure nuclei from the Walker tumour, Higashi *et al.* (1966) had to resort to very acid solutions of citric acid [2·5% (w/v) citric acid, pH 2·5–2·8] because of the difficulty of cell breakage. It is interesting that they used the zonal ultracentrifuge to separate RNA from their preparations.

Most of the above methods were primarily developed to isolate nuclei from normal mammalian liver, usually from the rat. Space does not allow a detailed comparison of methods for isolating nuclei from different tissues, and of course it would be most incorrect to imagine that those that are successful with rat liver would be of use with, for example, muscle or ascites-tumour cells. In general, the chief difficulty in isolating nuclei is in the cell breakage. Also, each tissue brings its special problems of contamination of the nuclear fraction. Nevertheless, the methods described above cover the broad span of techniques currently available for isolating

Chart 15. Method of Blobel and Potter (1966)

Rat liver homogenised in 0·25 M sucrose, 0·05 M tris-HCl, pH 7·5, 0·025 M KCl, 5 mM $MgCl_2$ (0·25 M sucrose, TKM). Homogenate filtered through cheesecloth. Filtrate mixed with 2 vol of 2·3 M sucrose, TKM to give final concn. of 1·62 M sucrose and layered over 2·3 M sucrose, TKM

Centrifuged
124 000*g* for 30 min

Pellet:

Nuclear fraction

Supernatant rejected

Chart 16. Method of Higashi *et al.* (1966)

Walker tumour fragmented in tissue press and homogenised in 2·5% (w/v) citric acid to give pH 2·5–2·8

Centrifuged
600*g* for 10 min

Pellet
resuspended in 0·25 M sucrose, 1·5% (w/v) citric acid and layered over 0·88 M sucrose, 1·5% (w/v) citric acid

Supernatant rejected

Centrifuged
900*g* for 10 min

Pellet:

Nuclear fraction

Supernatant rejected

nuclei, and special adaptations for different tissues generally exploit one or more of the findings with mammalian liver. However, since recent years have seen a widening of interest in the range of cell types that may be analysed by subcellular fractionation, it may be useful to discuss a few techniques for the isolation of nuclei from tissues other than mammalian liver.

For those interested in tumour nuclei (see *Chart 16* above), a method for preparing nuclei from HeLa cells with the use of detergents (Vaughan *et al.*, 1967) is outlined in *Chart 17*.

The flow-sheet given in *Chart 18* describes a method for the isolation of purified nuclei from mouse L cells grown in tissue culture in Eagle's medium (Schildkraut and Maio, 1968). Wray and Stubblefield (1970) have recently described the isolation of nuclei from mammalian fibroblasts using buffered 0·5 M hexylene glycol, 1 mM $CaCl_2$ as an isolation medium. Lipke *et al.* (1969) have

Chart 17. Method of Vaughan *et al.* (1967)

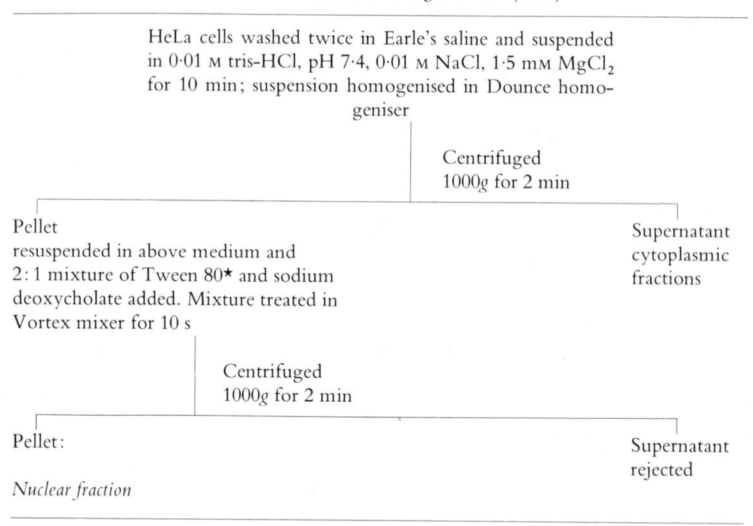

* polyoxyethylene sorbitol monopalmitate

described a rather complicated method for the isolation of nuclei from granulocytes. Fluorescein mercuric acetate was added to 'stabilise' the nuclei. The cells were then disrupted by sonication, followed by blending with glass beads. The nuclei were then washed successively with acetate buffer at pH 3·9 and cetylpyridinium chloride. This paper illustrates rather vividly the technical problems that may arise in nuclear isolation when one no longer uses rat or mouse liver as a source of nuclei.

Subcellular fractionation of brain presents considerable technical difficulty, mainly because of the heterogeneity of cell type in that tissue. Thus Kato and Kurokawa (1967) found it necessary to carry out a complex series of centrifugations, using a variety of sucrose and Ficol gradients, in order to rid nuclei from mammalian cerebral cortex of fragments of myelin, capillaries, erythrocytes and various cytoplasmic fragments. The range of cellular material and different nuclear types obtained during fractionation of brain homogenates is well illustrated in the method described by Løvtrup-Rein and McEwen (1966) for the isolation of rat-brain nuclei.

Finally, it should be mentioned that there is now increased interest in the isolation of nuclei from eukaryotic micro-organisms. Examples of some of the techniques used are as follows. Duffus (1969) described the isolation of nuclei from the yeast *Schizosaccharomyces pombe* using the Eaton press to disrupt the cells, and a sorbitol gradient for fractionation. The main problem with micro-organisms is to find a sufficiently mild, but efficient, method for

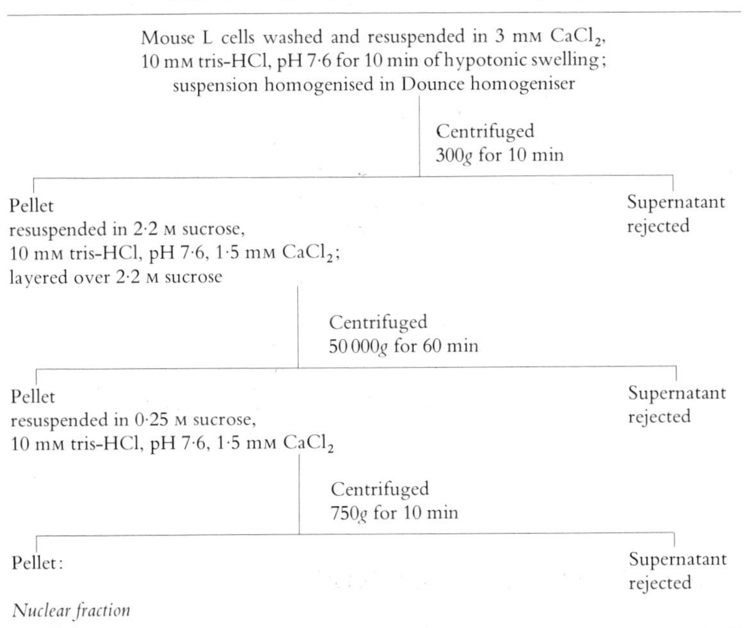

Chart 18. Method of Schildkraut and Maio (1968)

breaking the cells without damaging the nucleus. For example, Parenti *et al.* (1969) used a variety of methods to disrupt *Euglena gracilis*, including the French press and digestion with proteolytic enzymes, in order to isolate undamaged nuclei from this organism. Finally, Gorovsky (1970) made a detailed study of techniques for the isolation of both macro-nuclei and micro-nuclei from *Tetrahymena pyriformis*. The cells were disrupted with the Waring

Blendor in dilute gum arabic followed by the addition of 0·63% *n*-octanol. Using electron microscopy, Gorovsky examined the effect of different homogenising conditions, Ca^{2+} and Mg^{2+} ions, various buffers and a range of concentrations of *n*-octanol and gum arabic. These experiments illustrate the fact that there is still a great deal of work to be done in the development of suitable techniques for nuclear isolation from lower organisms.

There is little doubt that the zonal ultracentrifuge developed by N. G. Anderson and his associates will be of increasing use for isolating nuclei. Various fractionation schemes published during the development of this instrument showed clear separations between nuclei and various cytoplasmic components (e.g. Fisher and Cline, 1963). El-Aaser *et al.* (1966), in some pioneer studies, described the fractionation of a crude nuclear preparation from rat liver with an A-type zonal rotor. They were able to separate it into four fractions consisting of small vesicles, swollen mitochondria, sheets of plasma membrane and nuclei, respectively. The nuclear fraction, however, still contained some membranous material. The zonal ultracentrifuge has been used extensively in this Department to study nuclei from a range of tissues (see, for example, Johnston *et al.*, 1968; Haines *et al.*, 1969; Mathias *et al.*, 1969). The matter is discussed more fully by Johnston and Mathias in *Chapter 3* of this volume. There are now a variety of commercially available zonal rotors, and one may expect increasing use of this elegant technique for nuclear isolation. Not all the rotors, however, were specifically designed for isolation of nuclei, and Elrod *et al.* (1969) describe a plastic gradient-reorientating rotor (rotor A-XVI) that is particularly useful for isolation of nuclei. Nevertheless, the main problem with zonal centrifugation is still one of scale, since present instruments still require relatively large volumes of isolation medium and would be difficult to apply to the fractionation of milligramme amounts of tissue.

SUBFRACTIONATION OF NUCLEI

Although this article is mainly concerned with methods for isolating nuclei, some mention should be made of the increasing number of fractionations in which the nuclei, once isolated, are disrupted and further subfractionated. In particular, there has been increased emphasis recently on the isolation of nucleoli, and of the nuclear membrane system.

The role of the nucleolus in RNA synthesis, particularly of ribosomal RNA, has become increasingly clear (see, for example,

Vesco and Penman, 1968; Perry, 1969; Penman *et al.*, 1969; Hurlbert *et al.*, 1969). The nucleolus also appears to be considerably enriched in satellite DNA (Schildkraut and Maio, 1968; Jones, 1970). A variety of techniques have been employed for the isolation of nucleoli, and a representative method of nuclear fractionation is given in *Chart 19*. Unfortunately, the isolation of nucleoli is complicated by the fact that it appears that most tissues yield a heterogeneous population of nucleoli. For example, Kaufmann *et al.* (1968) found considerable variation in nucleolar size in normal and regenerating rat liver. Investigating the matter

Chart 19. Method of Schildkraut and Maio (1968) for nucleoli

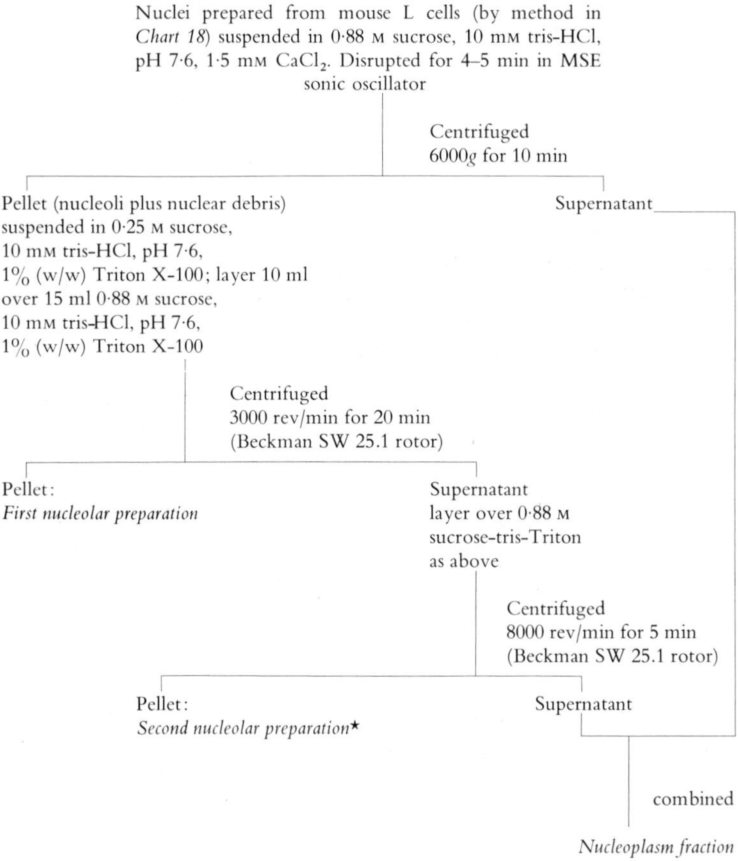

Nuclei prepared from mouse L cells (by method in *Chart 18*) suspended in 0·88 M sucrose, 10 mM tris-HCl, pH 7·6, 1·5 mM $CaCl_2$. Disrupted for 4–5 min in MSE sonic oscillator

Centrifuged 6000*g* for 10 min

Pellet (nucleoli plus nuclear debris) suspended in 0·25 M sucrose, 10 mM tris-HCl, pH 7·6, 1% (w/w) Triton X-100; layer 10 ml over 15 ml 0·88 M sucrose, 10 mM tris-HCl, pH 7·6, 1% (w/w) Triton X-100

Supernatant

Centrifuged 3000 rev/min for 20 min (Beckman SW 25.1 rotor)

Pellet: *First nucleolar preparation*

Supernatant layer over 0·88 M sucrose-tris-Triton as above

Centrifuged 8000 rev/min for 5 min (Beckman SW 25.1 rotor)

Pellet: *Second nucleolar preparation**

Supernatant

combined

Nucleoplasm fraction

* sometimes contaminated with fibrous chromatin

further, Johnston *et al.* (1969) isolated rat-liver nuclei of different ploidy by the zonal ultracentrifuge, and analysed the nucleoli obtained from them on sucrose gradients. They found considerable differences in the properties of the different classes of nucleoli, the most striking being that nucleoli from stromal cells had little labelled RNA after labelling *in vivo*, whereas tetraploid nuclei from the parenchymal cells yielded nucleoli whose RNA was most actively labelled. It is clear, therefore, that one must interpret with caution biochemical findings on nucleoli prepared from a mixture of cell types.

There have been several recent reports of the isolation of nuclear membranes from preparations of nuclei. Kashnig and Kasper (1969) first isolated rat-liver nuclei by the method by Blobel and Potter (1966) (see *Chart 15*), disrupted them ultrasonically and isolated nuclear-membrane and nucleolar fractions by centrifuging in a discontinuous sucrose gradient. Zbarsky *et al.* (1969) used similar methods, and observed that the nuclear membranes (identified by their characteristic pores) banded at a sucrose concentration having a density of 1·18 to 1·19 g/cm^3. They observed considerable oxidative activity in the nuclear membrane fraction, and suggested that it may be the site of nuclear oxidative phosphorylation. Zbarsky and co-workers also prepared nuclear membranes by extraction of intact nuclei with 0·2 M phosphate buffer at pH 7·2. Franke *et al.* (1970) used sonic vibrations to disrupt the nuclei and removed adhering nucleoproteins by washing with salt. (It is interesting that, in their characterisation of the nuclear membrane fraction, they noticed many similarities to endoplasmic reticulum membranes). Finally, Berezney *et al.* (1970) prepared nuclear membranes by prolonged digestion of bovine liver nuclei with deoxyribonuclease. In support of the findings of Zbarsky and co-workers, they observed cytochrome oxidase and NADH oxidase activity in the isolated membrane fraction.

CONCLUSIONS

I have attempted to survey the various methods available for nuclear isolation and given some indication of current work on subfractionation of nuclei. The account is not intended to be comprehensive and I apologise to those workers whose methods have been ignored. To a certain extent my choice of methods has been based on personal prejudice, but this is perhaps inevitable in a field as confused as this. Because of the difficulties outlined at the beginning, I have avoided a detailed assessment of the relative merits of the different methods. However, the main findings in Roodyn (1963) are probably still valid. It is to be hoped that future investigations will provide us with much greater information with which to assess the merits of the various techniques. Perhaps it is true to say that there is no perfect method, and the choice will depend on the reason for isolating the nuclei. Thus, if hypertonic media must be avoided and metabolic studies are to be performed, the techniques of Hogeboom *et al.* (1952) or Stirpe and Aldridge (1961) would be indicated. If metabolic activity is not under investigation but purity and ease of preparation are important, the citric acid procedures may still be relevant. Dense sucrose methods have been used to isolate nuclei for metabolic studies, but these are perhaps best suited to the study of nucleic acids or other chemical constituents of the nucleus. If the preservation of morphology is essential, more complex salts media, as in the method of Maver *et al.* (1952), may be required. Studies on soluble components may call for the use of organic solvents. Zonal centrifugation is probably essential for a rigorous analysis of a heterogeneous population of nuclei. Ideally, of course, there should be one method for isolating intact, uncontaminated nuclei suitable for metabolic and chemical studies. Until this is developed, we will have to select from the current methods those approaches which are most applicable to our particular problems. Since this will no doubt result in the publication of even more methods, we can probably expect more confusion in the subject before it is finally clarified. The confusion would be less, however, if adequate criteria for the assessment of the purity and structural and metabolic integrity of the isolated nuclei could be developed and generally applied.

APPENDIX

Details of three typical bulk methods for isolating rat-liver nuclei (see this Volume, Chapter 3 for details of the use of the zonal centrifuge).

Details of three methods for isolating rat-liver nuclei

I. *'Isotonic' sucrose-$CaCl_2$ technique* (Hogeboom *et al.*, 1952)—This technique gives somewhat variable results, depending on the degree of cell breakage. However, it is probably the most satisfactory for studies on enzyme systems that are inactivated by the high sucrose concentrations in the Chauveau method, since only 0·25–0·34 M sucrose is used.

1. Kill the rat and perfuse the liver with 0·15 M NaCl, either directly through the portal vein or by back-perfusion through the inferior vena cava. (The removal of most of the blood by careful perfusion is essential, otherwise serious contamination with

erythrocytes results.) Perfuse with homogenising medium for a short time.

2. Excise the liver, free it of gross connective tissue and blot it dry. Force it through a tissue press (or Latapie mincer) containing a mesh with 1 mm diam. holes and homogenise the mince in 9 vol of 0·25 M sucrose, 1·8 mM $CaCl_2$ for about 30 s in a Potter–Elvehjem homogeniser. (The optimum tightness of fit depends somewhat on the age of the liver, the amount of fibre and the power of the motor. It is best determined empirically with a series of homogenisers.)
3. Filter through one layer of sterile plain surgical lint and layer 10 ml of filtrate over 10 ml of 0·34 M sucrose, 0·18 mM $CaCl_2$. (It is most convenient to use centrifuge tubes made of precision-bore glass into which the head of the homogeniser can fit. 'Veridia' tubing (17 mm internal bore) is suitable for this purpose.)
4. Centrifuge for 10 min at $600g_{av}$, taking care to accelerate and decelerate the centrifuge as slowly as possible.
5. Reject the entire supernatant fluid to within about 2 mm of the pellet, preferably with a suction device. Resuspend in 3 ml of 0·25 M sucrose, 0·18 mM $CaCl_2$ by gentle homogenisation, layer over 10 ml of 0·34 M sucrose, 0·18 mM $CaCl_2$ and centrifuge again at $600g_{av}$ for 10 min.
6. Repeat the process of resuspension, layering and centrifuging as given in 5 once more, and resuspend the final pellet of nuclei in a suitable volume of 0·25 M sucrose, 0·18 mM $CaCl_2$.

II. *2·2 M sucrose technique* (Chauveau *et al.*, 1956)—The homogenate is prepared in 2·2 M sucrose, and on high-speed centrifuging the nuclei sediment and all the cytoplasmic contaminants float. The method is simple and gives nuclei with a good degree of purity.

1. Homogenise rat liver for 60–80 s in 2·2 M sucrose to give a 5% (w/v) homogenate using a Potter–Elvehjem homogeniser.
2. Centrifuge at $40\,000g_{max}$ for 1 h (Beckman No. 40 rotor, 21 000 rev/min).
3. Reject the supernatant fluid and resuspend the sediment of nuclei in a suitable volume of 2·2 M sucrose.

III. *2·2 M sucrose-$CaCl_2$ technique* (Maggio *et al.*, 1963)—This is a modification of the Chauveau technique and yields nuclei of excellent purity and with less clumping and aggregation.

1. As in the Chauveau method, perfusion is not essential, although there is no harm in using perfused liver. The liver is forced through a tissue press with 1 mm diam. holes and homogenised in 0·88 M sucrose, 1·5 mM $CaCl_2$ with a loose fitting Potter–Elvehjem homogeniser (0·009 in clearance) operating at 500 rev/min. Use 4 up and down movements.
2. Dilute to a 5% (w/v) homogenate and filter through 4 layers of cheesecloth.
3. Layer 20 ml of the filtrate over 10 ml of 2·2 M sucrose, 0·5 mM $CaCl_2$ in each of 3 centrifuge tubes of the Beckman SW 25.1 rotor. Centrifuge at 23 000 rev/min ($53\,000g_{av}$) for 90 min.
4. After centrifugation, most of the material will have remained above or at the interface between the 0·88 M and 2·2 M sucrose. A light translucent pellet of nuclei is at the bottom of the tube. Remove all the supernatant material by suction, wipe the inside of the centrifuge tube with cotton wool and resuspend the nuclei in 4 ml of 0·88 M sucrose per 3 g of liver originally taken.
5. If washing of the nuclei is required, layer 1·5–2·0 ml of the resuspended nuclei over a discontinuous gradient (1·5 ml of 1·5 M sucrose over 2 ml of 2·2 M sucrose) in 5 ml tubes of the Beckman SW 39L rotor. Centrifuge at 18 000 rev/min ($25\,000g_{av}$) for 45 min. Resuspend the final nuclear pellet in a suitable volume.

REFERENCES

ACKERMANN, D. (1904). *Hoppe-Seyler's Z. physiol. Chem.*, **43**, 299

ALBRECHT, C. F. (1968). *Expl Cell Res.*, **49**, 373

ALBRECHT, C. F. (1969). *Expl Cell Res.*, **56**, 44

ALLFREY, V. G., LITTAU, V. C. and MIRSKY, A. E. (1964). *J. Cell Biol.*, **21**, 213

ALLFREY, V. G., MIRSKY, A. E. and OSAWA, S. (1957). *J. gen. Physiol.*, **40**, 451

BALLAL, N. R., COLLINS, M. S., HALPERN, R. M. and SMITH, R. A. (1970). *Biochem. biophys. Res. Comm.*, **40**, 1201

BASHIRELAHI, N. and VILLEE, C. A. (1970). *Biochim. biophys. Acta*, **202**, 192

BEHRENS, M. (1932). *Hoppe-Seyler's Z physiol. Chem.*, **209**, 509

BEHRENS, M. and TAUBERT, M. (1952). *Hoppe-Seyler's Z. physiol. Chem.*, **291**, 213

BEREZNEY, R., FUNK, L. K. and CRANE, F. L. (1970). *Biochim. biophys. Res. Comm.*, **38**, 93

BETEL, I. (1969). *Arch. Biochem. Biophys.*, **134**, 271

BETEL, I. and KLOUWEN, H. M. (1967). *Biochim. biophys. Acta*, **131**, 453

BLOBEL, G. and POTTER, V. R. (1966). *Science, N.Y.*, **154**, 1662

BOLUND, L., RINGERTZ, N. R. and HARRIS, H. (1969). *J. Cell Sci.*, **4**, 71

BONDY, S. C., ROBERTS, S. and MORELOS, B. S. (1970). *Biochem. J.*, **119**, 665

BURNSTOCK, N. and PHILPOT, J. ST. L. (1959). *Expl Cell Res.*, **16**, 657

CHAUVEAU, J., MOULÉ, Y. and ROUILLER, C. H. (1956). *Expl Cell Res.*, **11**, 317

CHESTERTON, C. J. and BUTTERWORTH, P. H. W. (1971). *Eur. J. Biochem.*, **19**, 232

CROSSMAN, G. (1937). *Science, N.Y.*, **85**, 250

CURRIE, W. D., DAVIDIAN, N. MCC., ELLIOTT, W. B., RODMAN, N. F. and PENNIALL, R. (1966). *Archs Biochem. Biophys.*, **113**, 156

DE DUVE, C., BERTHET, J. and BEAUFAY, H. (1959). *Prog. Biophys. biophys. Chem.*, **9**, 325

DOUNCE, A. L. (1943). *J. biol. Chem.*, **147**, 685

DOUNCE, A. L. (1948). *Ann. N.Y. Acad Sci.*, **50**, 982

DOUNCE, A. L., TISHKOFF, G. H., BARNETT, S. R. and FREER, R. M. (1950). *J. gen. Physiol.*, **33**, 629

DOUNCE, A. L. and ICKOWICZ, R. (1969). *Arch. Biochem. Biophys.*, **131**, 210

DOUNCE, A. L. and ICKOWICZ, R. (1970). *Arch. Biochem. Biophys.*, **137**, 143

DOUNCE, A. L., SEAMAN, F. and MACKAY, M. (1966). *Arch. Biochem. Biophys.*, **117**, 550

DUFFUS, J. H. (1969). *Biochim. biophys. Acta*, **195**, 230

EL-AASER, A. A., FITZSIMONS, J. T. R., HINTON, R. H., REID, E., KLUCIS, E. and ALEXANDER, P. (1966). *Biochim. biophys. Acta*, **127**, 553

ELROD, L. H., PATRICK, L. G. and ANDERSON, N. G. (1969). *Analyt. Biochem.*, **30**, 230

FALZONE, J. A., BARROWS, C. H. and YIENGST, M. J. (1962). *Expl Cell Res.*, **26**, 552

FISHER, W. D. and CLINE, G. B. (1963). *Biochim. biophys. Acta*, **68**, 640

FRANKE, W. W., DEUMLING, B., ERMEN, B., JARASCH, E.-D., and KLEINIG, H. (1970). *J. Cell Biol.*, **46**, 379

FRIEDMAN, D. L. (1970). *Biochem. biophys. Res. Comm.*, **39**, 100

FUJIMURA, S., HASEGAWA, S., SHIMIZU, Y. and SUGIMURA, T. (1967). *Biochim. biophys. Acta*, **145**, 247

GALLWITZ, D. (1970). *Biochem. biophys. Res. Comm.*, **40**, 236

GEORGIEV, G. P. (1967). In *Enzyme Cytology* (Ed. Roodyn, D. B.), p. 27, London (Academic Press)

GILL, D. M. (1965). *J. Cell biol.*, **24**, 157

GOLDSTEIN, L. and TRESCOTT, O. H. (1970). *Proc. natn Acad. Sci., U.S.*, **67**, 1367

GOROVSKY, M. A. (1970). *J. Cell Biol.*, **47**, 619

HAINES, M. E., JOHNSTON, I. R. and MATHIAS, A. P. (1970). *FEBS Lett.*, **10**, 113

HAINES, M. E., JOHNSTON, I. R., MATHIAS, A. P. and RIDGE, D. (1969). *Biochem. J.*, **115**, 881

HASEGAWA, S., FUJIMURA, S., SHIMIZU, Y. and SUGIMURA, T. (1967). *Biochim. biophys. Acta*, **149**, 369

HIGASHI, K., NARAYANAN, K. S., ADAMS, H. R. and BUSCH, H. (1966). *Cancer Res.*, **26**, 1582

HOGEBOOM, G. H. and SCHNEIDER, W. C. (1952). *J. biol. Chem.*, **197**, 611

HOGEBOOM, G. H., SCHNEIDER, W. C. and STRIEBICH, M. J. (1952). *J. biol. Chem.*, **196**, 111

HOLTZMAN, E., SMITH, I. and PENMAN, S. (1966). *J. molec. Biol.*, **17**, 131

HURLBERT, R. B., MILLER, E. G. and VAUGHAN, C. L. (1969). In *Advances in Enzyme Regulation* (Ed. Weber, G.), vol. 7, p. 219, Oxford (Pergamon)

HUBERT, M. T., FAVARD, P., CARASSO, N., ROZENCWAJ, G. R. and ZALTA, J. P. (1962). *J. Microscopie*, **1**, 435

HYMER, W. C. and KUFF, E. L. (1964). *J. Histochem. Cytochem.*, **12**, 359

JOHNSTON, I. R., MATHIAS, A. P., PENNINGTON, F. and RIDGE, D. (1968). *Biochem. J.*, **109**, 127

JOHNSTON, I. R., MATHIAS, A. P., PENNINGTON, F. and RIDGE, D. (1969). *Biochim. biophys. Acta*, **195**, 563

JONES, K. W. (1970). *Nature, Lond.*, **225**, 912

KASHNIG, D. M. and KASPER, C. B. (1969). *J. biol. Chem.*, **244**, 3786

KATO, T. and KUROKAWA, M. (1967). *J. Cell Biol.*, **32**, 649

KAUFMANN, E., TRAUB, A. and TEITZ, Y. (1968). *Expl Cell Res.*, **49**, 215

KIRSCH, W. M., LEITNER, J. W., GAINEY, M., SCHULZ, D. W., LASHER, R. and NAKANE, P. (1970). *Science, N.Y.*, **168**, 1592

KODAMA, R. M. and TEDESCHI, H. (1963). *J. Cell Biol.*, **18**, 541

LADDA, R. L. and ESTENSEN, R. D. (1970). *Proc. Natn. Acad. Sci., U.S.*, **67**, 1528

LIAO, S., SAGHER, D. and FANG, S. M. (1968). *Nature, Lond.*, **220**, 1336

LINDSAY, J. G., BERRYMAN, S. and ADAMS, R. L. P. (1970). *Biochem. J.*, **119**, 839

LIPKE, H., HESTER, R., GROFF, W. and PETRALI, J. (1969). *J. cell. Physiol.*, **73**, 93

LØVTRUP-REIN, H. (1970). *Brain Res.*, **19**, 433

LØVTRUP-REIN, H. and MCEWEN, B. S. (1966). *J. Cell Biol.*, **30**, 405

LYNCH, W. E., BROWN, R. F., UMEDA, T., LANGRETH, S. G. and LIEBERMAN, I. (1970). *J. biol. Chem.*, **245**, 3911

MAGGIO, R., SIEKEVITZ, P. and PALADE, G. E. (1963). *J. Cell Biol.*, **18**, 267

MARSHAK, A. (1941). *J. gen. Physiol.*, **25**, 275

MATHIAS, A. P., RIDGE, D. and TREZONA, N. ST. G. (1969). *Biochem. J.*, **111**, 583

MAVER, M. E., GRECO, E. A., LOVTRUP, E. and DALTON, A. J. (1952). *J. Natn Cancer Inst.*, **13**, 687

MCBRIDE, O. W. and PETERSON, E. A. (1970). *J. Cell Biol.* **47**, 132

MIESCHER, F. (1871). *Medizinische-Chemische Untersuchungen*, vol. 4, p. 441, Berlin (Hirschwald)

MOULÉ, Y. (1970). *Eur. J. Biochem.*, **17**, 544

NISHIZUKA, Y., UEDA, K., NAKAZAWA, K. and HAYAISHI, O. (1967). *J. biol. Chem.*, **242**, 3164

ORD, M. J. and BELL, L. G. E. (1970). *Nature, Lond.*, **226**, 854

PARENTI, F., BRAWERMAN, G., PRESTON, J. F. and EISENSTADT, J. M. (1969). *Biochim. biophys. Acta*, **195**, 235

PENMAN, S., VESCO, C., WEINBERG, R. and ZYLBER, E. (1969). *Cold Spring Harbor Symp. Quant. Biol.*, **34**, 535

PERRY, R. P. (1969). In *Handbook of Molecular Cytology* (Ed. Lima-de-Faria, A.), p. 620, Amsterdam (North Holland)

POGO, A. O., LITTAU, V. C., ALLFREY, V. G. and MIRSKY, A. E. (1967). *Proc. natn. Acad. Sci., U.S.*, **57**, 743

POGO, B. G. T., POGO, A. O., ALLFREY, V. G. and MIRSKY, A. E. (1968). *Proc. natn. Acad. Sci., U.S.*, **59**, 1337

PRICE, C. E. (1970). *Cytobios*, **2**, 119

PLOZ (1871). *Medizinische-Chemische Untersuchungen*, vol. 4, p. 461, Berlin (Hirschwald)

RAPPOPORT, D. A., FRITZ, R. R. and MORACZEWISKI, A. (1963). *Biochim. biophys. Acta*, **74**, 42

READ, R. S. D. and MAURITZEN, C. M. (1970). *Can. J. Biochem.*, **48**, 559

REEDER, R. H., UEDA, K., HONJO, T., NISHIZUKA, Y. and HAYAISHI, O. (1967). *J. biol. Chem.*, **242**, 3172

RINGERTZ, N. R. (1969). In *Handbook of Molecular Cytology* (Ed. Lima-de-Faria, A.), p. 656, Amsterdam (North Holland)
ROEDER, R. G. and RUTTER, W. J. (1970). *Proc. natn. Acad. Sci., U.S.*, **65**, 675
ROODYN, D. B. (1956). *Biochem. J.*, **64**, 361
ROODYN, D. B. (1959). *Int. Rev. Cytol.*, **8**, 279
ROODYN, D. B. (1963). *Biochem. Soc. Symp.*, **23**, 20
ROODYN, D. B. (1965). *Int. Rev. Cytol.*, **18**, 99
ROODYN, D. B. (1967). *Enzyme Cytology*, London (Academic Press)
SCHILDKRAUT, C. L. and MAIO, J. J. (1968). *Biochim. biophys. Acta*, **161**, 76
SCHNEIDER, R. M. and PETERMANN, M. L. (1950). *Cancer Res.*, **10**, 751
SCHNEIDER, W. C. (1948). *J. biol. Chem.*, **176**, 259
SEE, Y. P. and FITT, P. S. (1970). *Biochem. J.*, **119**, 517
SIMARD, R. (1970). *Int. Rev. Cytol.*, **28**, 169
SMITH, F. G. (1856). Appendix to 1st American ed. of Carpenter, W. B. *The Microscope*, Philadelphia (Blanchard & Lea)
SNEIDER, T. W., BUSHNELL, D. E. and POTTER, V. R. (1970). *Cancer Res.*, **30**, 1867
STIRPE, F. and ALDRIDGE, W. N. (1961). *Biochem. J.*, **80**, 481
STONEBURG, C. A. (1939). *J. biol. Chem.*, **129**, 189
SWIFT, H. (1969). *Genetics*, **61** (Suppl.), 439
SZARSKI, H. (1970). *Nature, Lond.*, **226**, 651
VAUGHAN, M. H., WARNER, J. R. and DARNELL, J. E. (1967). *J. molec. Biol.*, **25**, 253
VESCO, C. and PENMAN, S. (1968). *Biochim. biophys. Acta*, **169**, 188
WAGAR, M. A., BURGOYNE, L. A. and ATKINSON, M. R. (1971). *Biochem. J.*, **121**, 803
WARBURG, O. (1910). *Hoppe-Seyler's Z. physiol. Chem.*, **70**, 413
WESER, U. (1970). *Z. Klin. Chem.*, **8**, 137
WIDNELL, C. C. and TATA, J. R. (1964). *Biochem. J.*, **92**, 313
WILBUR, K. M. and ANDERSON, N. G. (1951). *Expl Cell Res.*, **2**, 47
WILHELM, J. A., SPELSBERG, T. C. and HNILICA, L. S. (1971). *Sub-cell. Biochem.*, **1**, 39
WRAY, W. and STUBBLEFIELD, E. (1970). *Expl Cell Res.*, **59**, 469
ZBARSKY, I. B., PEREVOSHCHIKOVA, N., DELEKTORSKAYA, L. N. and DELEKTORSKY, V. V. (1969). *Nature, Lond.*, **221**, 257
ZENTGRAF, H., DEUMLING, B. and FRANKE, W. W. (1969). *Expl Cell Res.*, **56**, 333
ZIGMOND, R. E. and MCEWEN, B. S. (1970). *J. Neurochem.*, **17**, 889

3

THE BIOCHEMICAL PROPERTIES OF NUCLEI FRACTIONATED BY ZONAL CENTRIFUGATION

I. R. Johnston and A. P. Mathias

The function of the nucleus in the determination of heredity and its role in protein synthesis explain the intense interest in this organelle. Many biochemists in their studies of the nucleus have used rat liver, and often they have glossed over the complex cellular structure of this organ. It consists of a continuous mass of parenchymal cells through which tunnel a network of capillaries (the sinusoids) (Elias and Shemick, 1969). These convey predominantly venous blood from the gastrointestinal tract to the heart. The capillaries are formed chiefly of flattened overlapping littoral cells, the Kupffer cells, active in phagocytosis, and some other cells including fat storage cells. Connective tissue cells are located in the pericapillary space. The parenchymal cells are organised in a system of inter-connected walls or plates one cell thick—known as a muralium. The bile canaliculi have walls which are formed of grooves in adjacent liver cells. Along the margins of these grooves the paren-chymal cells are in tight contact. The canaliculi form a continuous net of polygonal meshes, each mesh embracing a parenchymal cell. They converge into ducts of gradually increasing size lined with bile duct cells. Finally there are the cells of the walls of main blood vessels. In adult rat liver the parenchyma comprises 90% of the mass but only 60% of the cell population (Fukuda and Sibatani, 1953).

Whilst the non-parenchymal or stromal cells are diploid, the parenchyma of most rodents is polyploid and in adult mice hexadecaploids are quite common. An appreciable proportion of the parenchymal cells are binucleate, which may represent a stage in the development of polyploidy (Carriere, 1969). It is important to assess the effects of polyploidy and the cellular origin on the biochemical activities of liver nuclei. Accordingly, we have sought methods for the separation of nuclei which differ in their content of DNA and in their morphology. Because of the requirement for adequate amounts of material for biochemical analyses, we have concentrated on the use of zonal centrifuges which offer high capacity, convenience and speed of working (Anderson, 1966).

ZONAL CENTRIFUGATION

CONDITIONS FOR CENTRIFUGATION

Our preliminary experiments revealed that good resolution of the nuclei in crude homogenates or from intermediate stages in the purification could not be achieved. Satisfactory separations necessitate the use of nuclei that have been centrifuged through 2·2 M sucrose. Normally we have used nuclei isolated by the modification of the procedure of Chaveau devised by Widnell and Tata (1964), which yields preparations that retain good biochemical activity. It is important to control the pH and the concentration of divalent metal ions in the solutions of sucrose used both for the purification and the zonal centrifugation of the nuclei. A pH of 7·4 and 1 mM $MgCl_2$ give nuclei that are well rounded and show a minimal tendency to aggregation. At lower pH values the nuclei become crinkled and tend to aggregate. In the absence of Mg^{2+}, the nuclei swell and many nuclei are ruptured. Losses, amounting to 40% or more of the cellular DNA, occur in the purification of the nuclei. These losses arise mostly from incomplete breakage of cells and trapping in supernatant plugs, but they are apparently not selective from any particular type of cell.

The standard conditions for nuclear separations that we employ (Johnston *et al.*, 1968) involve a gradient of sucrose (1000 ml) ranging from 20 to 50% (w/w) containing 1 mM $MgCl_2$ and adjusted to pH 7·4 with $NaHCO_3$ (final concentration approx. 1 mM). Higher concentrations of $MgCl_2$ cause aggregation and lead to blurring of the separations. This gradient was chosen so that it would be sufficiently concentrated to prevent excessive sedimentation of the nuclei during the displacement from the rotor at the end of the run, and sufficiently steep to ensure stability of the zones. The

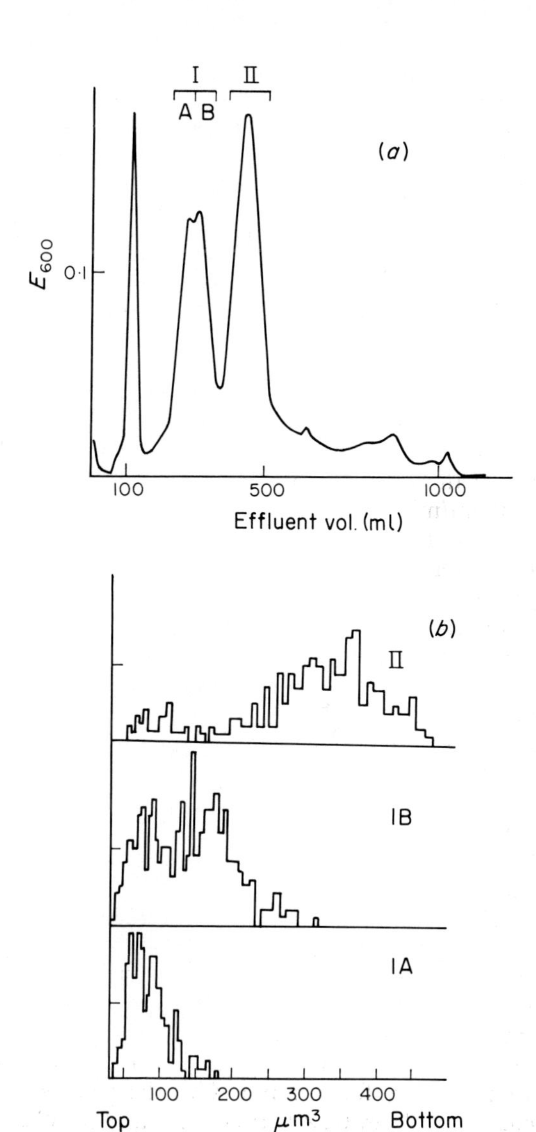

Figure 1. (a) *Elution profile of rat-liver nuclei isolated from Norwegian hooded rats weighing 200 g. After centrifugation in the AXII zonal rotor using the conditions described in the text the contents of the rotor were displaced through a flow cell (1 cm path length) mounted in a Gilford recording spectrophotometer and monitored at 600 nm. After collection, the fractions were pooled as indicated.* (b) *Histograms of nuclear volumes calculated from measurements of the nuclear diameters made with a phase-contrast microscope fitted with a micrometer eyepiece. The bar on the ordinate represents ten nuclei*

gradient is linear with respect to volume and is introduced into the AXII rotor at 50 ml/min, whilst it is spinning at 600 rev/min in an MSE Mistral 6L centrifuge. The underlay is 55% (w/w) sucrose and the temperature is maintained at 5°C. The nuclei equivalent to 6 g wet weight of liver are suspended in 10 to 20 ml of 15% buffered sucrose containing 1 mM $MgCl_2$. Care must be taken to ensure that the nuclei are completely drained of the 2·2 M sucrose used in the last stage of their purification before preparing the sample suspension otherwise the density of the latter may exceed that of the top of the gradient and cause instability. If the weight of sample is much increased, the concentration of sucrose used to suspend the nuclei must be reduced to 10% (w/w). The sample is followed by an overlay of 10% (w/w) sucrose which is pumped in until the sample zone is 6 cm from the axis of rotation. After 1 h at 600 rev/min, the contents of the rotor are displaced by pumping 55% (w/w) sucrose at 50 ml/min to the edge of the rotor. The effluent is monitored at 600 nm using a flow cell mounted in a recording spectrophotometer. A typical profile is shown in *Figure 1*. The peak on the extreme left is caused mainly by the interface between the overlay and the sample. In addition, there is a small amount of material, mainly membranous, that remains at the interface and marks the initial position of the sample. Apart from this, two major peaks are seen. The slower-moving is a double peak. It is often seen as single peak, either with a pronounced shoulder or, less frequently, with a dip at the peak, as in *Figure 1*. If a washed pellet which has been sedimented at 700*g* is used as the sample, instead of the purified nuclei, poor resolution is achieved and cell debris and unbroken cells are strewn throughout the gradient.

The nuclei are collected manually in fractions of 40 ml, or automatically in fractions of 12·5 ml in a fraction cutter modified to eliminate spillage during the change over from one tube to the next. When necessary, the smaller fractions can be combined in appropriate groups as indicated by the optical density trace. The nuclei are recovered from the fractions by diluting them with an equal volume of 0·32 M sucrose, 3 mM $MgCl_2$, pH 7·4. After sedimentation at 1000*g* for 5 min at 4°C, the pelleted nuclei are resuspended, normally in 0·32 M sucrose, 1 mM $MgCl_2$, or some other solution suitable for a particular purpose.

CHARACTERISATION OF NUCLEI

The characterisation of the nuclei from the zones shown in *Figure 1* requires both microscopical examination and chemical analyses.

The numbers of nuclei in each fraction are determined after suitable dilution of the fractions with saline, or sucrose–saline, using either a Coulter particle counter with a 100 μm orifice, or a haemocytometer. Diameters of nuclei may be measured by the use of a micrometer eyepiece under phase contrast, with an oil immersion lens at an overall magnification of 1000 ×. Stromal nuclei are distinguished from parenchymal nuclei as they are smaller, and many are ellipsoidal. Their volumes (V) are calculated from the formula for prolate spheroids:

$$V = \frac{\pi}{6} ab^2$$

where the lengths of the major and minor axes are a and b, respectively.

The content of DNA and RNA are measured most conveniently by adding $HClO_4$, to a final concentration of 10% (w/v), directly to a suspension of a known number of nuclei in dilute sucrose. The precipitate is collected and analysed by standard procedures. Protein may be determined in the same way. The results are expressed in picogrammes per nucleus.

The analyses for liver nuclei prepared from Norwegian hooded rats weighing 200 g, the fractionation of which is shown in *Figure 1*, are summarised in *Table 1*. It is obvious that zone 1 contains diploid nuclei and zone 2 tetraploid nuclei. Partial resolution of the stromal from the parenchymal diploid nuclei is achieved, the slower-moving part of zone 1 (1A) containing a

Table 1 CHARACTERISATION OF FRACTIONS OF RAT-LIVER NUCLEI OBTAINED BY ZONAL CENTRIFUGATION*

	Zone 1A	*Zone 1B*	*Zone 2*
Mean nuclear volume (μm^3)	80	180	350
DNA per nucleus (pg)†	7·5	7·5	14·7
Corrected value (pg)‡	7·5	7·5	15·7

* Fractionation shown in *Figure 1*.
† DNA was measured by the method of Burton as described by Widnell and Tata (1964).
‡ This makes allowance for contamination of zone 2 by diploid nuclei.

considerable proportion of ellipsoidal nuclei. Nuclei from rat kidney, which are known to be virtually exclusively diploid, yield a single peak corresponding in position to zone 1 of the liver nuclei. This behaviour, together with analyses for DNA and the ratios of the nuclear volumes, confirms the identification of zone 1 and zone 2 in *Figure 1* as corresponding to diploid and tetraploid nuclei, respectively.

The relative proportion of diploid and tetraploid nuclei changes with age. Nuclei from livers of new-born rats are nearly all diploid. The tetraploid peak becomes distinguishable between the fourth and fifth week after birth. The proportion of tetraploid nuclei increases progressively with age and very old rats have appreciable proportions of octaploid nuclei (Johnston *et al.*, 1968*a*).

When nuclei from the livers of mice of the N.I.H. strain were examined, three main zones and one minor one were observed (*Figure 2*). We have shown that these correspond to a diploid zone, in which partial separation of stromal and parenchymal nuclei occurs, a tetraploid zone, an octaploid zone and finally a zone in

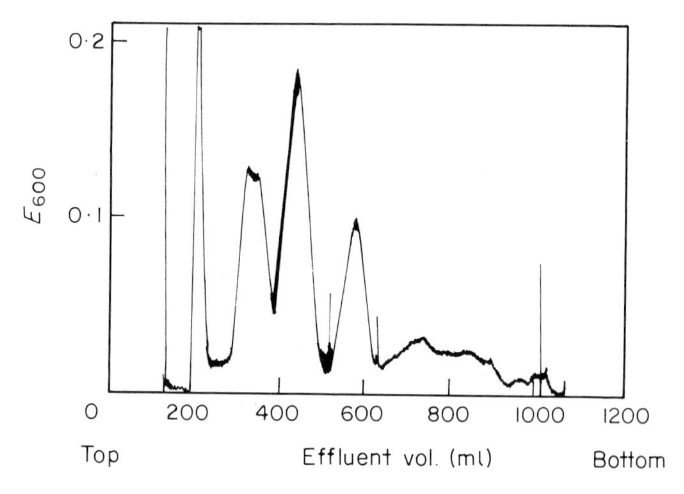

Figure 2. Profile of liver nuclei from adult albino mice of the N.I.H. strain. Twelve animals (average weight, 27 g) gave 13·6 g of liver

which many nuclei of the size expected of hexadecaploid nuclei are found. It is evident that zonal centrifugation affords an excellent method for fractionating nuclei according to their degree of ploidy. The contamination of the faster-moving zone by nuclei of the next lowest level of ploidy may be explained by the tendency of the larger nuclei, as they start to form a discrete zone, to carry along with them some of the smaller nuclei. Possibly this is assisted by a reversible association of nuclei. Also, nuclei which possess the same amount of DNA but differ in shape may be separated partially. The degree of separation is influenced by the pH (*Figure 3*).

Lowering the pH improves the resolution of the diploid stromal from the diploid parenchymal nuclei. Raising the concentration of Mg^{2+} in the gradient to 3 mM has the same effect. However, this gain must be set against the disadvantage that many nuclei are lost by increased aggregation at low pH or high concentration of Mg^{2+}.

The upper and lower faces of the A-series of zonal rotors are made of perspex, thus permitting direct observation during centrifugation of the movement and positions of the migrating zones. This

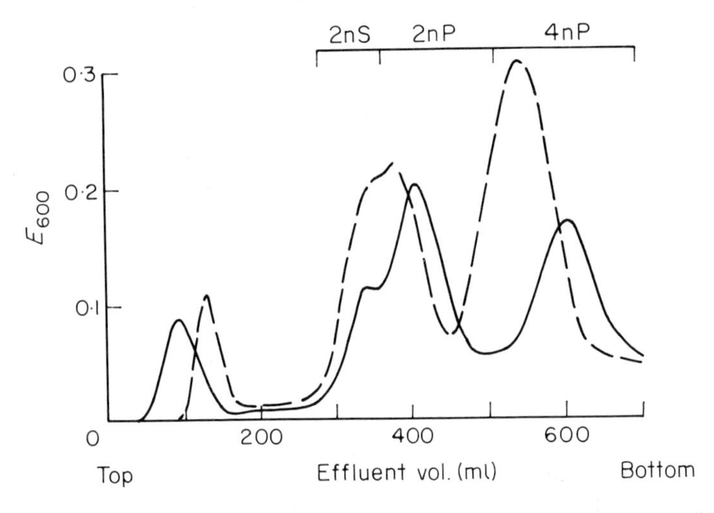

Figure 3. Comparison of the elution profiles of rat-liver nuclei centrifuged in a gradient buffered at pH 6 with 0·001 M sodium cacodylate–HCl (———) and pH 7·4 (— — —); 2nS, diploid stromal; 2nP, diploid parenchymal; 4nP, tetraploid parenchymal nuclei

enables measurements to be made of the rate of movement of the particles at various positions in the gradient. As a consequence, the zonal rotor may be used to provide analytical information. An example of this is the determination of the density of rat-liver nuclei. Because of their high density, it is not practicable to do this directly by an isopycnic experiment using sucrose solutions as the medium. However, by carrying out runs in a variety of gradients, including 50 to 69% (w/w) sucrose at 2500 rev/min for 6·5 h, and observing their rate of migration, it is possible to compute the density of the nuclei at different points of known sucrose concentration (equivalent to density). The density of the nuclei varies with the density of the medium, presumably because they are permeable to sucrose. Extrapolation of a plot of $\rho_p - \rho_m$ against ρ_m (*Figure 4*) to zero provides an estimate of 1·35 g/cm^3 as the buoyant density of rat-

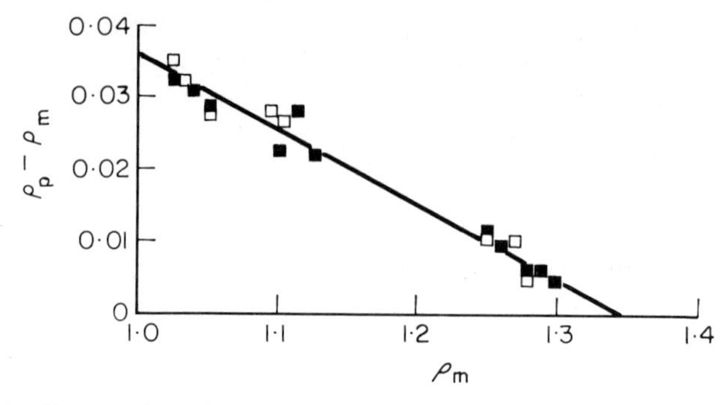

Figure 4. Variation of particle density (ρ_p) with density of the medium (ρ_m) for diploid nuclei (□) and tetraploid nuclei (■) at 5°C

liver nuclei (both diploid and tetraploid) in sucrose. Calculations suggest that approximately 10% of the volume of the nuclei is impermeable to sucrose.

BIOCHEMICAL PROPERTIES

The results and the advantage, if any, of polyploidy are largely a matter of conjecture. The availability of a technique for fractionating nuclei enables us to investigate the biochemical consequences within the nucleus of doubling its content of DNA. Also, we can discover to what extent nuclei of different cells within the same tissue vary in their enzymic composition. Another question that is now opened to study is the homogeneity or otherwise of a given type of nucleus. We have examined the distribution of a number of nuclear enzymes within the zonal profiles; our findings are summarised in the following sections.

SYNTHESIS OF DNA

The synthesis of a macromolecule within a subcellular organelle may be studied in two ways. One method is to investigate the activities *in vitro* of the enzymes supposedly implicated in the synthesis and determine their specific activities. The alternative is to establish the actual capacity for synthesis within the intact organ. The two approaches do not necessarily lead to the same conclusions. This discrepancy is marked in the consideration of the synthesis of DNA.

A standard procedure for examining the synthesis of DNA

within the tissues is to inject [^{3}H]thymidine and follow its incorporation. When rats are given an intraperitoneal dose of labelled thymidine and killed within two h, that is, a period considerably shorter than the duration of S phase, the nuclei are labelled to an extent that depends on the age of the animal. The labelled nuclei may then be examined by zonal centrifugation. At the end of the experiment, the effluent from the zonal rotor is collected in fractions and, after removal of a sample for the determination of the numbers of nuclei, the radioactivity in each fraction is estimated by liquid scintillation counting of the recovered nuclei. The results are expressed either directly as counts/min in each fraction, or as specific activities calculated as incorporation per 10^6 or 10^9 nuclei.

In liver nuclei from rats weighing 25 g a zone of nuclei which have incorporated [^{3}H]thymidine is seen; however, this zone sediments faster than the main zone of nuclei (*Figure 5*). With older

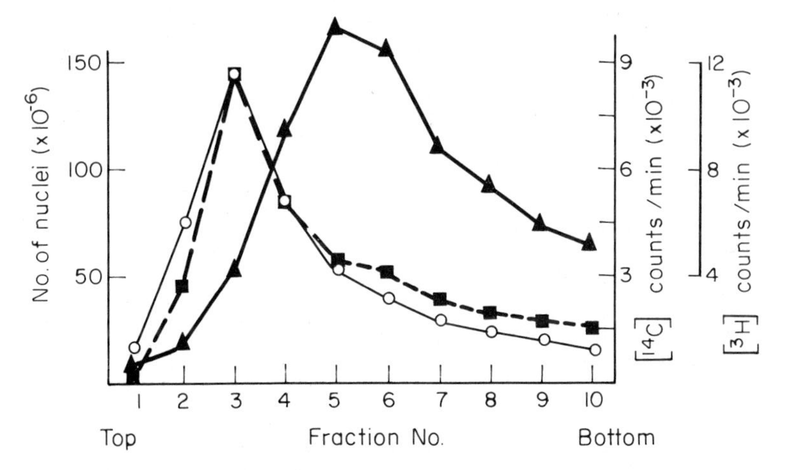

Figure 5. Synthesis of DNA and RNA in vivo *in the livers of rats weighing 25 g. Five animals each received 1 μCi of [6–^{14}C]orotic acid and 30 μCi of [^{3}H]thymidine intraperitoneally 70 min before death.* (○), *Number of nuclei;* (■), *incorporation of [^{14}C]orotic acid;* (▲), *incorporation of [^{3}H]thymidine*

rats (*Figure 6*), in which some of the parenchymal cells of the liver are tetraploid, there are two peaks of incorporation of [^{3}H]thymidine, one between the diploid and tetraploid zones and the second sedimenting ahead of the tetraploid nuclei (Johnston *et al.*, 1968*b*; Haines *et al.*, 1969). Liver nuclei from adult mice, because of their greater degree of polyploidy, give a more complex pattern (*Figure 7*). These patterns of DNA synthesis *in vivo* can be understood when it is realised that nuclei engaged in the replication of their chromatin will have amounts of DNA and protein intermediate between that of the class of nuclei from which they are derived and the next highest ploidy class. Zonal centrifugation offers a means of greatly concentrating these nuclei, although comparatively few of the nuclei found in the zones of incorporation of [^{3}H]thymidine are actually in S phase.

The synthesis of DNA has also been investigated by examining the distribution of DNA polymerase among fractionated nuclei

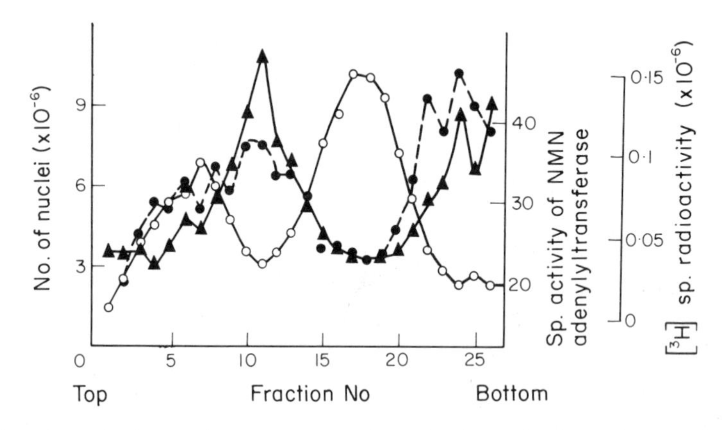

Figure 6. Distribution of NMN adenyltransferase (NAD pyrophosphorylase) activity and synthesis of DNA in vivo *in liver nuclei in rats weighing 200 g.* (●), *NMN adenylyltransferase activity (μmoles of NAD synthesised/h per 10^9 nuclei);* (▲), *[^{3}H]thymidine incorporation (counts/min per 10^9 nuclei);* (○)*; number of nuclei*

(Haines *et al.*, 1970). For this purpose, the conditions of zonal centrifugation were modified to favour the retention of maximum enzyme activity in the nuclei after recovery from the rotor. The gradient consisted of 26 to 66% (w/w) glycerol containing 1 mM $MgCl_2$, 5 mM mercaptoethanol and 5 mM tris-HCl, pH 7·4, in an HS–A zonal rotor. Centrifugation was carried out at 4°C in an MSE High-speed 18 centrifuge (fitted with a speed controller) for 15 min at 1200 rev/min. The higher running speed and the fast deceleration of the HS–A rotor make it possible to keep to a minimum the time of centrifugation. *Figure 8* shows the distribution of DNA polymerase activity among the different classes of nuclei, assayed in the presence and absence of exogenous native rat-liver DNA. There are substantial variations of enzyme activity through the gradient. Activity is associated with the parenchymal rather than the stromal nuclei. The tetraploid nuclei in the region of the apex of the peak have approximately twice the activity of the diploid parenchymal nuclei. What is most remarkable is that the specific activity of the DNA polymerase decreases substantially in both of the regions of the gradient where nuclei involved *in vivo* in DNA synthesis are found (cf. *Figure 6*). The data in *Figure 8* may be used to calculate the ratio of the specific activity of DNA polymerase with and without exogenous template DNA (*Figure 9*). From this it is obvious that the addition of exogenous DNA produces the greatest stimulation in nuclei involved in DNA synthesis *in vivo*, though it might have been anticipated that the DNA polymerase in nuclei active in DNA synthesis would be in the form of a complex with the endogenous DNA, and hence less accessible to added template. Sonication of the nuclei before assay

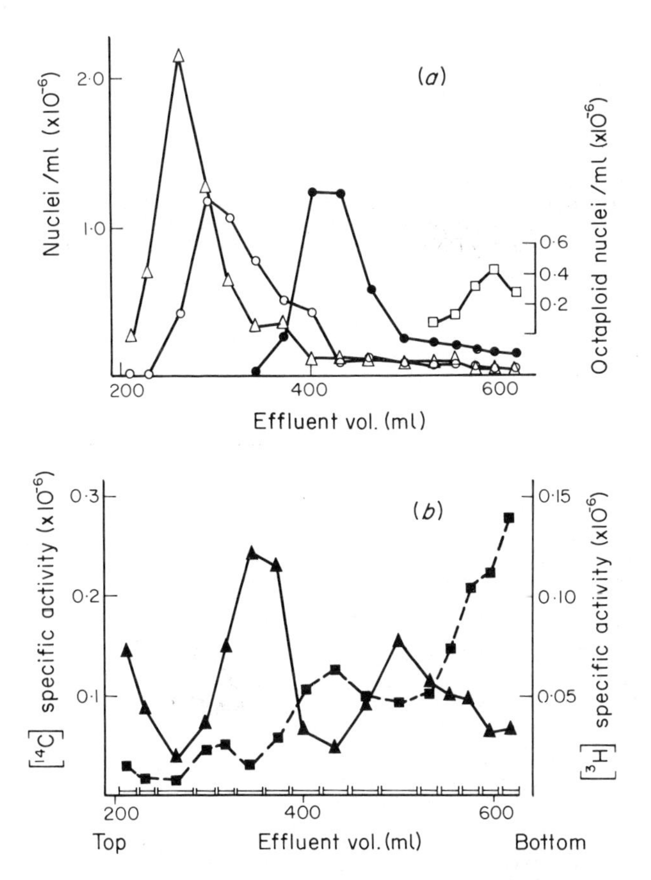

Figure 7. Synthesis of DNA and RNA in liver nuclei from mice weighing 25 g. (a) *Numbers of nuclei of various classes in each fraction.* (△), *Stromal diploid nuclei;* (○), *parenchymal diploid nuclei;* (●), *tetraploid nuclei;* (□), *octaploid nuclei.* (b) *Incorporation of* [6–^{14}C]*orotic acid* (■) *and* [^{3}H]*thymidine* (▲), *expressed as counts/min per* 10^9 *nuclei. Twelve male mice each received 2* μ*Ci of* [^{14}C]*orotic acid and 20* μ*Ci of* [^{3}H]*thymidine 75 min before death*

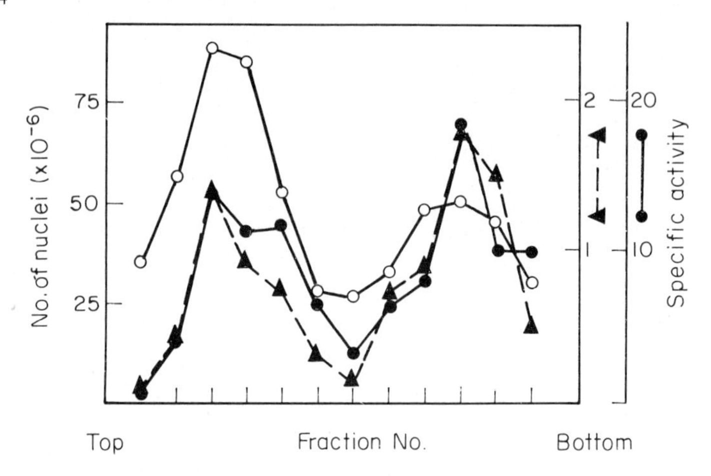

Figure 8. Specific activities of DNA polymerase, expressed as pmoles of [^3H]thymidylic acid incorporated per 10^6 nuclei, in liver nuclei separated by zonal centrifugation from Wistar rats weighing 120 g. (○), Nuclei per fraction; (●), DNA polymerase assayed in the presence of exogenous native rat-liver DNA; (▲), activity measured in the absence of exogenous DNA

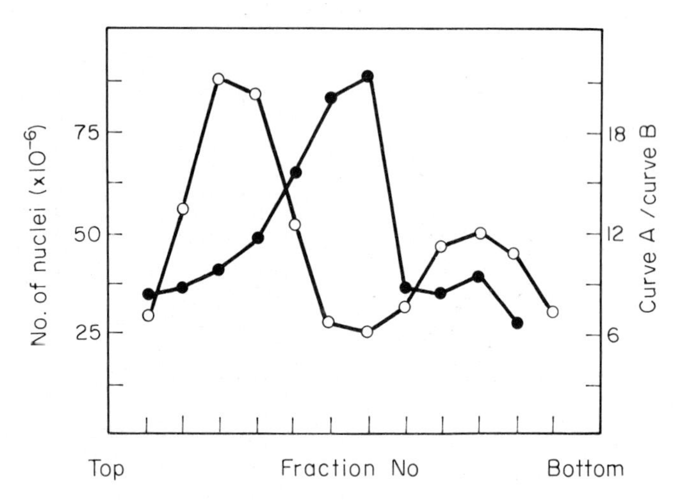

Figure 9. Ratio of specific activity of DNA polymerase assayed with and without added rat-liver DNA. (○), Numbers of nuclei per fraction; (●), ratio of activity with added DNA to activity without added DNA

produces no enhancement of the enzyme activity in the zonal fractions, thus confirming that both substrates and template can reach the enzyme in intact nuclei.

The conditions employed for the assay *in vitro* of enzymes for such complex processes as the synthesis of DNA are inevitably arbitrary. However, considerable alteration of the conditions from those used for the assays shown in *Figure 8* do not alter the distribution of the enzyme activity, nor does the substitution of calf-thymus for rat-liver DNA. If assays of the zonal fractions are performed to compare the template activities of native and denatured DNA, no change in preference is found at any point in the gradient, nor, within the limits of the somewhat insensitive experiments, have we been able to detect any variation in the K_m values for the deoxyribonucleotide triphosphates or substantial deoxyribonuclease activity in any zonal fraction.

It is not easy to resolve the paradox that the nuclei participating in DNA synthesis *in vivo* have the lowest levels of DNA polymerase detectable *in vitro*. Possibly the enzyme is most readily lost from these nuclei during their isolation. Alternatively, the enzyme may be more labile when forming part of a complex which is active in DNA synthesis, than it is in nuclei in G_1 or G_2 phase. Another possibility is that nuclear DNA polymerase may be required for repair rather than for synthesis of DNA. It will be necessary to examine the distribution of other enzymes required for the synthesis of DNA, such as the ligase and specific endonucleases.

SYNTHESIS OF RNA

Next to the replication of its DNA the most important function of the nucleus is the synthesis of RNA. If rats are pulse-labelled with [^{14}C]orotic acid and the liver nuclei fractionated by zonal centrifugation, we may estimate the capacity for synthesis of RNA *in vivo* by determining the extent of incorporation. The results of such an experiment for rats weighing 160 g are shown in *Figure 10*. Little synthesis of RNA occurs in the stromal nuclei. The specific incorporation reaches a plateau as we pass across the parenchymal section of the diploid peak, often shows a slight fall in the region containing the nuclei involved in DNA synthesis, and rises in the tetraploid peak to a value approximately twice that of the parenchymal diploid plateau. The relatively low amount of incorporation into the stromal nuclei is also observed in younger rats (*Figure 5*).

RNA polymerase probably exists in several different forms in

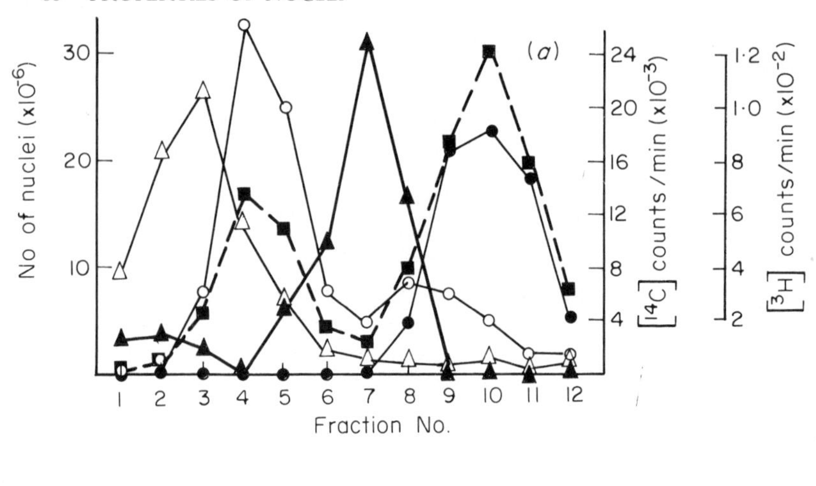

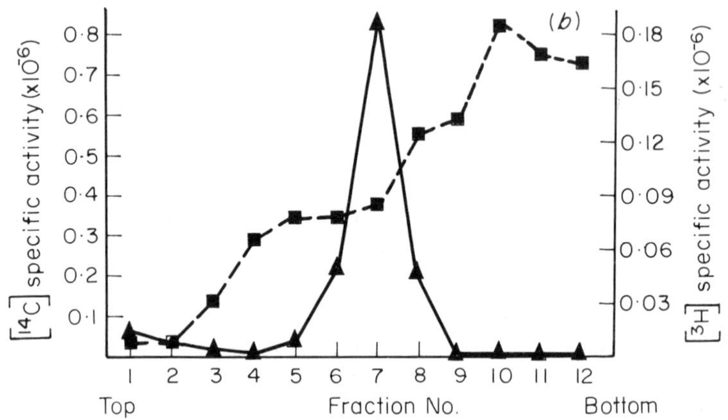

Figure 10. The synthesis in vivo *of DNA and RNA by liver nuclei from rats weighing 160 g. Three rats each received 10 μCi of* [6–^{14}C]*orotic acid (60 μCi/μmole) and 50 μCi of* [^{3}H]*thymidine (18·6 μCi/μmole) intraperitoneally 1 h before death.* (a) *Numbers of nuclei in each fraction:* (△), *stromal diploid nuclei;* (○), *parenchymal diploid nuclei;* (●), *tetraploid nuclei. Incorporation of* [^{14}C]*orotic acid* (■) *and* [^{3}H]*thymidine* (▲). (b) *Incorporation of* [^{14}C]*orotic acid* (■) *and* [^{3}H]*thymidine* (▲) *expressed as counts/min per* 10^9 *nuclei*

mammalian nuclei. However, estimates of its activity may be made by measuring the activity in the presence of Mg^{2+} or Mn^{2+} and ammonium sulphate. The distribution of the activity in young and old rats is shown in *Table 2*. The activities measured in the presence of Mn^{2+} are much higher, but the same shift of activity from diploid to tetraploid nuclei as the animals grow older is evident. A more detailed analysis of the distribution of RNA polymerase is shown in *Figure 11*. The specific activity is very low in the stromal region of the diploid zone. The activity falls in the region of DNA synthesis and then increases as the proportion of tetraploid nuclei rises. The problem of the relative specific activities of the diploid and tetraploid parenchymal nuclei has been studied in a series of experiments. In rats weighing 170 to 190 g the activities of the diploid stromal, diploid parenchymal and tetraploid parenchymal

Table 2. DISTRIBUTION OF RNA POLYMERASE ACTIVITY IN THE DIPLOID AND TETRAPLOID ZONES OF RAT-LIVER NUCLEI

Weight of rats	*Diploid nuclei*	*Tetraploid nuclei*
25 g	2295*	—
200 g	424*	3987*

* pmoles of [^{3}H]UMP incorporated from [^{3}H]UTP per 10^9 nuclei in presence of Mg^{2+} and ATP, GTP and CTP (Widnell and Tata, 1966).

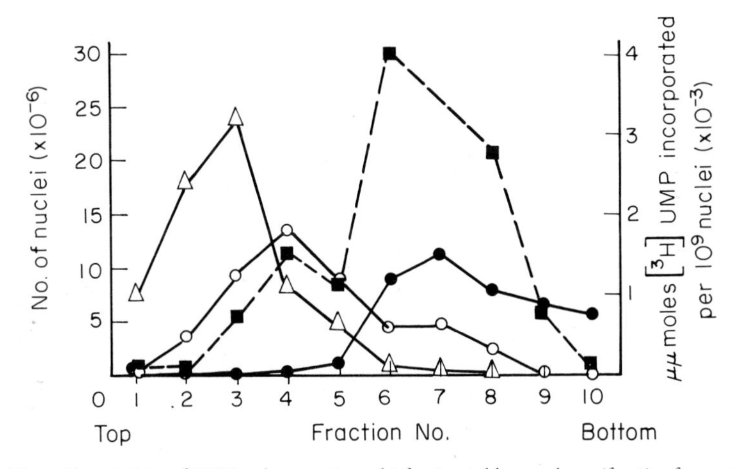

Figure 11. Activity of RNA polymerase in nuclei fractionated by zonal centrifugation from rats weighing 190 g. (△), Numbers of stromal diploid nuclei in fractions; (○), numbers of parenchymal diploid nuclei; (●), numbers of tetraploid nuclei; (■), activity of RNA polymerase expressed as pmoles of [^{3}H]UMP incorporated per 10^9 nuclei

nuclei are in the ratio 0:1·0:2·8. For older rats the ratio of parenchymal diploid to tetraploid activity approaches 1:4. Thus the activity of RNA polymerase per tetraploid nucleus is more than twice that in the diploid parenchymal nucleus. In contrast, the extent of incorporation of precursors into nuclear RNA *in vivo* would appear to indicate a ratio of almost exactly 2:1 for tetraploid as compared to diploid nuclei (*Figure 10*). These findings emphasise again the need to assess the measurements of enzyme activity *in vitro* in the light of the effective activity *in vivo*.

In nuclei from mice, the regions of RNA synthesis show a pattern relative to DNA synthesis that is similar to that in rats (*Figure 7*). The relative specific activities of the various classes of nuclei are given in *Table 3*.

Much of the nuclear synthesis of RNA, and probably all that of ribosomal RNA, occurs within the nucleolus. Nucleoli may be

Table 3. RELATIVE SPECIFIC ACTIVITIES OF THE VARIOUS CLASSES OF MOUSE-LIVER NUCLEI

Nuclei	RNA *polymerase*★ *measured in presence of*		[^{14}C]*orotic acid*★† *incorporation in vivo*
	Mg^{2+}	$Mn^{2+}+(NH_4)_2SO_4$	
Diploid stromal	0·2	0·3	0·25
Diploid parenchymal	1·0	1·0	1·0
Tetraploid parenchymal	2·2	1·8	2·1
Octaploid parenchymal	5·9	8·6	4·2

★ Activity of diploid parenchymal nuclei is taken as 1·0.
† Incorporation of [^{14}C]orotic acid into nuclear RNA is calculated from the data shown in *Figure 7*.

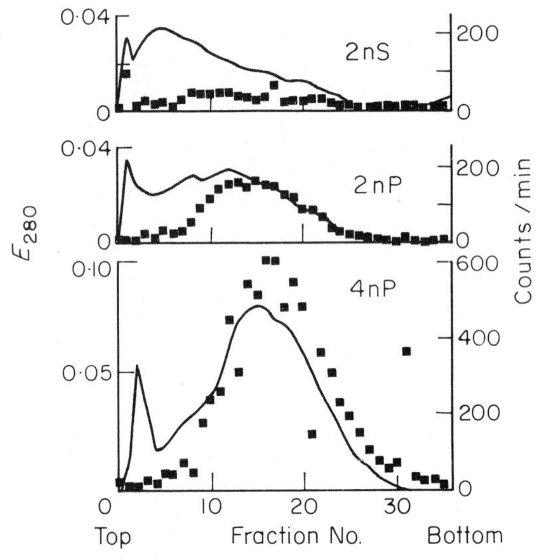

Figure 12. Centrifugation of nucleoli from diploid stromal (2nS), diploid parenchymal (2nP) and tetraploid parenchymal (4nP) nuclei. Rats (150 g) were injected with [^{14}C]orotic acid (5 μCi/animal) 2 h before death. The nuclei were isolated and separated at pH 6 (see Figure 3*). Nucleoli were prepared from the separate classes of nuclei by sonication in unbuffered 0·32* M *sucrose at 0°C for a total of 5 min, using the MSE 100-W ultrasonic disintegrator with titanium probe. Centrifugation of the nucleoli was for 15 min at 4300 rev/min in an SW 25.1 rotor at 4°C.* ——— E_{280nm}; (■) *incorporation of* [^{14}C]*orotic acid*

prepared from nuclei fractionated by zonal centrifugation (Johnston *et al.*, 1969). They form a broad band when centrifuged in a swinging-out rotor in a 30 to 60% (w/v) sucrose gradient (*Figure 12*). The stromal nucleoli are the smallest and slowest moving. The nucleoli from tetraploid nuclei have the bulk of the newly-synthesised RNA. Microscopic observation shows them to have the greater mean size.

METABOLISM OF NAD

The vital role of the nucleus in anabolic and catabolic reaction of NAD is apparent from *Figure 13* (Haines *et al.*, 1969). We have studied the distribution of NAD pyrophosphorylase (NMN-adenylyl-transferase) and poly(ADP-ribose) polymerase. In mature

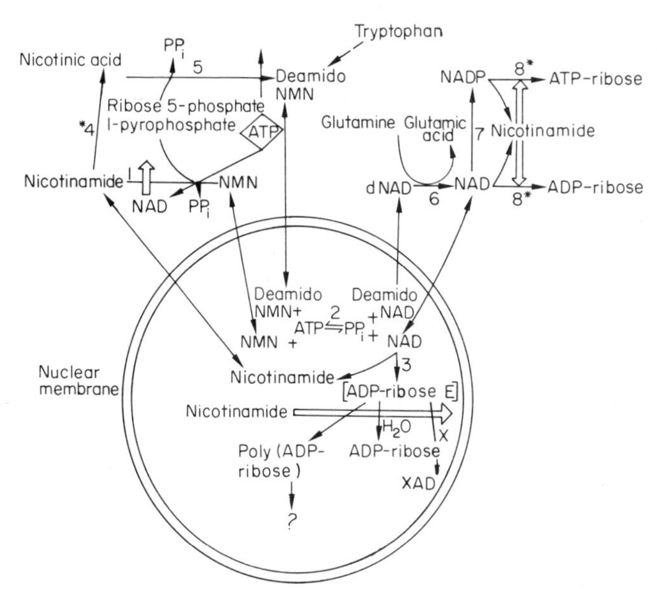

Figure 13. Pathways of NAD metabolism in liver tissue. The enzymes involved are: (1) *NMN pyrophosphorylase (NMN–pyrophosphate phosphoribosyltransferase, EC 2.4.2.12);* (2) *deamido-NAD pyrophosphorylase (ATP–nicotinate mononucleotide adenylyltransferase, EC 2.7.7.18) and NMN adenylyltransferase (ATP–NMN adenylyl-transferase, EC 2.7.7.1);* (3) *poly (ADP-ribose) polymerase or nuclear NAD glycohydrolase (EC 3.2.2.5), or both;* (4) *nicotinamide deamidase;* (5) *nicotinate mononucleotide pyrophosphorylase (nicotinate mononucleotide–pyrophosphate phosphoribosyltransferase, EC 2.4.2.11);* (6) *NAD synthetase [deamido-NAD–L-glutamine amidoligase (AMP), EC 6.3.5.1];* (7) *NAD kinase (ATP–NAD 2′-phosphotransferase, EC 2.7.1.23);* (8) *microsomal NAD glycohydrolase (EC 3.2.2.5).* ⟷ *represents transport across the nuclear membrane,* ◇→ *allosteric effectors,* ⟹ *inhibitors. Microsomal enzymes are indicated by an asterisk. X indicates a pyridine base*

rats (200 g) the first of these enzymes has peaks of specific activity in the regions of DNA synthesis ahead of the zones of diploid and the tetraploid nuclei (*Figure 6*). The same correlation of high levels of NAD pyrophosphorylase with activity in the synthesis of DNA is also apparent in very young rats. However, in rats of about 60 g, which are passing through a state where the transition of the liver from a diploid to a predominantly tetraploid level is in its most rapid phase, a more complex pattern is observed. Control experiments demonstrated that the variations in specific enzyme activity were not due to changes in the amounts of NAD glycohydrolase.

The highest activities of the poly(ADP-ribose) polymerase are found in the tetraploid nuclei, in the fractions with maximum synthesis of RNA *in vivo*. The poly(ADP-ribose) polymerase activity falls off steeply on either side but rises again to give a double maximum in the diploid peak (*Figure 14*). The sharp decline of the specific activity of the nuclei in the regions of DNA synthesis is striking.

We have demonstrated that the nuclei from diploid cells in S phase lie between the diploid and tetraploid zones. As the nuclei

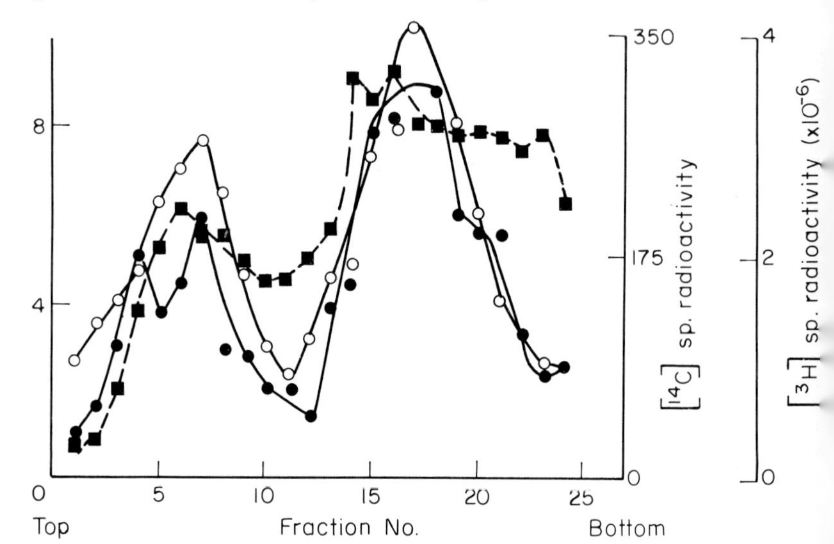

Figure 14. Distribution of poly (ADP-ribose) polymerase activity and synthesis of RNA in vivo *in various classes of liver nuclei in rats weighing 150 g. Three animals each received 25 μCi of [^{3}H]orotic acid (1 Ci/mmole) 1 h before death.* (●), *activity of poly (ADP-ribose) polymerase expressed as nmoles of [^{14}C]ADP-ribose incorporated per 15 min per 10^9 nuclei;* (■), *incorporation of [^{3}H]UMP (counts/min per 10^9 nuclei);* (○), *number of nuclei in 4 ml of each fraction. Two 4 ml samples of each fraction from the zonal centrifugation were taken, one for the enzyme assay and the other for measurement of [^{3}H] incorporation. The remainder was used to determine the number of nuclei with the Coulter Counter*

pass from S into G_2, they will be found further into the trailing edge of the tetraploid zone. After mitosis, the nuclei will move back into the diploid zone. On this basis, we can deduce that the changes in the specific activities of the nuclear enzymes of NAD metabolism may result in a lowered concentration of NAD at the time immediately preceding mitosis. This could be advantageous in view of the apparent inverse correlation between the concentration of nicotinamide and mitosis (Morton, 1961).

NUCLEI OF AVIAN RED CELLS

Avian red cells may be fractionated in an AXII zonal rotor using a 20 to 50% (w/w) gradient of sucrose buffered with 1 mM tris-HCl pH 7·2 (Mathias *et al.*, 1969) (*Figure 15*). The trailing edge of the main peak has a much higher proportion of reticulocytes, which are relatively rich in RNA. They have a higher ratio of nuclear to

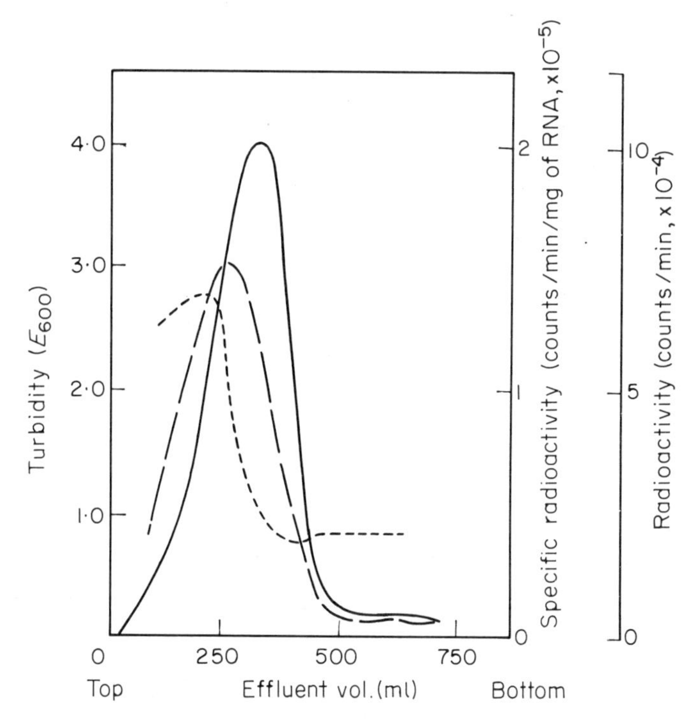

Figure 15. Elution profile of pigeon reticulocytes labelled in vivo *with [^{32}P]phosphate. ———, Turbidity (E_{600nm}); – – – – –, radioactivity; . . . . . , specific radioactivity*

cytoplasmic RNA and are more active in the synthesis of RNA than the mature erythrocytes, which advance further into the gradient. The nuclei from the reticulocytes sediment more slowly than those from the erythrocytes (*Figure 16*), despite the fact that they have a mean volume of 35 μm^3 compared to 25 μm^3 for the erythrocyte nuclei. This implies that the reticulocyte nuclei are less

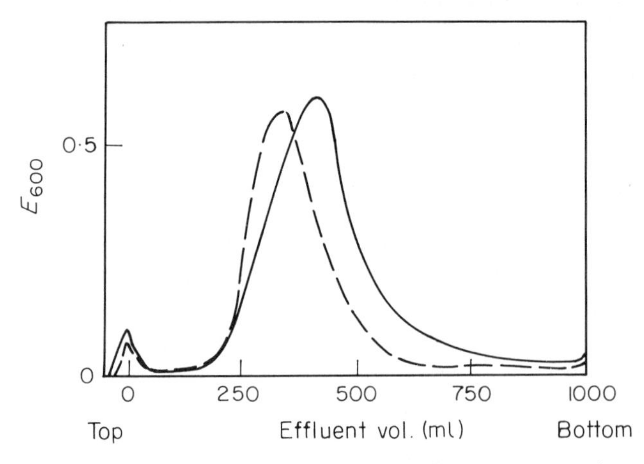

Figure 16. Elution profile of nuclei isolated from pigeon erythrocytes (———) and reticulocytes (– – – – –) using the standard conditions for centrifugation employed for liver nuclei but prolonging the time of centrifugation to 4 h at 600 rev/min

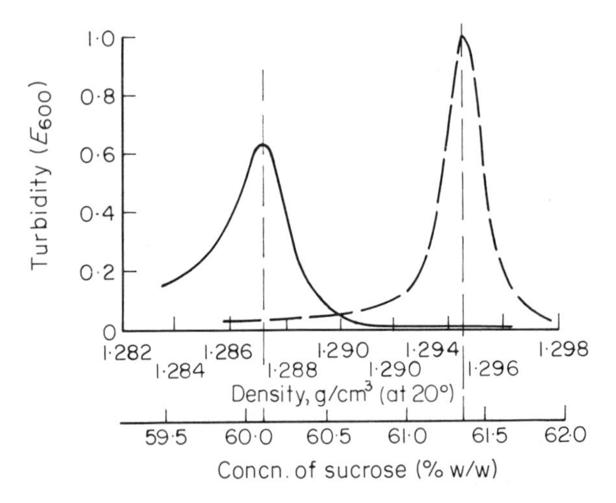

Figure 17. Isopycnic zonal centrifugation of nuclei from pigeon reticulocytes (———) and erythrocytes (– – – – –)

dense than nuclei from erythrocytes. This hypothesis was confirmed in an experiment in which the zonal rotor was used in an analytical mode to determine the isopycnic points of the two classes of nuclei (*Figure 17*). Centrifugation was carried out for 10 h at 3200 rev/min in a gradient of 53 to 63% (w/w) sucrose containing 1 mM $MgCl_2$, pH 7·2, by which time the nuclei had formed sharp immobile bands. The reticulocyte nuclei had a banding density of 1·287 g/cm^3 whilst those from erythrocytes banded at 1·294 g/cm^3.

CONCLUSIONS

Some of the biochemical properties of liver nuclei from young adult rats are summarised in *Figure 18* and *Table 4*, which enable us

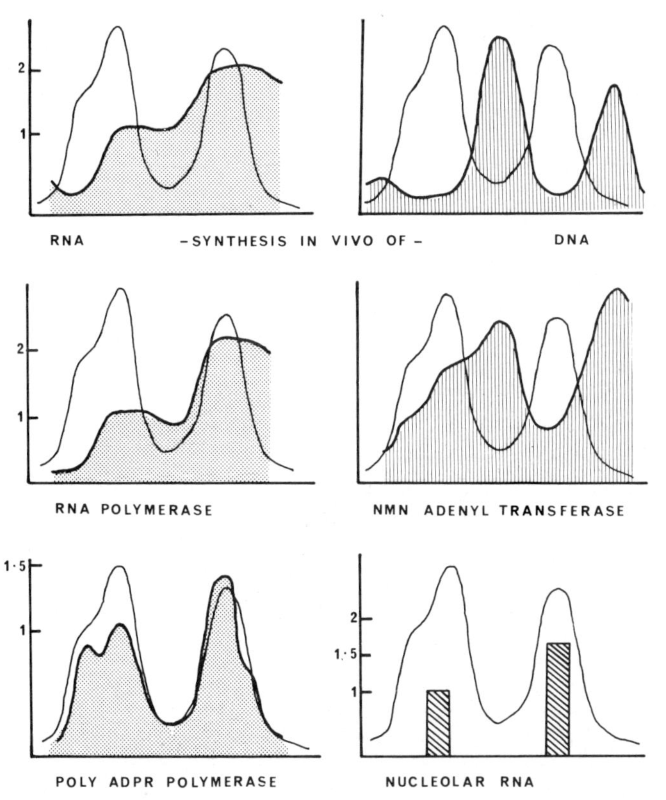

Figure 18. Biochemical properties of nuclei from young adult rats fractionated by rate-zonal sedimentation. The specific activity of the parenchymal diploid nuclei in G_1 (or G_0) is arbitrarily set at 1 in each case

to discern some of the factors that influence the activity of the nucleus. The relative inertness of the stromal nucleus is evident. Some of the biochemical properties of the parenchymal nucleus appear to be dictated by gene dosage, for example the synthesis *in vivo* of RNA and possibly the levels of RNA polymerase. Others show a closer correlation with the ratio of surface area to volume which, when comparing diploid to tetraploid, is 1:1·59. DNA polymerase, poly(ADP-ribosome) polymerase and nucleolar RNA synthesis may fit into this category. Lastly, we may note the variation of enzyme activity with position in the cell cycle

Table 4. RELATIVE ACTIVITIES OF VARIOUS CLASSES OF LIVER NUCLEI IN YOUNG ADULT RATS

Activity	2nS*	2nP*†	4nP*
Synthesis of RNA *in vivo*	0	1	2
RNA polymerase	0	1	2
DNA polymerase‡	0	1	1·5–1·8
NAD pyrophosphorylase‡	0·66	1	0·5
Poly(ADP-ribose) polymerase	0·8	1	1·4
Synthesis of nucleolar RNA	0	1	1·66

* 2nS, 2nP and 4nP are diploid stromal, diploid parenchymal and tetraploid parenchymal nuclei, resp.
† Activity of diploid parenchymal nuclei is taken as 1·0.
‡ Estimated.

(for example, NAD pyrophosphorylase). Some enzyme distributions do not fit readily into any of these patterns. The importance of comparing measurements *in vitro* of enzyme levels with actual activity *in vivo* requires strong emphasis.

These studies have revealed the degree of heterogeneity that exists in isolated nuclei. This heterogeneity can originate in the type of cell from which the nucleus was derived, the position of the cell in the cell cycle and the level of ploidy. The enzymic properties are influenced by gene dosage and the surface area of the nucleus. The zonal centrifuge provides a versatile and valuable technique to examine and probe this heterogeneity. Apart from its preparative application, it can be exploited to provide analytical information.

Acknowledgments. We thank our colleagues, and in particular M. E. Haines, F. Pennington and D. Ridge, for their contribution to these studies, which were supported by grants from the Science Research Council and the Wellcome Trust. We are grateful to the Editors of the *Biochemical Journal*, *Biochimica et Biophysica Acta*, *FEBS Letters* and *Nature* for permission to reproduce various Figures.

REFERENCES

ANDERSON, N. G. (1966). *Science, N.Y.*, **154**, 103

CARRIERE, R. (1969). *International Review of Cytology*, **25**, 201

ELIAS, H. and SHEMICK, J. (1969). *The Morphology of the Liver*, New York and London (Academic Press)

FUKUDA, M. and SIBATANI, A. (1953). *J. Biochem. Tokyo*, **40**, 95

HAINES, M. E., JOHNSTON, I. R., MATHIAS, A. P. and RIDGE, D. (1969). *Biochem. J.*, **115**, 881

HAINES, M. E., JOHNSTON, I. R. and RIDGE, D. (1970). *FEBS Letters*, **10**, 113

JOHNSTON, I. R., MATHIAS, A. P., PENNINGTON, F. and RIDGE, D. (1968). (*a*) *Biochem. J.*, **109**, 127; (*b*) *Nature, Lond.*, **220**, 668

JOHNSTON, I. R., MATHIAS, A. P., PENNINGTON, F. and RIDGE, D. (1969). *Biochim. biophys. Acta*, **195**, 563

MATHIAS, A. P., RIDGE, D. and TREZONA, N. ST. G. (1969). *Biochem. J.*, **111**, 583

MORTON, R. K. (1961). *Aust. J. Sci.*, **24**, 260

WIDNELL, C. C. and TATA, J. R. (1964). *Biochem. J.*, **92**, 313

WIDNELL, C. C. and TATA, J. R. (1966). *Biochim. biophys. Acta*, **123**, 478

4

PREPARATION OF MITOCHONDRIA FROM ANIMAL TISSUES AND YEASTS

J. B. Chappell and R. G. Hansford

The writing of this article has been made easier in one sense, and not so easy in another, by the publication of Volume 10 of *Methods in Enzymology*. In that volume, to which the reader is referred, the preparation of mitochondria from a wide variety of cells, tissues and organisms is described. We have decided, therefore, to describe only methods, or their modifications, which we know or believe to be superior to those described in that volume and of which we have personal experience. Criteria by which the morphological and biochemical integrity of the resulting mitochondria may be assessed are also set out.

PREPARATION OF MITOCHONDRIA: GENERAL PRINCIPLES

The isolation of mitochondria in quantity from tissues and cells involves, first, breakage of the cell membrane with minimal damage to the mitochondrial membrane by, for example, homogenisation and, second, isolation of the mitochondria from the homogenate by centrifugation. Some of the general principles involved are discussed below, and illustrated in the following descriptions of preparative techniques in use in the laboratory of biochemistry of Bristol University.

HOMOGENISATION

With soft, non-fibrous tissues homogenisation is best performed with either a manually-operated Potter–Elvehjem or a Dounce homogeniser. The tissue is chopped with scissors to form about 1 mm cubes and transferred to the mortar of a Dounce homogeniser. It is then broken up by raising and lowering the pestle, making 6 or so passes in all. At this stage a compromise must be made between yield and structural damage to mitochondria. The pestle is best made of Teflon and should have a diameter 0·25 to 0·33 mm less than the internal one of the mortar.

With fibrous or hard tissues, two approaches are possible. One is to apply mechanical forces of considerable magnitude which not only cause cell breakage but also damage a variable proportion of the mitochondrial population. The problem then is to separate the damaged from the undamaged mitochondria. This approach is used with beef heart in many laboratories (for example, Sanadi and Fluharty, 1963) where 'heavy' (undamaged) mitochondria are separated from 'light' (mechanically damaged). An alternative approach is to incubate the chopped tissue or the cell suspension with an enzyme or enzymes which cause removal of fibrous materials, such as collagen, or of hard cell walls, such as the polysaccharide wall of yeast. For the preparation of morphologically and biochemically intact mitochondria this method is to be preferred. After enzymatic digestion, conventional manual homogenisation releases the mitochondria, which may then be isolated by differential centrifugation.

CENTRIFUGATION

(*a*) *Differential centrifugation*—In this method, which is the most widely used for preparing mitochondria for biochemical studies, separation is achieved as a result of differences in size and density, and to some extent shape, of subcellular organelles. Centrifugal fields of sufficient magnitude and duration are employed to cause more or less complete sedimentation of the organelle which is to be isolated. Purification is achieved by resuspension and resedimentation and by pouring off loosely sedimented material.

(*b*) *Density-gradient centrifugation*—Here two methods may be distinguished. In the first, isopycnic centrifugation, the homogenate is layered on top of either a continuous or a discontinuous density gradient, and centrifugation is prolonged until the subcellular particles are in density equilibrium with the surrounding medium.

In the second method, rate-zonal centrifugation, sedimentation through a shallow gradient is used and centrifugation is stopped before it is complete, so that separation depends on sedimentation velocity.

In general, density-gradient centrifugation has not been used widely for the preparation of mitochondria, despite the fact that these methods give preparations of greater purity. The relatively long time taken to achieve separation and the effects of hypertonicity, if sucrose is used, lead to damage of mitochondria. The damaging effects of hypertonicity can be overcome by use of solutions of Ficoll, a polymer of sucrose (supplied by Pharmacia, Uppsala, Sweden), or other polysaccharides, which have density properties similar to that of sucrose solutions but do not make a significant contribution to the tonicity of the solution.

In some cases use of a density gradient is almost unavoidable. For example, with the method of Schatz *et al.* (1963) for disrupting yeast cells with a colloid mill, it is necessary to 'rescue' the remaining intact mitochondria, and this may be achieved by density-gradient centrifugation. Gray and Whittaker (1962) have used a discontinuous sucrose density gradient to separate brain mitochondria from the 'synaptosomes' which are produced from nerve endings on homogenisation of brain cortex.

PREPARATION OF RAT-LIVER MITOCHONDRIA

This method obviously owes much to classical techniques. It does differ from them, however, and attention is drawn to the use of a buffered preparation medium. The use of ethylene-glycol-bis-(aminoethyl)tetra-acetate (EGTA) which chelates Ca^{2+} ions specifically, is strongly advised. The more widely used chelating agent EDTA has the drawback that it extracts Mg^{2+} ions from the mitochondrial membrane, thereby making it leaky to K^+ ions.

The livers are removed as rapidly as possible from four or five rats killed by decapitation, and are rinsed with 30 ml of ice-cold preparation medium (0·25 M sucrose, 3·4 mM tris-HCl, pH 7·4 at 20°C, 1 mM EGTA). The medium is decanted, and the lobes of liver are finely chopped with scissors in another 30 ml of medium. Again the liquid is carefully decanted. This washing procedure serves to remove blood and is repeated once more. The chopped liver is transferred to a Dounce homogeniser which is then filled to the neck with medium, making 55 ml in all. The tissue is homogenised by five passes of the pestle. The homogenate is then poured into a 250 ml conical flask (in ice) and diluted with another 50 ml of medium, which is used to wash out any remaining homogenate from the homogeniser. The diluted homogenate is then equally divided between four 50 ml polythene centrifuge tubes and centrifuged at 2000 rev/min for 10 min in the 8 × 50 fixed-angle rotor of an MSE High-speed 18 centrifuge. This represents 4800*g*-min at the bottom of the tube and 2500*g*-min at the top. The supernatant fluid is carefully decanted into 4 clean centrifuge tubes and centrifuged at 9000 rev/min for 7 min. This represents 77 000*g*-min maximum and 40 000*g*-min minimum. For this high-speed spin the centrifuge is set to accelerate at its maximum rate and cut back when it reaches the desired speed. The supernatant fluid is decanted, and each pellet is gently resuspended by stirring with a 'cold finger'—a test tube filled with ice. A smooth resuspension can only be made if this is done *before* any fresh medium is added. The thick suspension so formed is then progressively diluted with medium. The four pellets are taken up in about 80 ml of preparation medium and poured into two MSE High-speed 18 centrifuge tubes. Another high-speed centrifugation as above, another resuspension, this time to fill one tube, and a third high-speed centrifugation follow. The single pellet from this final centrifugation is gently resuspended, and a little medium added, to give a total volume of about 3 ml.

It is important to pre-cool the medium and all apparatus, to avoid contamination with ice at any stage, and to make the preparation as rapidly as possible, especially the early stages. A good preparation takes no more than $1\frac{1}{2}$ h.

Special care must be taken when mitochondria are to be used for spectroscopic observations of the redox behaviour of the cytochromes to ensure that all red cells have been removed. This is particularly so when anaerobic-aerobic transitions of the cytochromes are to be followed or when cyanoferrate is added, since haemoglobin, oxyhaemoglobin and methaemoglobin all have considerable absorbancy in the regions of interest in the absorption spectra.

One method used to remove red cells in this laboratory is to intercalate a low-speed centrifugation before the final high-speed centrifugation described above. Another technique is to convert the haemoglobin to methaemoglobin at this stage, by incubating the mitochondrial suspension in the sucrose medium, to which cyanoferrate (final concentration, 5 mM) has been added, for 10 min at 0°C. With rat red cells this causes the conversion of (oxy)haemoglobin to methaemoglobin, which does not undergo spectroscopic changes on aerobic–anaerobic transitions. Caution must be employed with this technique when working with tissues other than rat tissues, because, for example, human red cells are impermeable to cyanoferrate. The safest approach, however, is to try to remove as many red cells as possible by careful centrifugation.

PREPARATION OF RAT-HEART MITOCHONDRIA

This method employs the proteolytic enzyme *nagarse*★, the use of which in the preparation of mitochondria from muscle tissue was developed by Chance and Hagihara (1961). The hearts from 4 or 5 rats are washed free of blood by lightly chopping with scissors in about 30 ml of preparation medium (0·21 M mannitol, 0·07 M sucrose, 5 mM tris-HCl, pH 7·4 at 20°C, 1 mM EGTA) and twice decanting the liquid. The pieces of heart are then transferred to a small beaker containing 7 to 8 mg of nagarse dissolved in 20 ml of preparation medium. After thorough chopping the material is transferred to a Dounce homogeniser, diluted to 55 ml and homogenised by hand. After this initial homogenisation, the tissue is left incubating with the enzyme for 15 min, strictly at 0°C. Homogenisation is then repeated, using five strokes of the pestle. The homogenate is transferred to two 50 ml centrifuge tubes and centrifuged at 9000 rev/min for 7 min as described for the rat-liver preparation. The pellet is resuspended using the 'cold-finger' technique and again centrifuged in two 50 ml tubes. The purpose of this centrifugation is the rapid removal of the majority of the proteolytic enzyme. The entire pellet is resuspended, again to the same volume, and cellular debris removed by centrifuging for 10 min at 1700 rev/min in an MSE High-speed 18 centrifuge (equivalent to 3500*g*-min maximum and 1800*g*-min minimum). The supernatant fluid is decanted very carefully, leaving any mobile layer behind, and is then centrifuged for 7 min at 9000 rev/min. The entire pellet is resuspended in 40 ml of medium and centrifuged again for 7 min at 7500 rev/min (53 000*g*-min maximum and 27 000*g*-min minimum). The final pellet comprises two layers—the lower is deep red-brown in colour, the upper is light buff. The latter is sloughed off as cleanly as possible, and discarded. The lower pellet is suspended carefully in 1 to 2 ml of medium to give the mitochondrial suspension. This latter fractionation appears to be critical.

★ *Nagarse* is a non-specific proteolytic enzyme of bacterial origin, which may be purchased from Hughes & Hughes (Enzymes) Ltd., 12A High Street, Brentwood, Essex.

PREPARATION OF BLOWFLY FLIGHT-MUSCLE MITOCHONDRIA

The heads and abdomens are removed from approximately 50 flies which have been immobilised by prior cooling to 4°C and the flight muscle is expressed by squeezing the thoraces. The muscle is placed in approximately 30 ml of 0·25 M sucrose, 5 mM tris-HCl, 1 mM EGTA (pH adjusted to 7·7 at 0°C) on ice. When all the muscles have been removed they are transferred to a Dounce homogeniser and 3·5 mg of nagarse, dissolved in about 5 ml of medium, is added. After two light passes of the pestle (0·25 mm clearance on the diameter) the tissue is allowed to digest for 10 min, strictly at 0°C. During this interval, further light homogenisation is carried out. The resulting suspension, which should be essentially free of intact muscle bundles, is filtered through four thicknesses of washed surgical gauze, diluted to 80 ml and distributed in two 50 ml MSE High-speed 18 centrifuge tubes. Centrifugation at 7500 rev/min for 3 min (24 000*g*-min maximum and 12 000*g*-min minimum) easily suffices to sediment the mitochondria and allows any upper, light-coloured, layer of partially-digested myofibrils to be removed by gentle shaking. The dark-red pellet is resuspended by stirring with a 'cold finger', diluted to 40 ml and centrifuged as before, in one tube. The final pellet is resuspended in 2 ml of preparation medium, yielding a mitochondrial suspension of approx. 20 mg of protein per ml.

The essence of this naïve-seeming preparation is the total digestion of the myofibrils, which is possible at the high ratio of enzyme to protein used. This obviates the necessity for a low-speed centrifugation, which is unsuccessful in preparing mitochondria from this tissue since these organelles also sediment, enmeshed in bundles of myofibrils. The method described is very convenient and is successful as judged by the biochemical criteria of integrity set down below. The electron microscope reveals a preparation of essentially-intact mitochondria, uncontaminated by other organelles, provided that the digestion is carefully controlled. This involves a prior adjustment of the pH of the *ice-cold* medium to 7·7. Excessive digestion results in a proportion of grossly-swollen mitochondria, readily seen in the electron microscope, but which affect the biochemical characteristics of the preparation rather little.

One should always be cautious in accepting results obtained only from preparations involving the use of nagarse, as it is always possible that some membrane properties may have been altered by the enzyme. Indeed, there is some evidence that preparations made in this way may have a lowered content of endogenous tricarboxylate cycle intermediates and may even have an inactivated inner membrane 'carrier'.

PREPARATION OF YEAST MITOCHONDRIA

The following method for preparing yeast mitochondria is due to Duell *et al.* (1964). Yeast grown in a chemostat is collected overnight in a vessel cooled in ice. Cells are harvested by centrifuging at 1000*g* for 10 min (2000 rev/min in an MSE serum centrifuge). Packed cells are resuspended in cold distilled water using a vortex mixer and are pelleted in four 50 ml MSE High-speed 18 centrifuge tubes (3000 rev/min for 7 min). The washing is repeated, and the packed cells are then weighed. For enzymic digestion, they are resuspended in 2 ml of 0·63 M sorbitol, 0·1 M citrate/phosphate buffer, pH 5·8, 0·4 mM EDTA per g of wet cells, and 6 mg of mercaptoethylamine and 0·3 ml of snail-gut enzyme are added per g of wet cells (the latter may be obtained from Industrie Biologique Francaise S.A., Gennevilliers, France). The suspension is incubated for 1 h at 30°C with shaking. The spheroplasts and cells are then pelleted by centrifuging at 5000 rev/min for 10 min (30 000*g*-min maximum and 15 400*g*-min minimum). The supernatant fluid containing the snail-gut enzyme is decanted and the spheroplasts are gently suspended in ice-cold 1 M sorbitol. They are then sedimented again and the washing procedure is repeated. After this final wash, care is taken to ensure that as much sorbitol as possible is decanted, since any remaining at this stage makes the subsequent osmotic lysis difficult.

The spheroplasts are suspended in 0·125 M sucrose containing 25 mM phosphate, pH 6·8, and 1 mM EGTA. All the pellets are pooled, the volume is made up to 300 ml and the spheroplasts are shaken in a large flask to ensure proper mixing and lysis. The resulting suspension is divided between eight 50 ml centrifuge tubes and centrifuged at 3000 rev/min for 10 min (11 000*g*-min maximum and 5500*g*-min minimum). The supernatant fluid is decanted very carefully leaving any loose material behind, and is then centrifuged at 9000 rev/min for 7 min. The pellets are resuspended in 0·25 M sucrose, 50 mM phosphate, pH 6·8, 1 mM EGTA, diluted to 40 ml and centrifuged again. The final pellet is resuspended as before. The yield is of the order of 2·5 mg of mitochondrial protein per g cells (wet weight) for *Candida utilis* grown in a chemostat on 0·5% glucose.

PREPARATION OF BRAIN MITOCHONDRIA

Brain mitochondria isolated by conventional methods (that is, those suitable for liver or kidney) exhibit no or, at best, relatively poor respiratory control when assayed by polarographic techniques suitable for the mitochondria of other organs. Published isolation procedures for brain mitochondria which do exhibit improved respiratory control ratios are relatively laborious and lengthy, may involve costly and exotic media, and generally result in low yields per g of original tissue. The following procedure was developed by J. Garbus during his stay in the biochemical laboratory of Bristol University. It is simple and rapid and leads to good yields of functionally-competent brain mitochondria, using respiratory control as the index of functional integrity. The individual steps in the procedure were designed to minimise factors which are known to be deleterious to brain mitochondria. Rapidity of operation is desirable because of the susceptibility of nervous tissue to anoxia. EDTA is included in the isolation and polarographic medium to chelate Mg^{2+}. A rapid centrifugation scheme was developed using 0·25 M sucrose to obviate the use of higher concentrations of sucrose or Ficoll media, both of which are known to depress respiratory control. Potassium salts are used throughout because of the enhancing effect of this cation upon respiration and to minimise the Na^{+}/K^{+}-stimulated ATPase.

The brain is quickly removed from a stunned 200 to 250 g rat. Oddly, it was found that the respiratory control ratios are not so good with either smaller or larger animals. The brain can be removed within 15 to 30 s with the following procedure. The animal is stunned by a sharp blow so as not to damage the skull or cervical region. The head is bent forward and an incision is made severing the spine between the cervical vertebrae. The skin is cut away from the skull and reflected forwards. The sharp point of a pair of well-sharpened scissors is inserted into the foramen and the skull is cut along its lateral sutures. The flap of bone is reflected forwards and the brain is scooped out into a beaker which is packed in ice and contains at least 15 vol of cold medium (0·25 M sucrose, 0·5 mM potassium EDTA, 10 mM tris-HCl, pH 7·4). The brain is rinsed in the cold medium, blotted on filter paper and dropped into a graduated cylinder packed in ice and containing a measured volume of cold medium. The volume of the brain is measured and additional medium is added to give a 10% (v/v) homogenate. Homogenisation is by hand for 15 up-and-down strokes after the pestle reaches the bottom of the tube. The homogenate is then rapidly centrifuged at 4°C to remove blood cells, nuclei, unbroken cells and debris. This is accomplished by bringing the rotor to 4300 rev/min (2000*g*) as rapidly as possible and immediately turning off the power and allowing the rotor to decelerate with braking. With the particular model of the MSE High-speed 18 centrifuge used, it required 40 s to reach the requisite speed (4300 rev/min) and 105 s to decelerate. The supernatant fluid is removed with a chilled pipette, care being taken not to include any of the packed material. To increase the yield, the pellet can be rehomogenised and centrifuged as before. The combined supernatant fractions are then centrifuged to 20 000*g* in an analogous manner, that is, the rotor is taken to the requisite speed (13 500 rev/min) as rapidly as possible and immediately allowed to decelerate. The total time for this centrifugation was found to be $5\frac{1}{2}$ min. The turbid supernatant fraction is decanted and any very mobile material discarded. The mitochondrial pellet is resuspended in a volume of medium equal to $\frac{1}{2}$ that originally used and subjected to another rapid centrifugation to 20 000*g*. The pellet is finally resuspended in 1 ml of 0·25 M sucrose per brain, giving an average yield of about 20 mg of protein per ml.

ASSESSMENT OF THE PREPARATION

Some of the criteria which can be used are listed in *Table 1*. Of these, the demonstration of a high respiratory control ratio is perhaps the least easy to satisfy. Acceptable ratios for liver mitochondria are 4 or more for succinate oxidation, 6 or more for pyridine-nucleotide-linked oxidations.

Formerly, intact mitochondria were most easily prepared from tissues like liver and kidney, but with the digestion technique it became possible to prepare muscle mitochondria showing respiratory control ratios at least as high. Examples are given in *Table 2*. The figures obtained for blowfly flight-muscle mitochondria pinpoint the problem of why a preparation coupled so tightly with respect to NAD-linked oxidations may show poor respiratory control with flavin-linked substrates. In part the answer may lie in a failure to realise optimal conditions for state 3 respiration. In part it may also reflect the fact that the two respiratory control ratios are measuring different things. Thus a small proportion of mitochondrial particles oxidising, say, glycerolphosphate in an uncontrolled fashion may greatly reduce the respiratory control with that substrate but not with pyruvate. The reason is that the particles will be incompetent at carrying out the reactions of the Krebs cycle through loss of soluble enzymes and coenzymes. The flavin-linked oxidation

Table 1. CRITERIA OF 'BIOCHEMICAL INTEGRITY' OF MITOCHONDRIA

Property	*Remarks*
High respiratory control ratio	To be expressed as ratio of maximum rate in presence of ADP to that obtained when the ADP has been used up (*not* that observed before ADP is added); checking that another addition of ADP causes stimulation of respiration also necessary
High P/O ratios	Not as sensitive an indicator of mitochondrial damage; care to be taken both in calibration of oxygen electrode system and standardisation of ADP (Chappell, 1964)
Low ATPase activity	Normally found when good respiratory control ratios are observed; uncoupling agents stimulate very markedly
H^+ efflux on addition of O_2 to anaerobic suspension	(1) Preparation contains Ca^{2+}; H^+ efflux occurs when Ca^{2+} accumulated; (2) mitochondrial membrane damaged (e.g. by use of EDTA) and permeable to K^+
Reversed electron transport	Intramitochondrial NAD(P) rapidly and extensively reduced by e.g. succinate or ascorbate-TMPD in an energy-dependent process; this does not occur in uncoupled mitochondria
Latency of mitochondrial enzymes	Many mitochondrial enzymes are far more active when assayed with broken than with 'intact' mitochondria (Chappell and Greville, 1963); the most probable cause of 'latency' or 'crypticity' is impermeability of the mitochondrial membrane to substrate, products or both
e.g.	
(i) Glutamate or malate dehydrogenase	At least two factors are involved: permeability to substrate (glutamate, malate, etc.) and to NAD(P); the latter most likely causes the low order of activity of these enzymes when assayed with *intact* mitochondria
(ii) Fumarate or aconitate hydratase	With fumarate hydratase, 'latency' most probably due to the feeble penetration of the mitochondrial membrane by fumarate (Chappell and Haarhoff, 1967); that of aconitate hydratase observed only with substrate-depleted mitochondria (Chappell, 1964) and in absence of *l*-malate which activates penetration of mitochondrial membrane by citrate, isocitrate and *cis*-aconitate
(iii) Thiosulphate transulphurase (rhodanese) ($HCN + S_2O_3^{2-} \rightleftharpoons CNS^- + SO_3^{2-} + H^+$)	Perhaps the simplest case of 'latency' of a mitochondrial enzyme since both reactants and products are either uncharged small molecules (HCN) or small anions; swelling of mitochondria causes marked increase in enzyme activity (Greville and Chappell, 1959)
Stimulation by added cytochrome *c* and NAD(P)	Reveals an appreciable fraction of broken mitochondria; added cytochrome *c* stimulates respiration of particles most if a KCl (as opposed to sucrose) medium is used

would be less affected. By contrast, a leakiness of the inner mitochondrial membrane might be expected to lead to a more general lowering of respiratory control.

It is possible to envisage how the two distinct forms of damage might occur. Mechanical damage might lead to a preparation containing some damaged particles but with a proportion of 'good' mitochondria. Such a preparation is that of beef-heart mitochondria, using a Waring blendor. These may show good respiratory control with glutamate and malate as substrate but poor with succinate, and oxidise exogenous NADH. On the other hand, on storing for a week at 0°C, there is a general loss of respiratory control,

Table 2. RESPIRATORY CONTROL RATIOS OF MITOCHONDRIA

Substrate	*Rate of oxygen consumption,* mμg atoms O_2/min/mg protein	*Respiratory control ratio*
	Rat-heart mitochondria, 23°C	
Glutamate	200	16
Pyruvate/malate	290	10
Palmitoyl Cn/malate	221	not measured
NADH	12	—
	Candida utilis mitochondria, 30°C	
Succinate/rotenone	80	not measured
2·5 mM citrate/malate	78	not measured
Isocitrate/malate	6	not measured
Oxoglutarate/malate	112	3·0
Pyruvate/malate	188	3·9
NADH	324	1·5
	Blowfly flight-muscle mitochondria	
Pyruvate/proline (25°C)	725	20
Glycerolphosphate (20°C)	860	3·5

coupled with an increased tendency to oxidise citrate, to which there is a complete latency in the fresh preparation. This suggests a leakiness of the mitochondrial membrane but not such disruption that the pyridine nucleotide is lost.

Clearly, such a heterogeneity in the mitochondrial preparation may not matter at all in some experiments, for example, studies of the oxidation of pyridine-nucleotide-linked substrates, but will be an acute embarrassment in others, such as the oxidation of added NADH by means of a 'shuttle'. The latter is indeed another criterion of integrity that can be used, since well-prepared mitochondria do not oxidise exogenous NADH (Lehninger, 1951). Preparations of muscle mitochondria often fail to meet this requirement—this

Table 3. MITOCHONDRIAL ARTEFACTS

Mitochondria do not	*Possible cause*	*Test to determine cause*	*Remarks*
Consume O_2 at reasonable rates	(i) contamination of apparatus with respiratory inhibitors, e.g. antimycin, rotenone	wash apparatus with ethanol	use less inhibitor in future
	(ii) too few mitochondria	increase number of mitochondria or add bovine plasma albumin	—
	(iii) due to damage during preparation have lost cytochrome *c* and/or NAD	add back these co-factors	improve method of preparation; avoid use of saline media
Oxidise succinate rapidly and linearly	oxaloacetate inhibition	add rotenone or amytal or a compound which removes oxaloacetate by transamination, e.g. glutamate or cysteine sulphinate	particularly with heart muscle preparations; especially marked in manometric experiments
Oxidise citrate, isocitrate or oxoglutarate	(i) no activator for carrier system present	add malate	—
	(ii) mitochondria do not have a carrier	try effect of malate	blowfly mitochondria do not appear to have a carrier
Oxidise pyruvate	no 'sparker' present	add malate or succinate	with blowfly mitochondria, these compounds do not penetrate; use either proline or $HCO_3^- + ATP$

Oxidise fatty acids or acyl-CoA compounds	(i) too high a concentration of substrate	reduce concentration and add bovine plasma albumin	test for detergent action by following light scattering
	(ii) no carnitine present	add carnitine or use acyl-carnitine as substrate	formation of an acyl-carnitine necessary before acyl groups can be transferred to the β-oxidation system
Show good respiratory control	(i) contamination of apparatus with uncoupling agents or oligomycin	wash apparatus with ethanol	use less uncoupler or one that does not adhere to glass and plastic surface, e.g. 2,4-dinitrophenol
	(ii) contamination of mitochondrial preparation with other subcellular organelles, e.g. myofibrils, microsomes, cell-membrane fragments containing ATPase	oligomycin should inhibit state 4 rate of respiration	improve separation procedure; also try leaving out Mg^{2+} and adding EGTA; most extra-mitochondrial ATPases are Mg^{2+}- or Ca^{2+}-activated
	(iii) presence of fatty acids liberated by phospholipase	add bovine plasma albumin	use bovine plasma albumin routinely in both preparation and assay
	(iv) sub-optimal concentrations of reactants	vary concentrations of phosphate and substrates	—
	(v) sub-optimal incubation temperature	vary temperature over range 15–37°C (perhaps lower with e.g. fish mitochondria)	—

again probably indicates the presence of a certain number of broken mitochondria. Such an oxidation of NADH is greatly reduced in heart mitochondria prepared as described here (see *Table 2*).

It is very important that the electron microscope should also be used in assessing mitochondrial preparations, if such facilities are available. This approach yields information regarding possible heterogeneity of preparations of organelles, as well as revealing damage and contamination with other cellular structures. *Figure 1* (reproduced by permission of Dr B. Sacktor) shows an electron

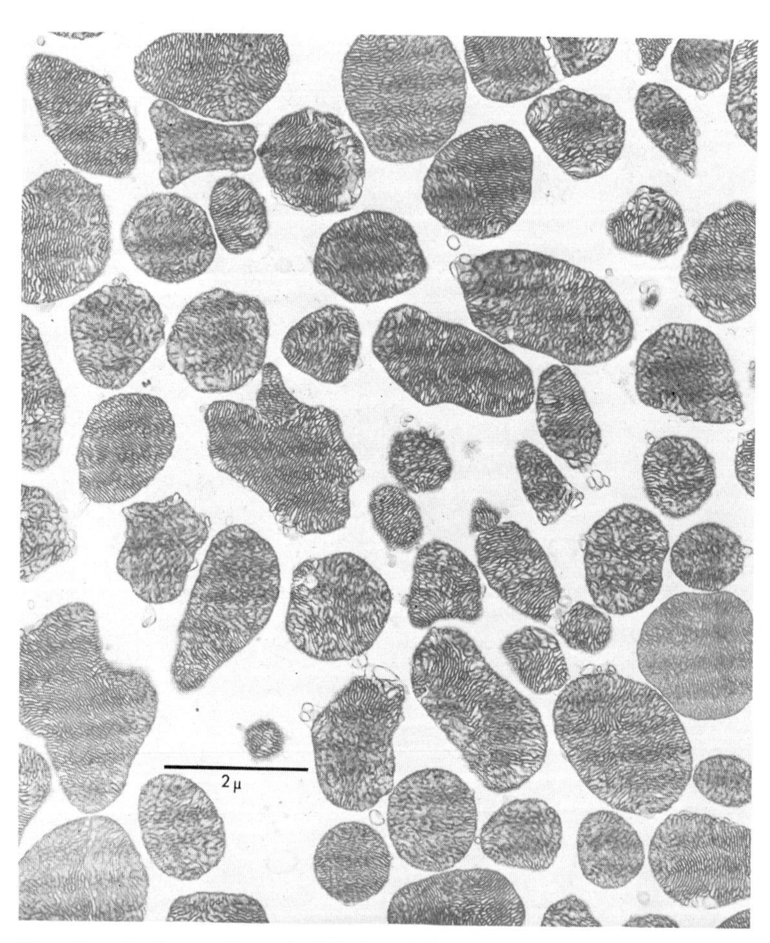

Figure 1. An electron micrograph of fly flight-muscle mitochondria isolated by the technique described in this paper. Contamination by other organelles is almost non-existent. The morphology of the mitochondria is almost unaltered from that of mitochondria in intact cells

micrograph of fly flight-muscle mitochondria prepared by one of us (R. G. H.). Mitochondria were fixed in suspension with 5% glutaraldehyde and then collected by centrifugation in a Coleman microfuge, to give a pellet not thicker than 0·2 mm. Following fixation with osmium tetroxide and embedding, sections were cut to investigate different areas of the pellet to minimise the risk that a selection of a proportion of the organelles had occurred. The field presented is entirely representative.

Finally, a number of artefacts are described (*Table 3*). This collection owes less to flippancy than to a realisation that the answers obtained are very dependent on the exact experimental procedures adopted.

REFERENCES

CHANCE, B. and HAGIHARA, B. (1961). *Proc. 5th int. Congr. Biochem., Moscow*, **5**, 3

CHAPPELL, J. B. (1964). *Biochem. J.*, **90**, 225

CHAPPELL, J. B. and GREVILLE, G. D. (1963). *Biochem. Soc. Symp.*, **23**, 39

CHAPPELL, J. B. and HAARHOFF, K. N. (1967). In *Biochemistry of Mitochondria* (ed. Slater, E. C., Kanuiga, Z. and Wojtczak, L.), p. 75, London (Academic Press)

DUELL, E. A., INOUE, S. and UTTER, M. F. (1964). *J. Bact.*, **88**, 1762

GRAY, E. G. and WHITTAKER, V. P. (1962). *J. Anat.*, **96**, 79

GREVILLE, G. D. and CHAPPELL, J. B. (1959). *Biochim. biophys. Acta*, **33**, 267

LEHNINGER, A. L. (1951). *J. biol. Chem.*, **190**, 345

Methods in Enzymology (1967). Vol. 10, (ed. Estabrook, R. W. and Pullman, M. E.), London (Academic Press)

SANADI, D. R. and FLUHARTY, A. L. (1963). *Biochemistry, N.Y.*, **2**, 523

SCHATZ, G., TUPPY, H. and KLIMA, J. (1963). *Z. Naturf.*, **186**, 1945

5

PREPARATION OF LYSOSOME-RICH FRACTIONS WITH OR WITHOUT PEROXISOMES

E. Reid

In this article an account is given of centrifugal methods for preparing lysosome-rich fractions from liver, preceded by a summary of relevant features of the cytoplasmic organelles involved. Centrifugal methods for preparing lysosomal fractions from other sources are then briefly surveyed, and non-centrifugal methods are touched on. Sources in which the lysosome field is reviewed more adequately include de Reuck and Cameron (1963*), de Duve and Baudhuin (1966), Straus (1967*) and Dingle and Fell (1969*).

The term differential centrifugation in this article connotes (perhaps rather restrictively) pelleting in tubes without a gradient, as one method for rate-sedimentation. An acceptable alternative is differential 'pelleting'. The term zonal centrifugation (as distinct from zonal rotor) has become ambiguous and is avoided. Equilibrium sedimentation and flotation are two forms of isopycnic centrifugation. Where purity is dubious, a term such as mitochondrial fraction rather than mitochondria is used.

DISCOVERY AND NATURE OF LYSOSOMES AND PEROXISOMES

The following sketch of the historical and biochemical background to these organelles may help in connection with the procedures to be described later.

* References marked with an asterisk have been placed at the end of the formal reference list under the book title.

LYSOSOMES

Hogeboom *et al.* (1948) pioneered the use of sucrose as a medium for differential centrifugation of liver homogenates. The distribution of various enzymic activities among the fractions obtained, namely, the crude nuclear, mitochondrial (large-granule), microsomal (small-granule) and supernatant fractions, was later summarised by Hogeboom *et al.* (1953). The mitochondrial fraction was rich in respiratory enzymes such as cytochrome oxidase (E.C.1.9.3.1), whereas the microsomal fraction was rich in the enzyme glucose-6-phosphatase (E.C.3.1.3.9), as studied extensively in de Duve's laboratory (Hers *et al.*, 1951). Each of the fractions was quite rich in uricase (urate oxidase, E.C.1.7.3.3). Mitochondria were believed, in agreement with Novikoff *et al.* (1953), to be the locus of acid phosphatase—an assumption accepted by Berthet and de Duve (1951)—and also of certain hepatic enzymes studied by other authors, for example, catalase (Greenfield and Price, 1956) and acid ribonuclease. Electron-microscopic techniques were not then available for morphological characterisation of the fractions, but dark-background light microscopy indicated that the large-granule fraction was indeed rich in mitochondria.

Meanwhile, the Louvain group led by C. de Duve was performing meticulous studies on the enzymology of cytoplasmic particles isolated by differential centrifugation in 0·25 M sucrose (Berthet and de Duve, 1951; Appelmans *et al.*, 1955; de Duve *et al.*, 1955). They were careful to validate their assay conditions, and shrewdly observed that the 'free' activity of acid phosphatase in fresh tissue preparations under isotonic conditions was low, and that latent activity became manifest with severe treatment such as freezing and thawing under hypotonic conditions. They inferred that the enzyme was situated in particles having a membrane which impeded access of substrate to the enzyme.

By 1955, as set down in a fascinating account by de Duve (1969*), this and other evidence had led them to propose that acid phosphatase is in a group of particles distinct from mitochondria, and designated lysosomes. As was argued in a paper that is now a classic (de Duve *et al.*, 1955), acid ribonuclease and certain other hydrolases are also present in these particles. In the experiments described in that paper, the Hogeboom–Schneider (1955) scheme was varied by preparing a small-bulk centrifugal fraction ('L') between the mitochondrial and microsomal fractions; this fraction was shown to be rich in hydrolases. This centrifugal evidence for a distinct organelle is more conclusive than evidence based on the release of enzyme activity *in vitro* compared with the release of mitochondrial enzyme activity, since, for example, the mitochondrial enzyme glutamate dehydrogenase is 'activated' as sluggishly as typical lysosomal enzymes (Bendall and de Duve, 1960). However, a recent approach, entailing the use of digitonin, furnishes enzyme activation data which enable mitochondria, lysosomes and peroxisomes to be distinguished as possible sites of enzyme activity (P. Jacques, personal communication).

Microscopy eventually came into line with biochemistry when peri-canalicular dense bodies which stain for acid phosphatase were shown to correspond to the postulated lysosomes (Novikoff, 1963*). Electron microscopy has established that these bodies are much scarcer than mitochondria and often somewhat smaller in size; they are bounded by a single membrane. Although acid phosphatase is the marker of choice for lysosomes, it is not axiomatic that any organelle which appears to stain for acid phosphatase in a tissue section must be a lysosome.

It is beyond the scope of this article to outline the present-day concept of primary and secondary lysosomes comprising an intracellular digestive system (Jacques, 1969*)—a concept that was taking shape some ten years after the discovery of lysosomes (de Reuck and Cameron, 1963*; Straus, 1967*). Suffice it to say that the biochemist must be aware of the probable existence in his material of secondary lysosomes which are possibly fragile and different in centrifugal behaviour from the parent lysosomes which serve as storage depots for the hydrolases.

For most lysosomal enzymes it is a normal finding that as much as 15% of the total homogenate activity (of the same order as the free activity of homogenates) appears in the final supernatant fraction (soluble phase), possibly reflecting metabolic activity in the cell (Reid and Nodes, 1959) whether or not disrupted secondary lysosomes are the source. There is some evidence that, with the Potter–Elvehjem homogeniser used under varying conditions, the appearance of such activity in the supernatant fraction is an all-or-none phenomenon (Reid and Nodes, 1963). However, de Duve's group have established that more drastic means of homogenisation, for example, a blendor, may cause biochemically disastrous solubilisation of lysosomal activity. 'Genuine' soluble cytoplasmic activity (possibly present *in vivo*) may conceivably be due partly to an isoenzyme different from the enzyme responsible for the corresponding activity in lysosomes, as happens with acid phosphatase in mouse liver if the substrate is *p*-nitrophenyl phosphate rather than β-glycerophosphate (Neil and Horner, 1965).

PEROXISOMES (de Duve and Baudhuin, 1966; Hogg and de Duve, 1970).

The soluble phase obtained from homogenates of liver in 0·25 M sucrose is notably rich in catalase (E.C.1.11.1.6), yet Greenfield and Price (1956) concluded (e.g. from trying polyvinylpyrrollidone in the medium) that the enzyme really resides in mitochondria. This view had to be modified when, as for acid phosphatase, evidence accumulated for a location in particles less readily sedimentable than mitochondria (Thomson and Klipfel, 1958). However, it turned out that the enzyme, together with certain others such as uricase and D-amino acid oxidase, is associated not with lysosomes but with so-called peroxisomes (also termed uricosomes). These correspond to the microbodies long known to electron microscopists. They are similar in size to lysosomes, and are hard to separate from them (see below), but differ in having a dense core of osmophilic material. They contain oxidoreductases rather than hydrolases (Roodyn, 1967*). Catalase is unusual in having definite latency like acid phosphatase, though differing in nature from the structure-linked latency of that enzyme (de Duve, 1965).

PRESENT-DAY PICTURE

Lysosomes and peroxisomes are now respectable members of the community of subcellular particles. They are known to occur in protozoa and invertebrates, and lysosomes occur in most tissues in vertebrates, as well as in plants. However, our knowledge of their biochemical properties stems mainly from work with rat liver. As with other organelles such as mitochondria (see Hartman and Hinton, 1971*), the early concept of a homogeneous population of particles has had to be modified in the case of lysosomes. Evidence from hormonal and other studies (e.g. Reid and Nodes, 1959; Slater *et al.*, 1963*; A. A. El-Aaser, unpublished experiments) showed that different lysosomal enzymes could vary in level independently. However, it is the centrifugal approach that has given conclusive evidence that the population is heterogeneous, possibly comprising two or more sub-classes rather than a continuous spectrum. This heterogeneity, for which evidence is mentioned later in this article, is not surprising in view of the functional progression of the primary lysosome within the cell, and of parallel morphological progression from buds (probably arising in the Golgi complex) to residual bodies. It should not, however, be taken for granted that the heterogeneity is solely at the intracellular level. There is already some evidence (Wattiaux *et al.*, 1963*) that Küpffer-cell lysosomes differ somewhat in enzyme complement from parenchymal-cell lysosomes. The problem of separating out particular types of cell (see Beaufay, 1969*; Dingle and Barrett, 1969*) in sufficient quantities to serve for preparing lysosomes must await the development of macro-techniques, possibly with use of zonal rotors (Reid, 1971*a*).

ISOLATION OF HEPATIC LYSOSOMES: OPERATIONAL ASPECTS

As is shown below, it is not difficult in principle to reproduce the conditions of de Duve *et al.* (1955) so as to obtain a lysosome-enriched pellet which will suffice for many purposes, although its lysosome content will be under 10% and it will also contain peroxisomes together with mitochondria and microsomes. It is much more difficult to obtain reasonably pure preparations of lysosomes free from peroxisomes, and *vice versa*, in fair yield. The source of the difficulty is the overlap in key properties, including size (approx. 1 μm), among mitochondria, lysosomes, peroxisomes, microsomes, and plasma-membrane fragments (Leighton *et al.*, 1968; Beaufay, 1969*; see also Hinton *et al.*, 1971*b**). Various values have been quoted for the respective densities, and indeed these may truly change with a change in the *in vitro* environment. For a hypertonic sucrose medium with no divalent cations added, median values of the following order are found: mitochondria, 1·19 g/cm^3 (1·16 g/cm^3 for fragments of the outer membrane); lysosomes containing acid phosphatase, 1·205 g/cm^3 (1·22 g/cm^3 for those containing acid ribonuclease); peroxisomes, 1·23 g/cm^3; smooth microsomes, 1·16 g/cm^3; plasma-membrane vesicles, 1·16 g/cm^3. In 0·25 M sucrose, however, mitochondria, lysosomes and peroxisomes all seem to have a median density of about 1·10 g/cm^3, and also the overlap in particle-size is unpropitious for rate-sedimentation separation. Moreover, in this medium, mitochondria become rounded rather than rod-like. It is fortunate that, because the sizes do not overlap exactly, hepatic lysosomes and peroxisomes do tend to sediment rather more slowly than mitochondria under differential-centrifugation conditions.

Discouraging though the picture is, an empirical approach may be rewarding if pursued with awareness of past failures and successes. Judicious choice of centrifugal conditions, coupled with expeditious working, is more efficacious than use of media other than sucrose. A particularly ingenious and effective approach is to inject the animals with an agent which lysosomes, manifesting their day-to-day function, ingest and thereby become changed in size and density (Wattiaux *et al.*, 1963★). The agent commonly used, as in examples given below, is the detergent Triton WR-1339 (supplied by Rohm and Hass).

PROCEDURES FOR EXAMINING FRACTIONS

For the purpose of examining the appearance of products obtained by centrifugation, standard electron-microscopic techniques are used, with fixation in glutaraldehyde and then in osmium tetroxide buffered at pH 7. Since the components of centrifugal pellets tend to be stratified, use should be made of a very small pellet, or of cryostat sections of a frozen suspension containing 1% gum acacia, or of an ultrafilter membrane on which the material is held (El-Aaser, 1971★). El-Aaser (1971) has described a useful variant of morphological examination, whereby the elements are stained for phosphatase activity with an appropriate substrate.

Assay of fractions, and of the homogenate, for marker enzymes is a laborious but essential step in subcellular work, calling for judgment and thoroughness (de Duve, 1967★; Reid, 1971*b*★). *Table 1* summarises suitable conditions for these enzymes. It is obviously advantageous to use automatic or, at least, work-simplified procedures wherever possible, with computer handling of the data (Hinton *et al.*, 1971*a*★). If sucrose is present in high concentration, corrections for its inhibitory actions must be made (Hinton *et al.*, 1969). If *p*-nitrophenyl phosphate (or phenyl phosphate) is chosen as the substrate for acid phosphatase for convenience, it must be remembered that the soluble phase of the cell contains an enzyme which attacks this substrate but not the recommended substrate, namely, β-glycerophosphate. If Triton X-100 is to be used in place of freezing and thawing to liberate latent activity, its concentration must be kept low (say 0·1% v/v) lest tissue protein be inadvertently carried through the acid-precipitation step and so sabotage the colorimetric determination of liberated phosphate.

CENTRIFUGAL METHODS FOR HEPATIC LYSOSOMES

PREPARATION OF ANIMALS AND TISSUE

For differential centrifugation (but not for zonal-rotor work) we routinely use rats fasted overnight, as in de Duve's laboratory; they may be killed by cervical fracture. Perfusion is usually of no

[illegible] SUITABLE CONDITIONS† FOR ASSAY OF MARKER ENZYMES

Enzyme	*Succinate dehydrogenase (mitochondria)*	*Acid phosphatase[a] (lysosomes)*	*Uricase (peroxisomes)*	*Glucose-6-phosphatase (microsomes)*	*5′-Nucleotidase (plasma membrane)*
Tissue equivalent (as fresh liver) per tube, for a lysosomal ('light-mitochondrial') fraction obtained by differential centrifugation	1 mg in 0·5 ml	(1) 10 mg[b] in 0·5 ml or (2) 2 mg in 0·5 ml	10 mg[c] in 1 ml, in silica cuvette	10 mg in 0·5 ml	10 mg in 0·5 ml
Buffer[d]	0·25 ml of 0·5 M phosphate, pH 7·4, cont. 0·25 mg INT[e]	0·4 ml of 0·3 M 3,3- dimethyl-glutarate, pH 5·0	1 ml of 5 mM phosphate, pH 7·4, cont. 0·2% (v/v) Triton X-100	0·4 ml of 0·3 M 3,3-dimethyl-glutarate, pH 6·4, cont. 25 mM EDTA	0·4 ml of 0·3 M tris-HCl, pH 7·8, cont. 12·5 mM $MgCl_2$
Substrate (pH adjusted; sometimes permissible to add it with the buffer); omit in blanks	0·25 ml of 0·3 M succinate (0·3 M malonate in blanks)	(1) 0·1 ml of 1 M β-glycero-phosphate or (2) 0·1 ml of 10 mM *p*-nitrophenyl-phosphate	1 ml of 10 mM phosphate, pH 7·4, cont. 86 μg of Na^+ urate[f]	0·1 ml of 50 mM glucose-6-phosphate (Ba^{2+} salt)	0·1 ml of 50 mM 5′-AMP or 5′-UMP
Incubation conditions (preferably with shaking except for uricase)	20 min at 37°C	30 min at 37°C (10 min for 'free' activity[b])	follow E_{292} for at least 5 min, preferably at 37°C	30 min at 37°C	30 min at 37°C
Final measurement on supernatant fluid after adding 1·5 ml of 6% trichloracetic acid at 0°C	E_{490} on ethyl acetate extract (4 ml)	estimate phosphate		estimate phosphate	estimate phosphate
Typical activity for whole homogenate, μmoles/g of liver/min	6	(1) 8 or (2) 20	4[c] at 20°C; 9[c] at 37°C	15	10

† The methods, as typically used for manual assays in the author's laboratory, are based largely on methods described by other authors, for example, de Duve *et al.* (1955). When protein has finally been removed, estimation of liberated phosphate can be done in a continuous-flow automatic analyser (with no dialysis step; ascorbic acid is a suitable agent for reducing the phosphomolybdate). Such an analyser can be used for the complete assay if it is *p*-nitrophenol that is liberated (Leighton *et al.*, 1968; Tappel, 1969*; Hinton *et al.*, 1971*a**).

a. Liberate latent activity by at least 5 freeze-thaws of the hypotonic tissue suspension.
b. For 'free' activity use the equivalent of 50 mg of liver, with 0·25 M sucrose present.
c. The amount of tissue needed is not well predictable (Baudhuin *et al.*, 1964).
d. Buffer pH values at temperature of incubation.
e. 2-(*p*-iodophenyl)-3-(*p*-nitrophenyl)-5-phenyltetrazolium chloride.
f. Solution must be fresh; it becomes inhibitory when stored.

advantage, although it should be remembered that both erythrocytes and plasma contain acid phosphatases. Perfusion may, in fact, lead to an increase in the sedimentation rate of cytoplasmic organelles (Jacques, 1958) and a decrease in density of plasma-membrane fragments (Coleman *et al.*, 1967).

A blendor should on no account be used for homogenising (Berthet and de Duve, 1951; cf. Sawant *et al.*, 1964). Homogenising devices suitable for lysosome isolation have been described by Dingle and Barrett (1969*). However, the conventional Potter–Elvehjem homogeniser works well for liver. A few strokes are sufficient, with the pestle, diameter 0·2 to 0·3 mm less than that of the vessel, rotating at 2000 rev/min. The proportion of the acid-phosphatase activity of the homogenate recovered in the final supernatant fraction after differential centrifugation should then, with normal rats, be under 15% and ideally as little as 5%. The homogenate should not be too concentrated, lest the subsequent separation be poor. Preferably at least 10 ml (minimum 5 ml) of medium should be used per g of liver. It is helpful to re-homogenise the low-speed pellet mentioned below in fresh medium and combine the supernatant fractions (de Duve *et al.*, 1955). Before this low-speed centrifugation, it is advantageous to remove any large pieces of connective tissue (along with adhering bile-duct cells) by filtration; a nylon tea-strainer works well.

MEDIA

The isotonic sucrose solution (0·25 M) that is normally used should, as a precaution, be buffered with tris-HCl (5 mM, pH 7·4) or bicarbonate (5 mM). Optionally, to ensure preservation of glucose-6-phosphatase, it may be supplemented with EDTA (1 mM). For concentrated solutions of sucrose, as used in zonal work, guidance on quantities and on gradient-making will be found in Reid (1971*a*); control of pH is recommended. The medium of Birbeck and Reid (1956), as an alternative to isotonic sucrose, consists of: raffinose, 0·23 M; dextran of high molecular weight (approx. 150 000; blood-substitute grade), 6%; and heparin, 4000 I.U. per 100 ml; EDTA (1 mM) is optional.

PELLETING IN CENTRIFUGE TUBES

Whether or not lysosomes are to be pelleted, conventional centrifugation should be done initially. If only nuclei, together with debris and unbroken cells, are to be removed, this is achieved by centrifuging at 600*g* for 10 min ($6 \times 10^3 g$-min); in the present context the conditions are not critical. Time can be saved, with some loss of yield, by omitting this centrifugation and relying on the next one (commonly in a fixed-angle rotor) to remove nuclei, etc. as well as mitochondria. The latter centrifugation, and the subsequent one to harvest the lysosomes (preceded and followed by washing centrifugations), are critical steps where conditions must be chosen by each worker and then faithfully followed for the sake of reproducibility. Values for the product of RCF and time (*g*-min) give a little help in enabling comparable conditions to be established with different centrifuges and rotors, particularly if the total value includes the acceleration and deceleration stages. Centrifuges can be fitted with devices which not only furnish an integral record of $\omega^2 t$ (ω = angular velocity) but also, as in de Duve's laboratory, regulate them. However, published *g*-min values commonly refer only to the steady-speed stage, and in any case it turns out that different rotors are better compared on the basis of angular velocity than nominal RCF (Appelmans *et al.*, 1955), even if this RCF indeed refers to the middle of the tube contents (g_{av}). The rationale of centrifugation is semi-empirical; fixed-angle rotors behave quite comparably to swing-out rotors (Schumaker, 1967; Anderson, 1968). Inescapably, each worker must establish his own conditions and abide by them.

The choice of centrifugal conditions for isolating a fraction enriched in lysosomes (and also in peroxisomes) must be based on literature for analytical differential centrifugation. Some loss of yield must obviously be accepted for the sake of minimising contamination with mitochondrial and microsomal material. However, in preparing lysosome-enriched pellets little is gained by departing from centrifugal conditions carefully chosen as the best compromise for analytical work (*Table 2* and *Figure 1*), since a tenfold purification is the best that can be hoped for. Notwithstanding thorough washing, the product is likely to be richer in peroxisomes than lysosomes (Beaufay, 1969★; cf. uricase values in *Table 2*), and will contain microsomes as judged by glucose-6-phosphatase activity (*Figure 1*) together with mitochondria and fragments thereof. There should, however, be little contamination with plasma membrane fragments, despite the apparent presence of 5′-nucleotidase activity when assayed in the presence of Mg^{2+}. This activity is low in the lysosomal fraction if attack on the nucleotide substrate by acid phosphatase is inhibited by tartrate, or obviated by keeping it latent, or corrected for by parallel assays in the presence of EDTA (*Figure 1*). These results illustrate how assays under ill-chosen conditions can be misleading.

Table 2. CENTRIFUGATION CONDITIONS FOR PREPARING LYSOSOME-RICH FRACTIONS†

Authors and notes	*Rotor used*	*Conditions to sediment*[a] *Heavy mitochondria*	*Lysosomes*	*Enzymes*[b] *in lysosomal fraction*	*Further treatment*	*Enzymes*[b] *in final lysosomal preparation*
de Duve *et al.* (1955) (integrated $\omega^2 t$)	Spinco no. 40	1: 12 500 max 2: see note *c* 3: 33	25 000 max see note *c* 250	AP: 40%; 5 AR: 33%; 5 Ur: 50%; 7		
Sawant *et al.* (1964) (blendor!)	Sorvall GSA	1: unstated 2: 10 min 3: 33	unstated 20 min 330	AP: 36%; 4 AR: 40%; 5	Resuspended and centrifuged 3 times; layering in second re-centrifugation	AP: 11%; '67' AR: 15%; '72' (but see text)
Rahman *et al.* (1967)	Conditions of de Duve *et al.* (1955)			AP: 36%; 4	Other treatments were alternatives	
El-Aaser and Reid (1969; and unpublished) (cf. Reid and Nodes, 1959)	MSE 8 × 50	1: 5000 2: 10 min 3: 30	12 000 15 min 225	AP: 55%; 7 AR: 31%; 5 Ur: 72%; 12		
Sloat and Allen (1969)	Sorvall SS-34	1: unstated 2: 10 min 3: 25	unstated 20 min 170	AP: 45%		

Leighton *et al.* (1968) and (flotation) Trouet (1964) (Triton WR-1339 injected)	Spinco no. 30	1: centri- 2: fugation 3: omitted	unstated see note *c* 340		Flotation in swing-out rotor[d]	AP: 20%; 39 see note *e*
		1: unstated 2: see note *c* 3: 24	unstated see note *c* 340		(i) Banding in Beaufay's rotor, then (ii) flotation in swing-out rotor[d]	(i) AP: 11%; 19 Ur: 0·2%; 0·4[f] (ii) AP: 8%; 49 see note *e*
Stahn *et al.* (1970; 1971★)	Conditions of de Duve *et al.* (1955) then 4 washing steps			AP: 6%; 8 AS: 9%; 14 see note *g*	Electrophoresis (see p. 114)	AP: 4%; 40[h] AS: 7%; 240[h]

† Normally in 0·25 M sucrose with a fixed-angle rotor.

a. Speed of rotor, rev/min (1); time (2); g-min $\times 10^{-3}$ (3).
b. Representative values for acid phosphatase (AP), acid ribonuclease (AR), uricase (Ur) and arylsulphatase (AS) given as per cent yield (left) and relative specific activity (right) compared with the homogenate; the pellets were usually washed well before assay.
c. Time depends on rates of acceleration and deceleration since integral $\omega^2 t$ used.
d. Two layers of sucrose (10 ml of 14·3% (w/w), density 1·06 g/cm^3, and 20 ml of 34·5% (w/w), density 1·155 g/cm^3) introduced into Spinco SW25.2 rotor tube; 25 ml (equivalent to 25 g of liver) of suspension of unwashed lysosomal-mitochondrial pellet in 45% (w/w) sucrose (density, 1·21 g/cm^3) layered underneath these, then 60·7% (w/w) sucrose (density, 1·30 g/cm^3) to fill the tube; tubes were centrifuged at 25 000 rev/min for 120 min then contents displaced with a fluorocarbon using a special pumping device; lysosomes recovered from interface between first two sucrose layers.
e. Fraction low in non-lysosomal elements including peroxisomes (catalase as marker).
f. For a peroxisome side-fraction at density 1·23 g/cm^3 (lysosomes near 1·12 g/cm^3), AP = 0·3%, 0·3; Ur = 18%, 50.
g. Relative specific activity of catalase, 3.
h. Relative specific activity given for fractions with highest activity.

To achieve a purification approaching ten-fold merely by conventional centrifugation is relatively easy. However, the step of removing the supernatant fluid from the pelleted mitochondrial fraction requires especial care. Removal of this supernatant fraction for the subsequent lysosome centrifugation is best done with a bulb-operated pipette with its tip bent by 90 degrees, so that there is minimal disturbance of the loose material overlying the pellet.

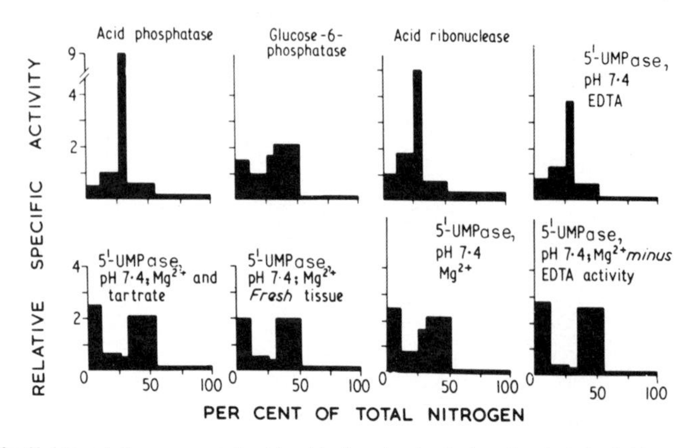

The five 'blocks' in each diagram represent, from left to right, the crude nuclear fraction, a 'heavy' mitochondrial fraction, a 'light' mitochondrial fraction (rich in lysosomes), the microsomal fraction, and the supernatant ('soluble') fraction, respectively. The height of each block represents the relative specific activity, that is, the ratio of the percentage of the total activity recovered to the percentage of total N recovered, so that the area of each block is proportional to the amount of activity in that fraction (de Duve et al., *1955). The conditions of centrifugation and assay were similar to those of de Duve* et al. *(1955), care having been taken to liberate latent activity except where 'fresh tissue' is specified. Succinate dehydrogenase (not illustrated) is mainly in the 'heavy' mitochondrial fraction. For 5'-nucleotidase assayed at pH 7·4 as illustrated, or at pH 5·0, the activity in the presence of 1 mM EDTA (without Mg^{2+}) was only 1/5 of that in the presence of 10 mM Mg^{2+}. Note that when the activity in presence of EDTA (due to acid phosphatase) is subtracted from that in presence of Mg^{2+}, the bimodal distribution of the latter is sharpened (bottom right). Similar sharpening is evident when acid phosphatase activity is minimised with 10 mM L-tartrate, or by avoiding freezing and thawing.*

Figure 1. Distribution of enzyme activities among fractions obtained from a liver homogenate by differential centrifugation (from an unpublished experiment, courtesy of Dr A. A. El-Aaser, similar to one shown in El-Aaser and Reid, 1969)

This 'fluffy layer' (Applemans *et al.*, 1955) contains microsomes and possibly some damaged mitochondria, but is low in acid phosphatase. Any fluffy-layer material carried over into the lysosomal pellet is readily removed when the latter is washed.

The raffinose–dextran–heparin medium of Birbeck and Reid (1956) may be marginally advantageous in reducing the microsomal contamination of the light-mitochondrial (lysosomal) pellet, as of the heavy-mitochondrial pellet. With this medium acid phosphatase is found mainly in the light-mitochondrial and microsomal pellets, whereas with sucrose it occurs mainly in the heavy- and light-mitochondrial pellets. It is not known whether the sedimentation behaviour of peroxisomes in this medium is similarly altered. In general, none of the various media tried in de

Duve's laboratory have proved to be much superior to sucrose (Leighton *et al.*, 1968; Beaufay, 1969★).

BANDING IN CENTRIFUGE TUBES

Various authors have attempted to purify crude lysosomal preparations by banding in tubes, usually isopycnically. The Louvain school has achieved notable success with the use of rats which have been injected with dextran, or, more commonly, Triton WR-1339, whereby primary lysosomes decrease in density (but change little in sedimentation rate). *Table 2* shows the striking enrichment obtainable, with a swing-out rotor, when the flotation method of Trouet (1964) is applied to a crude pellet rich in mitochondria as well as lysosomes. Beaufay (1969★) has questioned the claim from Tappel's laboratory that 67-fold enrichment in acid phosphatase is obtainable with normal liver when a lysosome-containing pellet is subjected to further differential centrifugation with one banding step (30 min at 9500*g*) in sucrose up to 0·7 M (*Table 2*; Tappel *et al.*, 1963★; Sawant *et al.*, 1964). Beaufay argues that the claim is improbable and not substantiated, and that lysosomes are likely to comprise little more than 20% of the product. As Beaufay (1969★) points out, the method mentioned in later papers from Tappel's laboratory (e.g. Ragab *et al.*, 1967) apparently gives an enrichment of only 10 to 27 times, and complete data for marker enzymes such as uricase are still lacking.

While zonal rotors are the tool of choice for purification of lysosomes by banding, mention should be made of tube-centrifugation studies other than those cited above. The limit of what can be achieved by isopycnic re-centrifugation of a mitochondrial-lysosomal fraction from normal liver, when layered over a sucrose gradient, is shown in *Figure 7(a)* of a paper by Beaufay *et al.* (1964). Gradients of sucrose in D_2O, or of glycogen in 0·5 M sucrose, offered some advantages in distinguishing between mitochondria, lysosomes and peroxisomes for the purpose of enzyme localisation. After centrifugation for 150 min at 39 000 rev/min in a Spinco SW-39L rotor, mitochondria were found at about density 1·19 g/cm^3, uricase at about 1·25 g/cm^3, partly separated from acid phosphatase, and this last diffusely spread with median density 1·20 g/cm^3. Acid deoxyribonuclease showed a shallow peak at density 1·23 g/cm^3.

Acid ribonuclease, which does not quite coincide with acid phosphatase in fractions separated by differential centrifugation (de Duve *et al.*, 1955; Reid and Nodes, 1959),,sedimented faster than acid phosphatase in sucrose gradients (Rahman and Cerny, 1969). The paper by Rahman and Cerny (1969) serves to confirm that lysosomes are heterogeneous, but is almost valueless for the purpose of devising a lysosomal separation. The study of Sloat and Allen (1969) (see *Table 2*) was mainly concerned with two forms of lysosomal acid phosphatase ('bound' and 'soluble'; typically α-naphthyl phosphate as substrate), but is of interest because density-gradient centrifugation, of whole cytoplasmic fractions, was done both isopycnically (Beaufay *et al.*, 1964; see above), giving acid phosphatase at density 1·20 g/cm^3, and by rate-sedimentation with 1 M KCl present, ultimately with pelleting of acid phosphatase but not of glucose-6-phosphatase. These experiments hardly provide a model for the design of separation conditions.

BANDING IN ZONAL ROTORS

If the limitation of running only one sample at a time is acceptable, then the method of choice for separating lysosomes and their sub-classes is centrifugation in a zonal rotor under carefully chosen conditions. A zonal rotor in the hands of an experienced operator allows material equivalent to approx. 10 g of liver to be well resolved, largely because of the design feature whereby the rotor is loaded and unloaded while it is rotating (see Anderson, 1966; Reid, 1971*a*).

Using a special rotor, which is not available commercially, Beaufay and collaborators (see *Table 2*) effectively separated lysosomes, virtually free of contaminating elements, by applying their profound experience of the behaviour of particles when centrifuged isopycnically in tubes. This work, now being followed up by Beaufay (personal communication) with a zonal rotor suitable for rate-sedimentation separations, is a challenge to zonal workers who have to make use of a commercial rotor. They should choose a suitable rotor, not a slow one like the AXII as perforce used by Rahman *et al.* (1967), since, in such rotors, attainment of a satisfactory RCF calls for a short path-length separation with restriction of the gradient to the edge of the rotor. The work with the AXII rotor (described with confusing mention of a 'cushion' of 0·25 M glucose) did provide confirmation of the view that the lysosomal population is heterogeneous, as judged by assays for acid phosphatase and acid ribonuclease (Rahman *et al.*, 1967). These authors handicapped themselves by performing the zonal separations with a whole homogenate or with a crude pellet spun down from the homogenate. Centrifugation was at 4000 rev/min for periods of up to 120 min, so that lysosomes could hardly have come completely to isopycnic equilibrium, nor could mitochondria, although this is unclear since the authors' graphs lack density values. Uricase banded similarly to acid phosphatase, and succinoxidase similarly to acid ribonuclease, yet the authors state that the banding was 'markedly different'. Livers from rats injected with Triton WR-1339 gave a poorer peak for acid ribonuclease, this peak and the acid phosphatase peak being almost unchanged in their positions.

In early studies with an AXII rotor, Schuel *et al.* (1968) centrifuged a whole homogenate, regrettably without a prior pelleting step. By rate sedimentation for 285 min they obtained a lysosomal fraction, the contaminants of which included microsomes. The enrichment factor was typically 5, if one exceptional value of 23 is disregarded. With an AXII rotor, tolerable separation of glucose-6-phosphatase, acid phosphatase, succinate dehydrogenase and 5′-nucleotidase (in that order from centre to edge of the rotor) can indeed be achieved, as shown by Hartman and Hinton (1971★), who centrifuged for 60 min a resuspended pellet containing the nuclei, mitochondria and lysosomes from 10 g of normal liver through a gradient which had a flattened middle portion around density 1·12 g/cm^3. However, Hartman and Hinton (1971★) were concerned with isolating mitochondria (and also nuclei) rather than lysosomes, as were other zonal workers cited by them. They used the raffinose–dextran–heparin medium of Birbeck and Reid (1956) for the homogenate (although not for the gradient), with apparent benefit in minimising microsomal contamination of the lysosomes. However, sucrose is not contra-indicated, and indeed was used throughout in rate-sedimentation work, with the AXII rotor, described by Hartman and Reid (1969). Here the zonal separation was for 75 min, but might advantageously have been slightly longer with a view to moving the mitochondria farther from the rotor centre, and away from the lysosomes.

Rate-sedimentation separations are best done with a somewhat faster rotor as described below, but preferably not with a B-type rotor since these are rotating too quickly during loading and have rather short path-lengths. However, B-type rotors can be used if no other is available. A method of questionable efficacy has been reported by Brown (1968), who centrifuged a whole homogenate through a Ficoll gradient in a BXV rotor for 15 min at 10 000 rev/min. Acid phosphatase (*p*-nitrophenyl phosphate as substrate) partly moved into the gradient, and a further centrifugation (possibly isopycnic) achieved some purification, though no estimates were given. Ficoll, even if advantageous, is rather expensive for zonal-rotor work. In the method briefly described by Withers *et al.*, (1968) using sucrose in a BXIV rotor, the authors claim to have purified lysosomes isopycnically. However, the peak was not properly characterised, and Beaufay (1969★) is rightly sceptical about the supposed separation from mitochondria. A particular difficulty that bedevils lysosomal separations is the possible presence of outer-membrane fragments derived from mitochondria. Monoamine oxidase assays with a suitable substrate are helpful in this connection (Baudhuin *et al.*, 1964; Hinton *et al.*, 1971*b*★).

After much of the foregoing work had been done, a faster version of the AXII rotor was marketed, designated 'HS' (Measuring and Scientific Equipment Ltd) or 'Z–XV' (International Equipment Co.). This rotor is particularly appropriate for isolating lysosomes. In common with the AXII rotor, its construction allows the mitochondrial band to be observed during the actual centrifugation. It gives an RCF of 4000 near the centre and almost 13 000 at the edge. Isopycnic banding of lysosomes could be slow with this rotor, but this hardly matters; the above-mentioned work with tubes shows the isopycnic approach to be usually unhelpful, notwithstanding the striking separations obtained isopycnically by Beaufay using his special rotor. In the following description of the techniques reported by Burge and Hinton (1971★), the principle is that with the chosen running time there is isopycnic banding of mitochondria but not of lysosomes or peroxisomes.

TECHNIQUE WITH AN HS ZONAL ROTOR

A mitochondrial-lysosomal pellet (11 500*g* for 15 min) from 9 g of liver is resuspended in about 25 ml of 0·25 M sucrose containing 5 mM tris-HCl, pH 7·4 (as for the gradient itself). An alternative starting material is the whole cytoplasmic fraction, but this must be equivalent to not more than 2 g of liver if microsomal contamination is to be minimised. The rotor is loaded, while rotating at about 1500 rev/min, with a pre-cooled (4°C) sucrose gradient which is steep initially and again when a density of 1·13 g/cm^3 is reached (cf. *Figure 2*). Use may be made of a commercial gradient former (e.g. Beckman or MSE) or of a device built and operated as described elsewhere (Reid, 1971*a*; for quantities, see Burge and Hinton, 1971★). If mitochondria are not required as a side product, a simple gradient linear with volume might be tried, ranging from 1·01 to 1·18 g/cm^3 in density, but then there is the risk that small lysosomes might still be close to the microsomes and that large lysosomes might sediment close to mitochondria. When the gradient (550 ml) has been introduced, the rotor is filled with 2 M sucrose as the 'cushion', some of which is then displaced when the sample is loaded from a syringe, which is operated with steady, but not excessive pressure lest a cross-leak develop. The usual overlay of 0·25 M sucrose is then introduced, and the feed-head detached. The rotor is accelerated and run for 45 min at 9000 rev/min. Finally its contents are displaced with 2·2 M sucrose, and collected in fractions, the volumes of which are either constant or subsequently measured. Before or after the actual collection the shape of the gradient is checked, suitably by an on-line or a manual refractometer. Fuller information on the relevant apparatus and procedures is given in Reid (1971*a*); a particular point to watch with the HS rotor is possible erratic running when the feed-head is in place, due to the drag which it exerts.

SEPARATIONS OBTAINED IN THE HS ZONAL ROTOR

Figure 2 shows an experiment with normal liver. Acid phosphatase is well separated from glucose-6-phosphatase, 5′-nucleotidase and succinate dehydrogenase, but not from acid ribonuclease. Of the acid phosphatase in the material loaded (64% of that in the original homogenate), 33% is recovered in the peak tubes (nos. 13 to 18), excluding the heavy lysosome region, with 15-fold enrichment on a protein basis as compared with the homogenate. The corresponding values for uricase (not plotted; 34% of the homogenate activity was loaded) are 43% and 11-fold, respectively. Since the amount of protein in tubes 13 to 18 is only 6% of that loaded (only 21% of the homogenate protein had been loaded), there is enrichment not only in acid phosphatase but also, regrettably, in glucose-6-phosphatase (four fold) and in 5′-nucleotidase (two fold), although not in succinate dehydrogenase. The amount of acid phosphatase in the sample zone is so low as to rule out the possibility that the lysosomes had suffered much damage.

Figure 3 shows an experiment with liver from rats which had been injected with Triton WR-1339. Acid phosphatase activity now sediments rather faster, overtaking, although still overlapping with, acid ribonuclease, and almost clear of uricase. While the relative specific activities at the peaks for acid phosphatase and uricase are hardly higher than those from normal liver, there is much less cross-contamination between peroxisomes and lysosomes or, more correctly, those lysosomes which have taken up the detergent to become secondary lysosomes. Break-up of

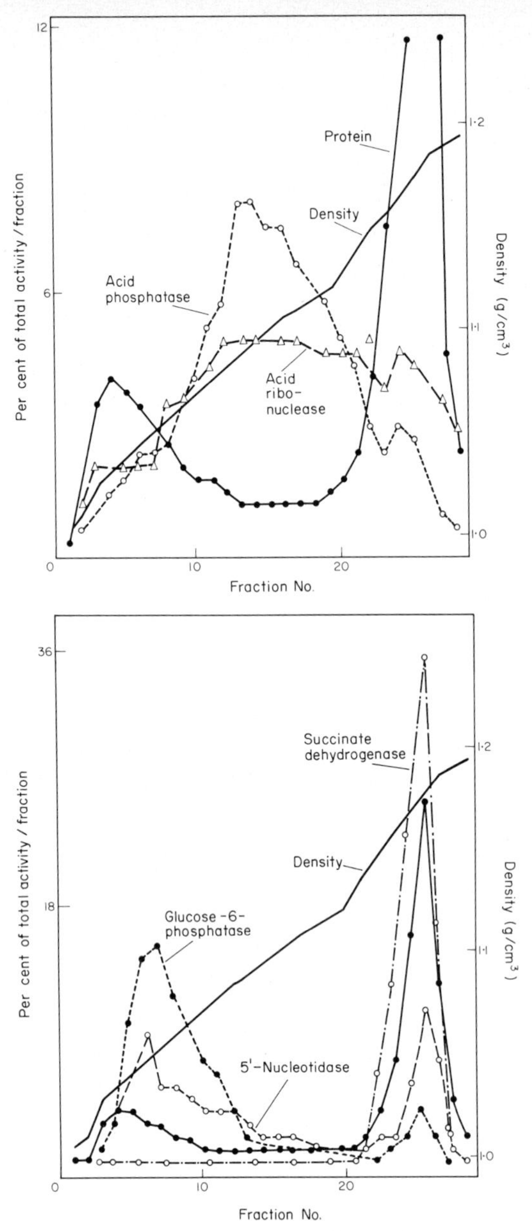

The fraction corresponds to a mixture of the 'heavy' and 'light' mitochondrial fractions of Figure 1. *It was prepared by resuspending the pellet obtained from a post-nuclear fraction from 9 g of liver centrifuged at 11 500g for 15 min* (i.e. *170 000g-min); centrifugation in the HS rotor was done as described in the text and 20 ml fractions were collected.*

———, *density gradient at 5°C (1·2 g/cm³ corresponds to 1·54* M *sucrose; the 2* M *sucrose cushion is beyond the last point shown);* ●——●, ○---○, △——△, ●---●, ○——○, *and* ○-·-○ *denote the percentage in each 20 ml fraction of the total protein, acid phosphatase, acid ribonuclease, glucose-6-phosphatase, 5'-nucleotidase and succinate dehydrogenase, respectively, recovered.*

Note (i) *protein has barely moved from the sample position;*
(ii) *heights of peaks do* not *denote specific activities (cf.* Figure 1*);*
(iii) *uricase (not illustrated) was similar in distribution to acid phosphatase, but the peak (fractions 13 to 18) was somewhat sharper.*

Figure 2. Distribution of enzyme activities and protein found after centrifuging a mitochondrial-lysosomal fraction from rat liver in an HS zonal rotor (from Burge and Hinton, 1971★)

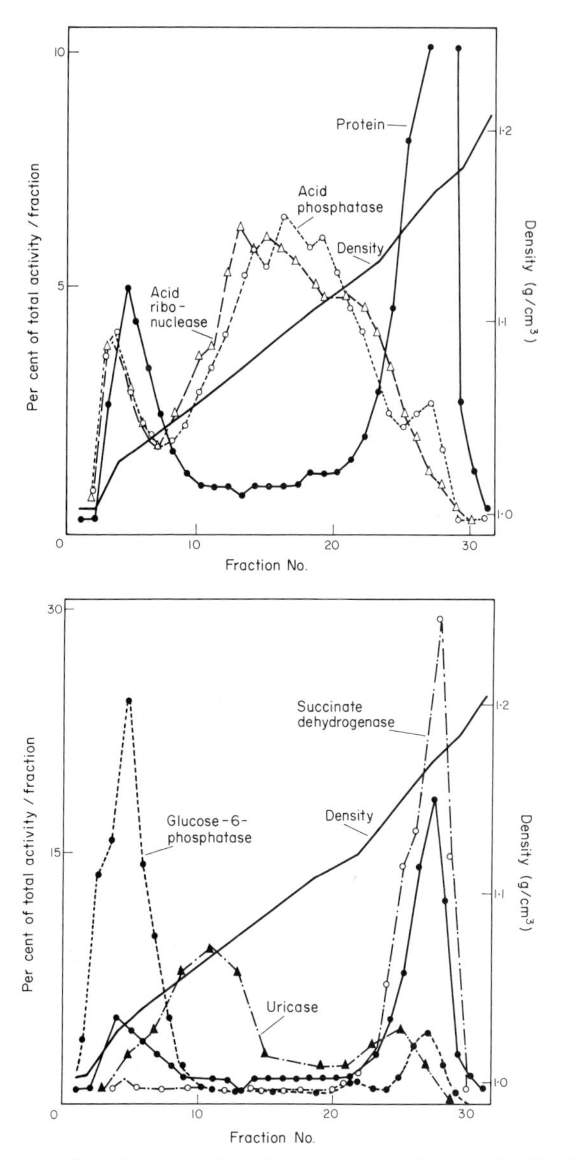

Preparation of sample and fractionation was as described for Figure 2; *the rats received one intraperitoneal injection of Triton WR-1339 (2·5 g/kg body weight) 3 days previously. Symbols are as for* Figure 2, *except* ▲— · —▲, *percentage in each fraction of total uricase activity recovered.*

Note: *acid phosphatase is shifted relative to uricase and acid ribonuclease (cf.* Figure 2*)*

Figure 3. Distribution of enzyme activities and protein found after centrifuging a mitochondrial-lysosomal fraction from livers of rats injected with Triton WR-1339 in an HS zonal rotor (from Burge and Hinton, 1971★)

some secondary lysosomes might account for the peak of acid phosphatase in the sample zone.

ISOPYCNIC RE-CENTRIFUGATION OF ZONAL FRACTIONS IN TUBES

In connection with their studies of acid ribonuclease, Burge and Hinton (1971★) subjected a sample of a zonal peak (fractions 13 to 18 of *Figure 2*) to equilibrium sedimentation in a sucrose gradient. Acid phosphatase and acid ribonuclease had coincident main peaks at density 1·20 g/cm^3, whereas uricase peaked at density 1·22 g/cm^3. The same behaviour was found with the 'heavy' edge of the zonal peak. With a zonal peak obtained after treatment with Triton WR-1339 (*Figure 3*), equilibrium sedimentation showed a shift to density 1·12 g/cm^3 for acid phosphatase but not for acid ribonuclease, although both enzymes now showed marked peaks in the soluble region. This is presumably because of damage to Triton-loaded lysosomes during pelleting and resuspension of the zonal material from the zonal fractions. (Membrane filtration might have enabled the lysosomes to be concentrated without damage.) In summary, isopycnic re-centrifugation (which could be done in a zonal rotor if sufficient material were available) gives, with Triton-altered lysosomes, an acid ribonuclease peak the leading edge of which is low in acid phosphatase although rich in uricase. The leading edge of the uricase peak is almost free of both hydrolases, whether or not Triton has been administered.

CENTRIFUGAL METHODS FOR TISSUES OTHER THAN LIVER

A partial survey of the literature is given in *Table 3*. In brief, fair success seems to have been achieved in purifying lysosomes from kidney, brain, adrenal medulla, and certain other sources. Much of the non-hepatic work is beset by two difficulties that are more serious than in work with liver. There is, first, the apparent heterogeneity of the lysosomal population, and second, the task of disrupting the cells without disrupting the lysosomes. Among other handicaps, one hepatic marker, uricase, is absent in some species, and glucose-6-phosphatase is absent from most non-hepatic tissues and cells†. Moreover, acid phosphatase is a questionable marker in the case of leucocytes (Beaufay, 1969★) and yeast (Matile, 1969★). It is a matter of opinion whether the term 'lysosome' should be applied to granules lacking acid phosphatase but

† The use of NAD glycohydrolase as a microsomal marker (Stagni and de Bernard, 1968) is of doubtful validity.

Table 3. SOME ATTEMPTS TO ISOLATE LYSOSOMAL FRACTIONS FROM TISSUES OTHER THAN LIVER

Author(s)	*Tissue*	*Comments*
Burge and Hinton (1971★)	Hepatoma transplants	Zonal rotor as for liver; mitochondria (small) poorly separated from lysosomes (cf. Hartman and Reid, 1969)
Straus (1956, 1963★, 1967★); *see also* Beaufay (1969★) *and* Goldstone *et al.* (1970)	Kidney, usually after injection of a foreign protein	Differential centrifugation to give several 'droplet' fractions with lysosomes of diverse size; fair purity, low yield
Shibko and Tappel (1965)	Kidney	Differential centrifugation, etc.; Beaufay (1969★) points out inconsistencies in the data
Wattiaux-De Coninck *et al.* (1965); de Duve and Baudhuin (1966); *see also* Beaufay (1969★)	Kidney	Equilibrium sedimentation (better than rate sedimentation) giving lysosomes + peroxisomes; Triton of limited help
Maunsbach (1969★); *see also* Beaufay (1969★)	Kidney	Rate and then equilibrium sedimentation; more enzymology desirable (good morphology)
Gitzelmann *et al.* (1964)	Intestine	Differential centrifugation
van Lancker and Holtzer (1959)	Pancreas	Differential centrifugation; lysosomes probably impure
Dingle and Barrett (1969★)	Kidney	Homogenate filtered through beads; lysosomes centrifuged into a Ficoll layer and then run in a zonal rotor
Koenig *et al.* (1964); Koenig (1969★, 1970)	Brain	Rate and then equilibrium sedimentation; evident heterogeneity of lysosomes
Spanner and Ansell (1971★)	Brain (cortex)	Crude pellet sedimented to equilibrium in zonal rotor; lysosomes 'of high purity'
Canonico and Bird (1970)	Skeletal muscle (Triton sometimes injected)	Zonal rotor, rate-sedimentation or isopycnic; 2 lysosomal populations —1 due to macrophages, etc.?
Vaes and Jacques (1965); Vaes (1969★)	Bone (from very young animals)	Rate sedimentation; cellular heterogeneity a problem—should perform a cell separation first?
Bowers and de Duve (1967); Bowers (1969★); *see also* Straus (1967★)	Lymphoid tissue, e.g. spleen, thymus	Rate or isopycnic sedimentation; evident heterogeneity of lysosomes; spleen and thymus behave similarly

Table 3 (*continued*)

Author(s)	*Tissue*	*Comments*
Schultz *et al.* (1965); cf. Baggiolini *et al.* (1970)	Leucocytes, platelets, etc.	Rate sedimentation; some enrichment in acid hydrolases (Baggiolini used zonal rotors for isopycnic centrifugation)
Wattiaux (1962); Munro *et al.* (1964)	Cultured cells	Fractions from HeLa cells and fibroblasts
Woessner (1965, 1969*); *see also* Slater *et al.* (1963*)	Uterus, prostrate, mammary gland	Rate sedimentation; handicap that tissue disruption may cause lysosomal disruption
Smith and Winkler (1969*)	Adrenal medulla	Rate and then equilibrium sedimentation; hydrolases were at density about 1·20 g/cm^3
Matile (1969*); Cartledge *et al.* (1971*)	Eukaryotic micro-organisms; higher plants (Matile, 1969*)	Rate and equilibrium sedimentation (zonal rotor) for lysosomes and peroxisomes; Matile also used flotation, and questions validity of acid phosphatase as marker (cf. Perlman and Mahler, 1970)

containing other latent hydrolases, with or without peroxidase. The literature evidently calls for critical scrutiny if an isolation is to be attempted, particularly since enzymic characterisation of fractions has often been poor, perhaps excusably so in the case of the peroxisomes, these organelles having fairly recently arrived on the biochemical scene.

METHODS BESIDES CENTRIFUGATION

Approaches other than centrifugation have been tried by various authors, for example, Sawant *et al.* (1964), usually with little success. There are, however, two that warrant mention. One, developed for kidney, is merely an alternative to the initial low-speed centrifugation usually performed to remove unbroken cells, etc. It consists of filtering the homogenate through layers of Ballotini beads varying in size, after which the filtrate is centrifuged through a double layer of sucrose–Ficoll, which prevents the lysosomes becoming impacted and damaged (Dingle and Barrett, 1969*). A filtration method has likewise been used to separate nuclei from enlarged renal lysosomes (Straus, 1967*). The other, developed by

Stahn *et al.* (1970; 1971*), entails continuous free-flow electrophoresis, in a special apparatus, of a lysosomal fraction prepared by differential centrifugation with several washing steps. The authors report overall yields of the order of 5%, with striking enrichment for lysosomal enzymes (*Table 2*). Arylsulphatase in a fast electrophoretic fraction was enriched 240-fold, and was evidently in a type of lysosome different from that containing acid phosphatase and β-glucuronidase. The lysosomes were almost free of peroxisomes and other elements. Stahn *et al.* (1971*) consider that the electrophoretic method is complementary, not an alternative, to centrifugation. A zonal-rotor separation could take the place of the differential centrifugation described. These striking results warrant trial of the electrophoretic technique in other laboratories.

CONCLUSIONS AND SUMMARY

Rate-sedimentation of a liver homogenate in 0·25 M sucrose without a gradient ('differential pelleting') can, with care, furnish a lysosomal–peroxisomal fraction in almost 50% yield with up to 10-fold enrichment in marker enzymes, albeit contaminated with mitochondria and microsomes. Partial separation of the organelles in a crude pellet, namely, lysosomes of at least two types, and peroxisomes, can be achieved by isopycnic sedimentation in tubes, preferably with pre-treatment of the rats (usually with Triton WR-1339) such that the lysosomes containing acid phosphatase become more readily separated from the other organelles. With a special rotor (not available commercially) for isopycnic sedimentation, a peroxisome-rich fraction has also been isolated, as well as a fraction enriched 19-fold in acid phosphatase. Enrichments approaching 50-fold have been obtained by isopycnic flotation in tubes, even without prior isopycnic sedimentation.

A commercial zonal rotor of the type 'HS' ('Z–XV') is advantageous for separations deliberately based on rate-sedimentation, and partly resolves peroxisomes and two classes of lysosomes if the rats are pre-treated with Triton WR-1339. The starting material should be a pelleted fraction, not a whole homogenate. Particularly striking separations have been reported for a novel electrophoretic technique applied to a centrifugal pellet.

Organelles of the lysosome type can be isolated from kidney and certain other sources, but with most non-hepatic tissues little progress has been made, because of heterogeneity in organelles and other technical difficulties, coupled, in some laboratories, with inadequate assays.

Acknowledgments. A word of gratitude is due to past and present members of my research group, particularly Dr A. A. El-Aaser and Dr R. H. Hinton. The group, which was at the Chester Beatty Research Institute till 1965, has had support from several sources including the Cancer Research Campaign, the Science Research Council and the Wellcome Trust. Valuable comments were received from Dr P. Jacques of the University of Louvain.

REFERENCES

ANDERSON, N. G. (1966). Ed., *Natn. Cancer Inst. Monog.*, **21**

ANDERSON, N. G. (1968). *Analyt. Biochem.*, **23**, 72

APPELMANS, F., WATTIAUX, R. and DE DUVE, C. (1955). *Biochem. J.*, **59**, 438

BAGGIOLINI, M., HIRSCH, J. G. and DE DUVE, C. (1970). *J. Cell Biol.*, **45**, 586

BAUDHUIN, P., BEAUFAY, H., RAHMAN-LI, Y., SELLINGER, O. Z., WATTIAUX, R., JACQUES, P. and DE DUVE, C. (1964). *Biochem. J.*, **92**, 179

BEAUFAY, H., JACQUES, P., BAUDHUIN, P., SELLINGER, O. Z., BERTHET, J. and DE DUVE, C. (1964). *Biochem. J.*, **92**, 184

BENDALL, D. S. and DE DUVE, C. (1960). *Biochem. J.*, **74**, 444

BERTHET, J. and DE DUVE, C. (1951). *Biochem. J.*, **50**, 174

BIRBECK, M. S. C. and REID, E. (1956). *J. biophys. biochem. Cytol.*, **2**, 609

BOWERS, W. E. and DE DUVE, C. (1967). *J. Cell Biol.*, **32**, 339

BROWN, D. H. (1968). *Biochim. biophys. Acta*, **162**, 152

CANONICO, P. G. and BIRD, J. W. C. (1970). *J. Cell. Biol.*, **45**, 321

COLEMAN, R., MICHEL, R. M., FINEAN, J. B. and HAWTHORNE, J. N. (1967). *Biochim. biophys. Acta*, **135**, 573

DE DUVE, C. (1965). *Harvey Lects.*, **59**, 49

DE DUVE, C. and BAUDHUIN, P. (1966). *Physiol. Revs.*, **46**, 323

DE DUVE, C., PRESSMAN, B. C., GIANETTO, R., WATTIAUX, R. and APPELMANS, F. (1955). *Biochem. J.*, **60**, 604

EL-AASER, A. A. and REID, E. (1969). *Histochem. J.*, **1**, 417

GITZELMAN, R., DAVIDSON, E. A. and OSINAK, J. (1964). *Biochim. biophys. Acta*, **85**, 69

GOLDSTONE, A., SZALO, E. and KOENIG, H. (1970). *Life Sci.*, **9**, 607

GREENFIELD, R. E. and PRICE, V. E. (1956). *J. biol. Chem.*, **220**, 607

HAGGIS, G. H. (1966). *The Electron Microscope in Molecular Biology*, London (Longmans)

HARTMAN, G. C. and REID, E. (1969). *FEBS Lett.*, **5**, 180

HERS, H. G., BERTHET, J., BERTHET, L. and DE DUVE, C. (1951). *Bull. Soc. Chim. Biol.*, **33**, 21

HINTON, R. H., BURGE, M. L. E. and HARTMAN, G. C. (1969). *Analyt. Biochem.*, **29**, 248

HOGEBOOM, G. H., SCHNEIDER, W. C. and PALADE, G. (1948). *J. biol. Chem.*, **172**, 619

HOGEBOOM, G. H., SCHNEIDER, W. C. and STRIEBICH, M. J. (1953). *Cancer Res.*, **13**, 617

HOGG, J. F. and DE DUVE, C. (1970). Eds., *The nature and function of peroxisomes, Ann. N.Y. Acad. Sci.*, **168**, 211–381

JACQUES, P. (1958). *Thesis*, University of Louvain, Belgium

KOENIG, H. (1970). In *Handbook of Neurochemistry* (ed. Lajtha, A.), vol. 2, p. 255, New York (Plenum Publishing Corp.)

KOENIG, H., GAINES, D., MCDONALD, T., GRAY, R. and SCOTT, J. (1964). *J. Neurochem.*, **11**, 729

LEIGHTON, F., POOLE, B., BEAUFAY, H., BAUDHUIN, P., COFFEY, J. W., FOWLER, S. and DE DUVE, C. (1968). *J. Cell Biol.*, **37**, 482

MUNRO, T. R., DANIEL, M. R. and DINGLE, J. T. (1964). *Exptl Cell Res.*, **35**, 515

NEIL, M. and HORNER, M. W. (1965). *Biochem. J.*, **92**, 217

NOVIKOFF, A. B., PODBER, E., RYAN, J. and NOE, E. (1953). *J. Histochem. Cytochem.*, **1**, 27

PERLMAN, P. S. and MAHLER, H. R. (1970). *Arch. Biochem. Biophys.*, **136**, 245

RAGAB, H., BECK, C., DILLARD, C. and TAPPEL, A. L. (1967). *Biochim. biophys. Acta*, **148**, 501

RAHMAN, Y. E. and CERNY, E. A. (1969). *Biochim. biophys. Acta*, **178**, 61

RAHMAN, Y. E., HOWE, J. F., NANCE, S. L. and THOMSON, J. F. (1967). *Biochim. biophys. Acta*, **146**, 484
REID, E. (1971*a*). Ed., *Separations with Zonal Rotors*, Guildford (Wolfson Bioanalytical Centre, University of Surrey)
REID, E. and NODES, J. T. (1959). *Ann. N.Y. Acad. Sci.*, **81**, 618
REID, E. and NODES, J. T. (1963). *Nature, Lond.*, **199**, 176
SAWANT, P. L., SHIBKO, S., KUMTA, U. S. and TAPPEL, A. L. (1964). *Biochim. biophys. Acta*, **85**, 82
SCHUEL, H., SCHUEL, R. and UNAKAR, N. J. (1968). *Analyt. Biochem.*, **25**, 146
SCHULTZ, J., CARLIN, R., ODDI, F., KAMINKER, K. and JONES, W. (1965). *Arch. Biochem. Biophys.*, **111**, 73
SHUMAKER, V. N. (1967). *Adv. biol. med. phys.*, **11**, 246
SHIBKO, S. and TAPPEL, A. (1965). *Biochem. J.*, **95**, 731
SLOAT, B. F. and ALLEN, J. M. (1969). *Ann. N.Y. Acad. Sci.*, **166**, 574
STAGNI, N. and DE BERNARD, B. (1968). *Biochim. biophys. Acta*, **170**, 129
STAHN, R., MAIER, K.-P. and HANNIG, K. (1970). *J. Cell Biol.*, **46**, 576
STRAUS, W. (1956). *J. biophys. biochem. Cytol.*, **2**, 513
THOMSON, J. F. and KLIPFEL, F. J. (1958). *Exptl Cell Res.*, **14**, 612
TROUET, A. (1964). *Arch. intern. Physiol. Biochim.*, **72**, 698
VAES, G. and JACQUES, P. (1965). *Biochem. J.*, **97**, 380, 389
VAN LANCKER, J. L. and HOLTZER, R. L. (1959). *J. biol. Chem.*, **234**, 2359
WATTIAUX, R. (1962). *Arch. int. Physiol.*, **70**, 765
WATTIAUX-DE CONINCK, S., RUTGEERTS, M. J. and WATTIAUX, R. (1965). *Biochim. biophys. Acta.*, **105**, 446
WITHERS, B., DAVIES, I. AB. I. and WYNN, C. H. (1968). *Biochem. biophys. res. Commun.*, **30**, 227
WOESSNER, J. F. (1965). *Intern. Rev. Connective Tissue Res.*, **3**, 221

References marked with an asterisk★ are to articles in the following books:
Ciba Symp. Lysosomes (ed. de Reuck, A. V. S. and Cameron, M. P.), London, 1963★ (J. and A. Churchill)
DE DUVE, C., p. 1
NOVIKOFF, A. B., p. 36
SLATER, T. F., GREENBAUM, A. L. and WANG, D. Y., p. 311
STRAUS, W., p. 151
TAPPEL, A. L., SAWANT, P. L. and SHIBKO, S., p. 78
WATTIAUX, R., WIBO, M. and BAUDHUIN, P., p. 176
Enzyme Cytology (ed. Roodyn, D. B.), London, 1967★ (Academic Press)
DE DUVE, C., p. 1
REID, E., p. 321
ROOYDN, D. B., p. 103
STRAUS, W., p. 239
Lysosomes in Biology and Pathology (ed. Dingle, J. T. and Fell, H. B.), Amsterdam, 1969★ (North-Holland)
BEAUFAY, H., vol. 2, p. 516
BOWERS, W. E., vol. 1, p. 167
DE DUVE, C., vol. 1, p. 3
DINGLE, J. T. and BARRETT, A. J., vol. 2, p. 555
DOTT, H. M., vol. 1, p. 330
JACQUES, P. J., vol. 2, p. 395
KOENIG, H., vol. 2, p. 111
MATILE, PH., vol. 1, p. 406
MAUNSBACH, A. B., vol. 1, p. 115
SMITH, A. D. and WINKLER, H., vol. 1, p. 155
TAPPEL, A. L., vol. 2, p. 547

VAES, G., vol. 1, p. 217
WOESSNER, J. F., vol. 1, p. 299
Separations with Zonal Rotors (ed. Reid, E.), Guildford, 1971★ (Wolfson Bioanalytical Centre, University of Surrey)
BURGE, M. L. E. and HINTON, R. H., p. S-5.1
CARTLEDGE, T. G., COOPER, R. A. and LLOYD, D., p. V-4.1
EL-AASER, A. A., p. B-5.1
HARTMAN, G. C. and HINTON, R. H., p. S-4.1
HINTON, R. H., p. Z-5.1
HINTON, R. H., MULLOCK, B. M. and REID, E., (*a*) p. B-4.1
HINTON, R. H., NORRIS, K. and REID, E., (*b*) p. S-2.1
REID, E., (*b*) p. B-3.1
SPANNER, S. and ANSELL, G. B., p. V-3.1
STAHN, R., MAIER, K. -P. and HANNIG, K., p. S-6.1

6

PURIFICATION OF PLASMA-MEMBRANE FRAGMENTS

R. H. Hinton

It is exceptionally difficult to discuss, in general terms, methods for the isolation of plasma-membrane fragments. While the aim of most procedures for the isolation of a particular cell component is to obtain that component intact and undamaged, this is clearly not possible in the case of the plasma membrane, for this must be broken in order to remove the cell contents. The nearest approach that can be obtained to an intact membrane preparation is the cell 'ghosts' which may be obtained from erythrocytes (Dodge *et al.*, 1962) and fat cells (Rodbell, 1967). When these cells are lysed under carefully controlled conditions, the membrane re-seals after the escape of the cell contents. Such membrane preparations are extremely useful for the study of transport phenomena, as the re-sealed membrane retains its polarity, and the composition of the internal space can be controlled in so far as the liquid in this space exchanges freely with the lysis medium during the escape of the cell contents. Such 'ghosts' are, however, by no means pure membrane preparations, since a proportion of the soluble cytoplasmic proteins and, in the case of the fat-cell ghosts, of the mitochondria, remain trapped inside the ghosts. In order to obtain cell-membrane preparations free from such contaminants the integrity of the membrane must be sacrificed.

Careful control of the homogenisation conditions is even more important for the isolation of plasma-membrane fragments than it is for the isolation of other cell components. Although each tissue presents its own particular problems, for the purposes of this discussion three groups can be distinguished. First, there are 'soft' tissues, such as liver and kidney, where there is little intercellular connective tissue, but the cells are bound together by strong junctional complexes. Second, there are related tissues in which such complexes are weak or non-existent. This class includes many tumours, and also cells grown in tissue culture. Finally, there are 'hard' tissues, such as muscle, which must be treated extremely vigorously in order to obtain any cell breakage. The methods developed for the isolation of plasma-membrane fragments have, in general, been concerned with tissues in the first two classes. In this article the problems in isolating plasma-membrane fragments from liver and hepatoma will be discussed in some detail to illustrate the difficulties which may be encountered with the first two types of tissue. Methods for other tissues, including those in the third class, will also be considered briefly.

LIVER AND HEPATOMA PLASMA-MEMBRANE FRAGMENTS

When liver is dispersed by use of a Potter–Elvehjem homogeniser, light-microscopic examination of the homogenate, or of fractions separated from it, suggests that the tissue is initially torn into extremely small scraps. Each consists of several attached cells (*Figure 1a*), which are successively broken and emptied of their contents (*Figure 1,b* and *c*) without any further rupture of the intercellular bonds. This results finally in the liberation of sheets of membrane which derive from several adjacent cells (*Figure 1d*). These large sheets of membrane sediment very rapidly and are normally recovered in the crude nuclear fraction prepared by differential centrifugation. When, however, the distribution of the plasma-membrane marker 5′-nucleotidase (Reid, 1966) is examined, it is clear that, even with the mildest homogenisation, only half of the enzyme activity is associated with the crude nuclear fraction, the remainder being associated with the microsomal fraction (El-Aaser and Reid, 1969). Increasing the speed of rotation of the homogeniser pestle, or the number of strokes, or decreasing the clearance between pestle and vessel, has little effect on the proportion of 5′-nucleotidase in the crude nuclear fraction. However, with a type of homogeniser in which release from high pressure plays a significant role in breaking the cells, for example the Chaikoff press (Emanuel and Chaikoff, 1957) or the nitrogen pressure homogeniser (Hunter and Commerford, 1961; see *Chapter*

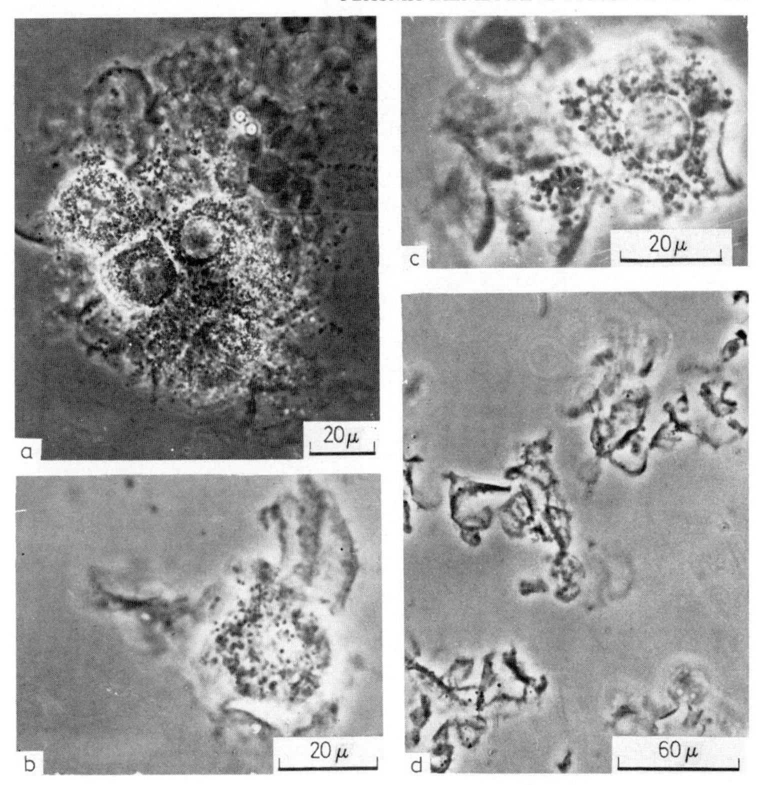

Figure 1. Liver cells at what appears to be different stages of fragmentation; (a) *small clump of cells, together with adhering nuclei;* (b) *whole cell, with adhering membranes of adjacent broken cells;* (c) *broken cell, with adhering membranes of adjacent cells;* (d) *clumps of membranes derived from a group of cells (*a, b *and* d *from Hinton* et al., *1970).*

Phase-contrast photomicrographs of material separated by centrifugation in an AXII zonal rotor; similar fragments can be seen in whole homogenates but the presence of small fragments makes photography difficult

1 of this volume), the plasma membrane becomes fragmented to vesicles (or sacs).

When, on the other hand, hepatoma tissue is disrupted in a Potter–Elvehjem homogeniser, the absence of any strong intercellular bonds results in the fragmentation of the tissue to single cells before any significant cell breakage occurs. The plasma-membrane fragments subsequently released are much smaller than those released from liver, but again fragments of two distinct sizes are found, namely sheets, in this case derived from the membrane of single cells, and vesicles.

The size distribution of the larger fragments of plasma membrane

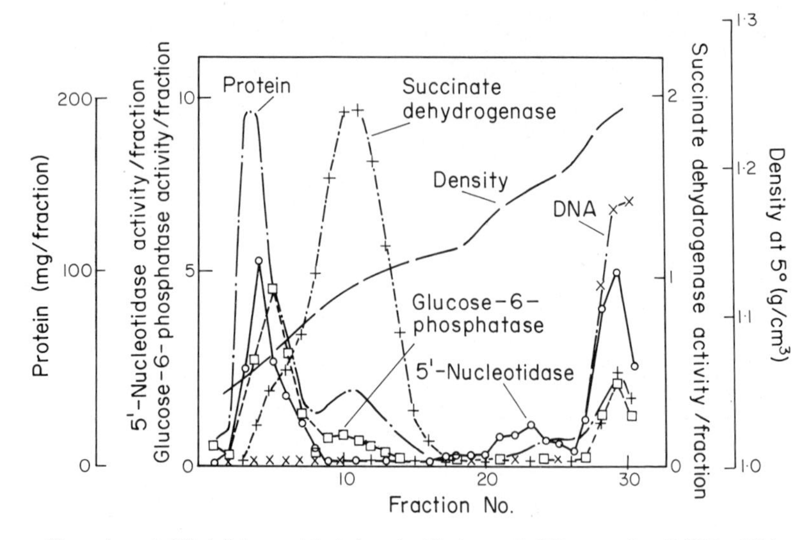

The sample comprised 20 ml of a homogenate (equivalent to 4 g of liver) prepared in 0·25 M *sucrose, 5* mM *$NaHCO_3$, pH 7·5, by 3 strokes in a Potter–Elvehjem homogeniser. The rotor was loaded with a sucrose gradient of form AII* (Table 5)*, containing 5* mM *$NaHCO_3$, the sample, and a 50 ml overlay of 0·08* M *sucrose; during acceleration a further 70 ml of overlay solution was taken up into the rotor; centrifugation was at 3700 rev/min for 60 min at 4°C*

Figure 2. Pattern obtained by centrifuging an homogenate of rat liver in an AXII zonal rotor

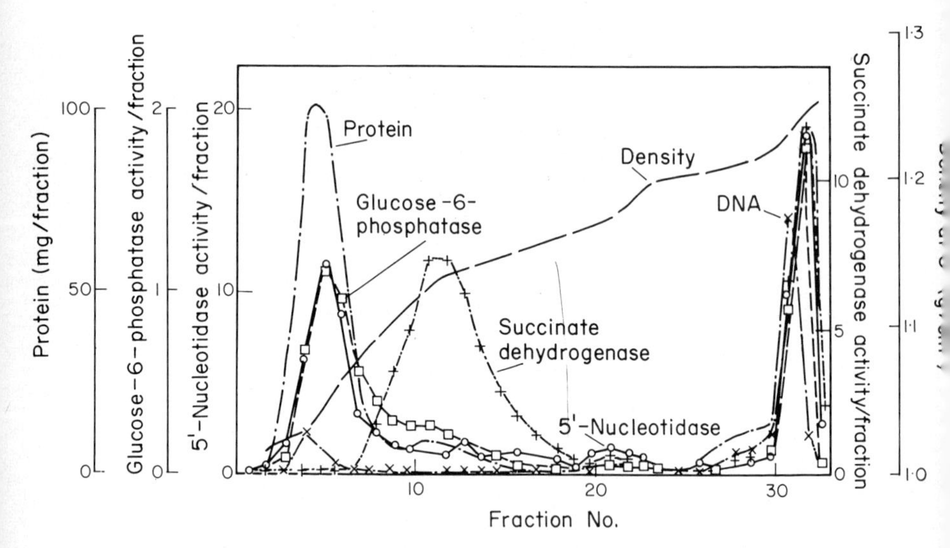

The sample (70 ml, equivalent to 12 g of hepatoma) was prepared, and the rotor loaded, as described for Figure 2*; centrifugation was at 3700 rev/min for 90 min at 4°C*

Figure 3. Pattern obtained by centrifuging an homogenate of rat hepatoma in an AXII zonal rotor

in a homogenate can be examined by centrifugation of the homogenate in an AXII zonal rotor. Results obtained with liver and hepatoma are shown in *Figures 2* and *3*, respectively. Fragments of microsomal size remain close to the original position of the sample, while whole cells and aggregates band with the nuclei at the dense end of the gradient. Free sheets of plasma membrane are found in the intermediate region. The large sheets of liver plasma membrane, detected by 5′-nucleotidase activity, sediment much faster than mitochondria and are banded at a density of 1·18 g/cm^3. The smaller sheets of hepatoma membrane sediment at a similar rate to the mitochondria, their broad distribution showing them to be heterogeneous in size.

Although a technique for the purification of liver plasma-membrane fragments by flotation in step gradients in small tubes was described as early as 1960 (Neville, 1960), before zonal rotors became generally available, methods for the purification of membrane fragments using zonal rotors can best be discussed first. The 'continuous scan' which they provide illustrates the principles of the separation and the sources of contamination better than the apparently arbitrary results obtained with step gradients in small tubes. Methods for separating membrane fragments from the microsomal fraction are discussed separately from those for separating fragments from the crude nuclear fraction. Background information on the use of zonal rotors will be found in Reid (1971).

SEPARATION OF MEMBRANE SHEETS

While small amounts of plasma-membrane sheets may be separated directly from liver homogenates by centrifugation in an AXII zonal rotor, only about 4 g of liver can be loaded on to the gradient. Attempts to increase the amount of material which can be handled in a single centrifugation step by performing a preliminary low-speed centrifugation, to collect the membrane fragments and remove small organelles, fail due to aggregation of the plasma-membrane fragments and the red blood cells which occurs on pelleting. This aggregation can be avoided by removal of the red blood cells by homogenisation in a hypotonic medium, or by perfusion of the liver *in situ* (Hinton *et al.*, 1970, 1971) or by homogenisation in a medium containing dextran, EDTA and heparin (Hartman and Hinton, 1971). Of these, perfusion is the preferred procedure as it is least likely to affect other cell organelles. Centrifugation of a crude nuclear fraction from perfused liver in an AXII zonal rotor with a suitable gradient results in the plasma-membrane fragments being collected in a narrow band at density 1·17 g/cm³, clearly separated from the neighbouring bands of mitochondria and red blood cells (*Figure 4*). These membranes are contaminated by other cell organelles trapped in the membrane sheets. The contaminants can be removed by rehomogenisation to break up the large sheets followed by flotation from sucrose of density 1·19 g/cm³.

A very similar method for the isolation of plasma-membrane fragments from rat or mouse liver using an AXII zonal rotor has

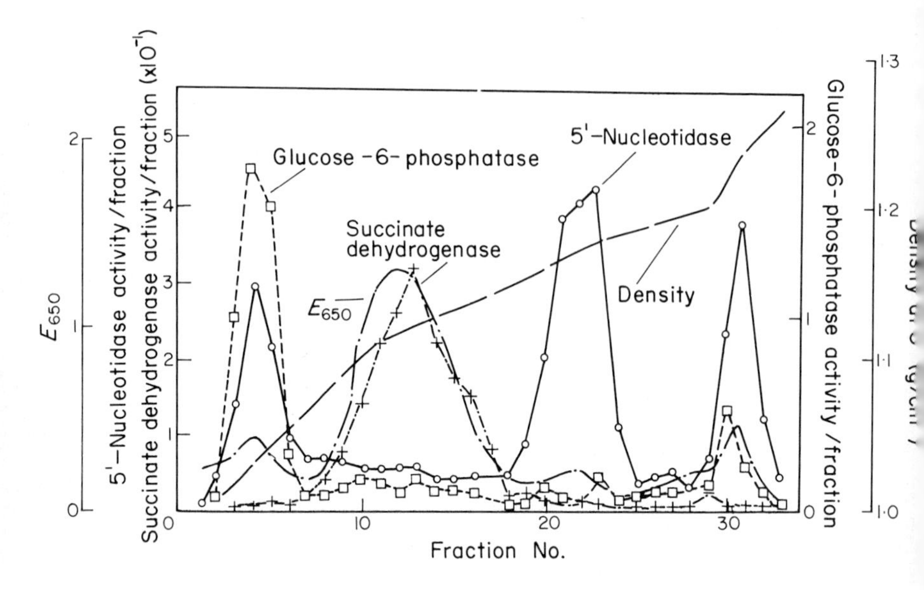

Liver was perfused with warm 0·25 M *sucrose, 5* mM *$NaHCO_3$, pH 7·5, then homogenised in ice-cold 0·25* M *sucrose, 5* mM *$NaHCO_3$, pH 7·5 by 3 strokes in a Potter–Elvehjem homogeniser; the homogenate was centrifuged at 400g for 10 min and the pellet (equivalent to 19 g of liver) was resuspended in homogenisation medium; the rotor was loaded as described for* Figure 2; *centrifugation was at 3700 rev/min for 60 min at 4°C*

Figure 4. Pattern obtained by centrifuging a crude nuclear fraction from perfused rat liver in an AXII zonal rotor

been described by Evans (1970, 1971). The red blood cells were, in this case, disrupted by homogenisation in hypotonic sucrose. Plasma membranes from up to 100 g of liver could be isolated in high purity in a single centrifugation step. Pfleger *et al.* (1968) and other groups (*Table 1*) have also described methods for the isolation of hepatic plasma membranes which make use of zonal rotors. However, these authors used B-series rotors in which the available path length for sedimenting particles is very much shorter than in

Table 1. SUMMARY OF METHODS FOR PURIFYING PLASMA-MEMBRANE FRAGMENTS FROM CRUDE NUCLEAR FRACTIONS OF LIVER USING ZONAL ROTORS

Pretreatment of tissue	*Type of homogeniser*	*Homogenisation medium*	*Summary of method*	*Appearance of fraction*	*Means of identification**	*Method described by*
None	Dounce	1 mM $NaHCO_3$	BXV rotor, short spin; remove front of gradient; prolonged spin	Sheets and vesicles	lm; em; enz	Pfleger *et al.* (1968); Anderson *et al.* (1968)
Perfuse with 0·08 M sucrose	Glass homogeniser, Teflon-coated pestle	0·08 M sucrose, 5 mM tris-HCl, pH 7·4	Rate zonal, BXV rotor; recycle plasma membrane-rich region	—	enz	Weaver and Boyle (1969)
None	Potter–Elvehjem	0·08 M sucrose, 5 mM $NaHCO_3$	Rate zonal, AXII rotor; isopycnic flotation	After AXII, large sheets; at end, small sheets	lm; em; enz	Hinton *et al.* (1970, 1971)
Perfuse with 0·25 M sucrose	Potter–Elvehjem	0·25 M sucrose, 5 mM $NaHCO_3$				
None	Dounce	1 mM $NaHCO_3$	Rate zonal, AXII rotor	Sheets	em, enz	Evans (1970, 1971)
Perfuse with 1 mM $NaHCO_3$	Dounce	1 mM $NaHCO_3$	Rate zonal, BXV rotor	—	enz	Newkirk and Waite (1971)

* lm, light microscopy; em, electron microscopy; enz, assay of specific enzymes.

the AXII rotor. They were, therefore, unable to resolve plasma-membrane fragments from smaller cell organelles such as mitochondria and denser, but similarly-sized, particles such as nuclei in a single centrifugation, but were forced to resort to a two-stage procedure.

While the methods using the AXII zonal rotor resolve plasma-membrane fragments directly from the crude nuclear fraction by taking advantage of a unique feature of these fragments—the combination of large size and low density—classical methods for the separation of membrane fragments are essentially two-stage (*Table 2*). First, small organelles are removed by a repeated series of washings, then the plasma membranes are floated away from the nuclei, usually through a step gradient which serves to remove any remaining mitochondria. It is essential that this final 'isopycnic' step be carried out by flotation, not sedimentation, for mitochondria, when separated by sedimentation, band at a density very similar to that of plasma-membrane fragments, but when separated by flotation they band at a significantly higher density (Beaufay *et al.*, 1964).

The problems with red blood cells which have been discussed earlier in connection with separations in zonal rotors are also evident in 'conventional' methods, and similar solutions are adopted. The earliest method for separating liver plasma-membrane fragments, that developed by Neville (1960), used a very hypotonic homogenisation medium (1 mM $NaHCO_3$). This method, as improved by Emmelot *et al.* (1964), has become the standard method for the isolation of liver plasma-membrane fragments. There are, however, indications from the work of Ray (1970) that the yield and purity of the product may be improved by the incorporation of 0·5 mM $CaCl_2$ in the homogenisation medium and also by extreme dilution of the homogenate before the initial preparation of the crude nuclear fraction. The second approach described in connection with the work on zonal rotors, of prior perfusion of the liver followed by homogenisation in an isotonic medium, had previously been used in a conventional separation scheme by Takeuchi and Terayama (1965), and subsequently by a number of other workers. A unique approach was taken by Coleman *et al.* (1967), who used a very vigorous rehomogenisation to break up aggregates formed during the initial pelleting of the crude nuclear fraction. They were rewarded by a high yield of membrane fragments, but the very low density (1·13 g/cm^3) at which the fragments were found suggests that the plasma membrane had been broken into two subfractions similar to those separated by Evans (1970), and that the fraction obtained is therefore not representative. A summary is given of the various published methods using conventional rotors for the isolation of plasma-membrane fragments in *Table 2*. Similar methods have also been used to prepare plasma-membrane fragments from hepatoma homogenates (*Table 3*), but the small size of such membrane fragments means that many will be lost if the ordinary low-speed pellet is collected as an initial step.

ISOLATION OF SMALL SHEETS OF PLASMA MEMBRANE

Provided that the centrifugal force during the initial pelleting and in the various washing steps is sufficiently high to pellet the small membrane sheets, the 'classical' methods described above can, in principle, be used to isolate membranes from hepatomas and similar tissues. Some problems arise, as will be discussed later, but the plasma-membrane sheets are sufficiently greater in size than the smooth endoplasmic reticulum vesicles, the density of which makes them the major potential contaminant, for the latter to be removed by differential centrifugation. A more radical recasting of zonal-rotor methods is necessary, as the plasma-membrane sheets can no longer be separated from the mitochondria by rate-sedimentation. However, an approach described below, developed in our laboratory by T. D. Prospero, gives highly satisfactory results and should be widely applicable.

An initial problem with hepatomas is that the red blood cells cannot be removed by perfusion and, moreover (at least with the tumours we have studied), the nuclei are so fragile that homogenisation in a hypotonic medium causes the formation of an intractable nucleoprotein gel. However, advantage may be taken of the relatively small size of the plasma-membrane fragments to remove red blood cells, together with aggregates and nuclei, by a very brief initial centrifugation. Mitochondria and membrane fragments may then be collected by centrifugation at a higher speed, resuspended, and loaded on to an HS zonal rotor. After centrifugation for a suitable time through an appropriate gradient, a band of plasma membranes is formed which, provided that a low concentration of Ca^{2+} ions (2 mM) is present in the homogenisation medium, is separated from mitochondria due to the difference in banding density and from membrane vesicles by virtue of the difference in sedimentation rate (*Figure 5*). If the Ca^{2+} ions are omitted from the homogenisation medium, the mitochondrial and plasma-membrane bands overlap. The major contaminant of the plasma-membrane fraction is amorphous material seen in

Table 2. SUMMARY OF METHODS FOR PURIFYING PLASMA-MEMBRANE FRAGMENTS FROM CRUDE NUCLEAR FRACTIONS OF LIVER USING CONVENTIONAL ROTORS

Pretreatment of tissue	*Type of homogeniser*	*Homogenisation medium*	*Summary of method*	*Appearance of fraction*	*Means of identification**	*Method described by*
(*i*) *Rat or Mouse*						
None	Dounce	1 mM $NaHCO_3$	Isopycnic flotation; rate sedimentation	Large sheets	lm	Neville (1968)
None	Potter–Elvehjem	1 mM $NaHCO_3$	Multiple washings; isopycnic flotation	Large sheets with intact bile canaliculi	lm; em; enz	Emmelot *et al.* (1964)
None	Dounce	1 mM $NaHCO_3$	Isopycnic flotation	Very rich in intact bile canaliculi	em; enz	Song *et al.* (1969)
None	Dounce	1 mM $NaHCO_3$, 0·5 mM $CaCl_2$	3 wash cycles; isopycnic flotation	Large sheets	em; enz	Ray (1970)
None	'Duall' homogeniser	0·25 M sucrose, 0·013 M phosphate, pH 7	Isopycnic flotation from KBr	Cell and nuclear membranes	imm	Herzenburg and Herzenburg (1961)
Perfuse with 0·3 M sucrose	Potter–Elvehjem	0·3 M sucrose, 5 mM $NaHCO_3$, pH 7·4	Wash 3 times, rehomogenise vigorously; isopycnic sedimentation	Vesicles and strips	enz	Coleman *et al.* (1967)
Perfuse with saline	Dounce	0·25 M sucrose, 5 mM $CaCl_2$	Wash twice; two cycles of isopycnic flotation	Large sheets	lm; em	Takeuchi and Terayama (1965)

None	Rubber pestle, glass vessel homogeniser	0·25 M sucrose	Float in sucrose containing 1 mM-$MgCl_2$; wash with EDTA-containing medium; float from sucrose containing 1 mM-EDTA	—	enz	Stein *et al.* (1968)
Perfuse with 0·9% (w/v) NaCl	Dounce	0·25 M sucrose, 0·5 mM $CaCl_2$, 5 mM tris-HCl, pH 7·4	Wash 7 times; isopycnic sedimentation; isopycnic flotation	Small sheets with tight junctions	lm; em; enz	Berman *et al.* (1969)
(ii) Ox						
None	Potter–Elvehjem, after mincing	1 mM $NaHCO_3$	as Neville (1968)	Large sheets with bile canaliculi	em; enz	Fleischer and Fleischer (1969)
(iii) Chick embryo						
None	Teflon–glass, hand-operated	0·25 M sucrose, 1 mM $MgCl_2$; *or* Krebs No. 3 buffer; *or* Hanks saline	Upper portion of pellet taken for isopycnic flotation	Membrane sheets	lm; em; enz	Rosenburg (1969)

* imm, immunology; others as *Table 1*.

Table 3. SUMMARY OF METHODS FOR THE PREPARATION OF PLASMA-MEMBRANE FRAGMENTS FROM RAT HEPATOMAS

Pretreatment of tissue	*Type of homogeniser*	*Homogenisation medium*	*Summary of method*	*Appearance of fraction*	*Means of identification★*	*Method described by*
None	Potter–Elvehjem	1 mM $NaHCO_3$, 2 mM $CaCl_2$	Low-speed pellet, wash; isopycnic flotation	Large sheets	em; enz	Emmelot and Bos (1970)
Prepare free cells	None; osmotic shock	1 mM $NaHCO_3$	As Takeuchi and Terayama (1965)	—	enz	Davydova (1968)
None	Potter–Elvehjem	0·25 M sucrose, 2 mM $CaCl_2$, 5 mM $NaHCO_3$, pH 7·4	Centrifuge mitochondrial-lysosomal fraction in HS rotor in presence of Ca^{2+}	—	enz	Prospero and Hinton (1972)

★ as *Table 1*.

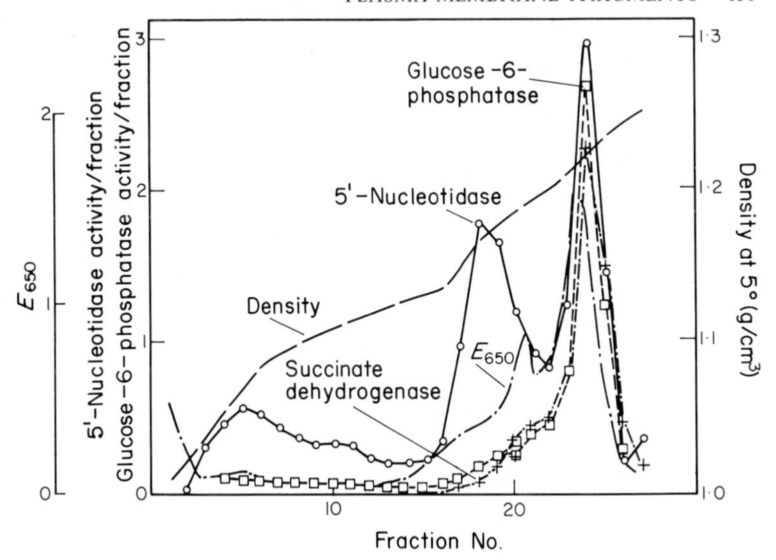

Hepatoma was homogenised in 0·25 M sucrose, 5mM $NaHCO_3$, 2 mM $CaCl_2$, pH 7·5; aggregates, nuclei and red blood cells were removed by centrifuging at 500g for 3 min; the supernatant fluid was centrifuged at 12 000g for 10 min and the pellet (mitochondrial-lysosomal fraction) was resuspended in homogenisation medium; the rotor contained a sucrose gradient of form HSII (Table 5) *and centrifugation was at 9000 rev/min for 90 min at 4°C*

Figure 5. Pattern obtained by centrifuging a mitochondrial-lysosomal fraction from rat hepatoma WD in an HS zonal rotor (T. D. Prospero, unpublished experiment)

Figure 5 as a broad protein band between density 1·14 and 1·20 g/cm³. This material is not enriched in any marker enzyme and may consist of elements from necrotic areas of the tumour. The plasma-membrane fraction so isolated may be further purified by flotation from dense sucrose. Analytical details on this preparation will be published elsewhere (Prospero and Hinton, 1972).

ISOLATION OF VESICLES DERIVED FROM THE PLASMA MEMBRANE

The unique combination of large size and low density makes plasma-membrane sheets relatively easy to isolate once methods have been developed to prevent their aggregation with other cell organelles. Plasma-membrane vesicles, on the other hand, have no such distinctive properties. There appears to be no significant difference in size between the different types of vesicle found in the micro-somal fraction. Plasma-membrane vesicles are, on average, slightly denser than vesicles derived from the Golgi apparatus (Fleischer and Fleischer, 1970) but possibly very slightly lighter than smooth endoplasmic reticulum (El-Aaser *et al.*, 1966). This latter difference can be accentuated if the membranes are separated by flotation rather than sedimentation (Wattiaux de Coninck and Wattiaux, 1969; Hinton *et al.*, 1971) but is still too small to be of any preparative use. The difference in density between plasma membrane-derived and endoplasmic reticulum-derived vesicles may be increased by inclusion of Mg^{2+} ions in the gradient (El-Aaser *et al.*, 1966) and so a considerable degree of separation between the two components obtained (*Figure 6*). Even so, the plasma-membrane region, as judged by the purification of 5′-nucleotidase, is still heavily contaminated by material which does not possess glucose-6-phosphatase activity, but probably includes fragments of the Golgi apparatus and of mitochondrial outer membranes (Hinton *et al.*, 1971). If however, the microsomal fraction is treated by sonication and with a low concentration of Pb^{2+} ions, a band of material highly enriched in the plasma-membrane marker 5′-nucleotidase can be separated by centrifugation in a BXIV zonal rotor (*Figure 7*). Although the purification factor for 5′-nucleotidase indicates that the material in this band is greatly enriched in plasma-membrane fragments (Hinton *et al.*, 1971), further experiments have indicated that it may represent a particular subfraction of the plasma membrane, for other 'marker' enzymes for the plasma membrane, notably L-leucyl β-napthyl amidase, are not concentrated in the band to the same degree (Norris *et al.*, 1972).

Another approach to the separation of plasma-membrane fragments from the hepatic microsomal fraction is isopycnic banding in gradients of Ficoll. This compound, unlike sucrose, does not penetrate vesicles to any extent, nor does it exert any appreciable osmotic pressure, and organelles do not necessarily show the same density relationships as they do in sucrose gradients (Amar-Costesec *et al.*, 1969). Kamat and Wallach (1965) developed a method for the separation of plasma-membrane vesicles from Ehrlich ascites cells following a systematic study of the properties of the plasma membrane and endoplasmic reticulum vesicles, especially in Ficoll-containing media (see Wallach, 1967). The method developed as a result of these studies would appear to be applicable to liver-membrane fragments as well, and yields preparations of plasma-membrane fragments whose purity, as judged by the specific activity of marker enzymes, compares well with that of preparations from the nuclear fraction (Graham *et al.*, 1968). House and Weidemann (1970) have described a method for

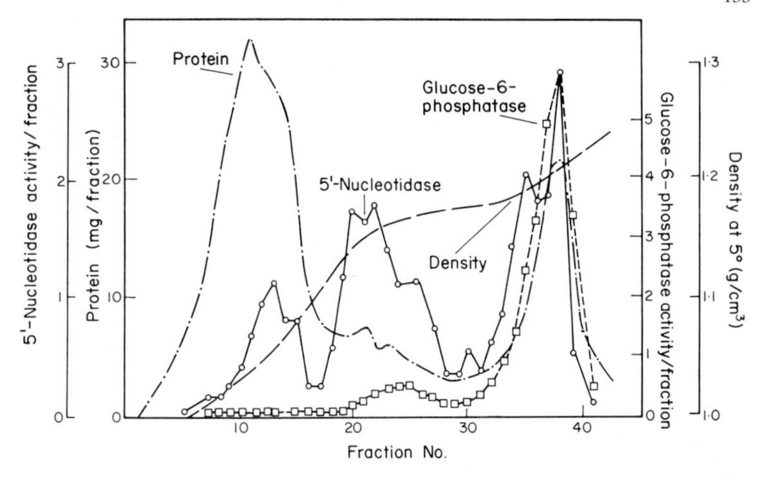

Liver (5 g) was homogenised in 0·25 M sucrose, 5 mM tris-HCl, 5 mM $MgCl_2$, pH 7·4, and the supernatant fraction prepared by centrifuging at 13 000g for 20 min; the rotor was loaded with a gradient (generated with a Beckman gradient engine) of sigmoid form, ranging from 0·25 M to 2 M sucrose with a plateau at 1·37 M sucrose (density, 1·175 g/cm³), the sample and a 200 ml overlay of water; the gradient solutions contained 5 mM tris-HCl, pH 7·4, and 5 mM $MgCl_2$; centrifugation was at 40 000 rev/min for 6 h at 4°C

Figure 6. Pattern obtained by centrifuging a post-mitochondrial supernatant fraction from rat liver in a BIV zonal rotor (from Fitzsimons, 1969)

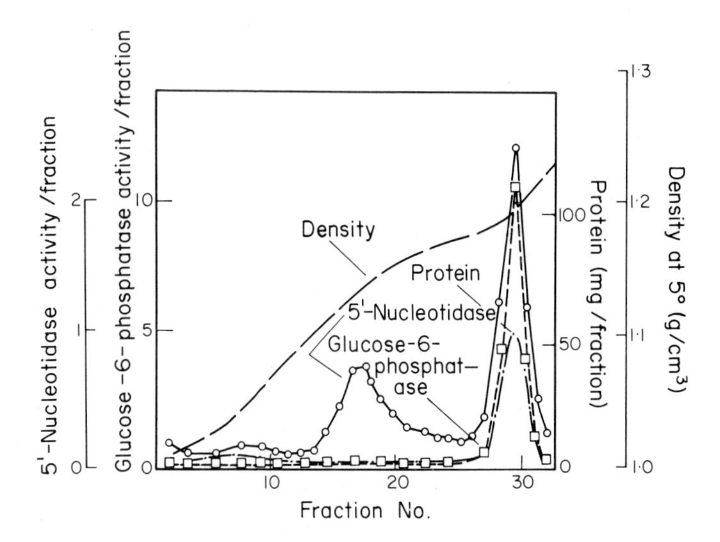

Rat-liver microsomes were prepared, sonicated and treated with 0·5 mM $Pb(NO_3)_2$ as described in the text; the rotor was loaded with a sucrose gradient of form B1 (Table 5), *the sample, and a 50 ml overlay of 0·08 M sucrose; centrifugation was at 47 000 rev/min for 150 min at 4°C*

Figure 7. Pattern obtained by centrifuging a rat-liver microsomal fraction, after sonication and treatment with $Pb(NO_3)_2$, in a BXIV zonal rotor (from Hinton et al.*, 1971)*

the isolation of plasma-membrane vesicles, based on that of Kamat and Wallach (1965), in which centrifugation times are drastically reduced. The plasma-membrane preparations so produced are of high purity, but in the absence of systematic 'scans' of gradients the principle underlying the separation achieved is unclear.

These methods are summarised in *Table 4*.

CHOICE OF METHODS FOR ISOLATING PLASMA-MEMBRANE FRAGMENTS FROM LIVER AND HEPATOMA

The classical method for isolating liver plasma-membrane fragments is undoubtedly that of Neville (1960), as modified by Emmelot *et al.* (1964). Use of this method will give results which are directly comparable both with the numerous results reported by Emmelot's group and with those of many other authors. However, there are a number of reasons for considering other techniques for separating plasma-membrane fragments.

A method for separating any cell organelle should ideally satisfy three criteria: it should provide a pure and representative product, the yield should be high and the technique should be as simple and rapid as possible. When these criteria are applied to the methods for preparing liver plasma-membrane fragments discussed in the previous section, it can be seen that, if 5′-nucleotidase is taken as a marker, there are fairly substantial differences in purity among the various preparations. The methods which use the AXII zonal rotor and the method of Ray (1970) give products with the greatest purity, while the 'yield' of the method of Ray is superior to any other method (see *Table 8*). The main distinction between the methods is, however, to be made on grounds of convenience. The advantage of being able to prepare 'control' fractions, which is a feature of methods in which the original homogenisation is carried out in isotonic sucrose, is important. There are also further disadvantages to hypotonic homogenisation media, notably the risk of nuclear lysis converting the preparation into an intractable jelly from which nothing can be extracted. This danger is especially pronounced in the 'classical' methods where several cycles of homogenisation and resuspension must be carried out to remove contaminating microsomes. The difficulties encountered by Marinetti and Gray (1967) and Davydova (1968) in preparing plasma membrane from perfused liver by the technique of

Table 4. SUMMARY OF METHODS FOR THE PREPARATION OF PLASMA-MEMBRANE FRAGMENTS FROM RAT-LIVER MICROSOMES

Pretreatment of tissue	*Type of homogeniser*	*Homogenisation medium*	*Summary of method*	*Appearance of fraction*	*Means of identification**	*Method described by*
None	Potter–Elvehjem	0·25 M sucrose, 0·2 mM $MgSO_4$	Isopycnic flotation, Ficoll gradient	—	enz	Graham *et al.* (1968)
None	'Teflon–glass'	0·25 M sucrose, 5 mM tris-HCl, pH 7·4	Short centrifugation, Ficoll gradient	—	enz	House and Weidemann (1970)
None	Potter–Elvehjem	0·25 M sucrose, 5 mM tris-HCl, pH 7·4	Sonicate, treat with Pb^{2+}, sonicate; isopycnic centrifugation	Vesicles	enz	Hinton *et al.* (1971); Norris *et al.* (1972)

* as *Table 1.*

Emmelot *et al.* (1964) are consistent with our observations that nuclei prepared from perfused liver are exceptionally fragile. The presence of Ca^{2+} ions in the homogenisation medium of Ray (1970) should help to prevent nuclear lysis. However, if an AXII zonal rotor is available then its use is preferable, for all washing steps can then be condensed into a single zonal centrifugation. If it is not available, the procedure of Emmelot *et al.* (1964) is to be preferred. The addition of $CaCl_2$ to the homogenisation medium, as suggested by Ray (1970), is to be recommended so long as the Ca^{2+} ions do not interfere with a subsequent part of the experiment.

In our experience nuclear lysis, which creates problems even with liver when hypotonic homogenisation media are used, is so pronounced in homogenates of hepatoma as to necessitate the use of an isotonic medium. Emmelot *et al.* (1964) reported that the problem can be overcome, even when very hypotonic media are used, by the addition of $CaCl_2$, and have described in detail the preparation of plasma-membrane fractions from various rat and mouse hepatomas (Emmelot and Bos, 1970). Unfortunately, the nuclei of different tumours vary in their sensitivity to osmotic stress, and we have found that with the hepatomas which we use the homogenisation must be carried out in an isotonic medium.

There is not yet sufficient data on plasma-membrane fragments prepared from the crude microsomal fraction to enable a more than preliminary judgement to be made of the three methods available. The separation on Ficoll gradients appears to provide a product of excellent purity in high yield. However, the fate of Golgi-body fragments, which in a comparable separation with Ehrlich ascites-cell microsomes band very close to the plasma-membrane fragments (Wallach, 1967), and of fragments of the outer mitochondrial membrane (as indicated by monoamine oxidase), which also have a similar banding density in Ficoll (Amar-Costesec *et al.*, 1969) is not clear, and will not be known until a systematic scan of a separation under these conditions is carried out. The method using Pb^{2+} ions and sonication is reliable and has been applied successfully to rat and dog liver and to a rat hepatoma (K. A. Norris, personal communication) but, as mentioned earlier, there is evidence that the 'plasma membrane' so prepared is not representative of all the fragments of the plasma membrane in the initial microsomal fraction. Careful examination of the results of House and Weidemann (1970), however, suggest that a similar fractionation may occur when Ficoll gradients are used. A full description of the method using treatment with Pb^{2+} ions and sonication (Hinton *et al.*, 1971) will be given, mainly because this is the method most familiar to the author.

DETAILS OF SELECTED PROCEDURES FOR THE PURIFICATION OF PLASMA-MEMBRANE FRAGMENTS FROM LIVER AND HEPATOMA

SEPARATION OF LARGE SHEETS OF LIVER PLASMA MEMBRANE IN THE AXII ZONAL ROTOR

The rats are killed by cervical fracture, care being taken not to tear the blood vessels in the neck. The body is opened, and the liver perfused with warm (37°C) 0·25 M sucrose containing 5 mM $NaHCO_3$(pH 7·5). In our studies, perfusion was carried out via the aorta, the inferior vena cava being slashed just above the diaphragm. A more direct route, for example, through the portal vein, would be desirable to avoid contamination by the gut bacteria which are released into the blood stream immediately after the death of the animal. As soon as the liver has blanched, it is extracted and immersed in ice-cold 0·25 M sucrose, 5 mM $NaHCO_3$. When about 20 g of liver has been collected it is blotted, weighed and transferred to fresh medium. Any liver which does not blanch rapidly during perfusion should be discarded.

The livers are then homogenised in about 9 vol of ice-cold 0·25 M sucrose, 5 mM $NaHCO_3$ (pH 7·5) by 3 strokes of a Potter–Elvehjem homogeniser using a loosely-fitting pestle (clearance about 0·33 mm) rotating at about 900 rev/min. After homogenisation, extra medium is added, if necessary, to bring the volume of the homogenate up to 10 ml per g of liver. A sample is removed, and the remainder of the homogenate is filtered through a coarse sieve to remove fragments of connective tissue. The filtered homogenate is then centrifuged for 10 min at 2000 rev/min (400*g*) in the 4 × 50 rotor of an MSE bench centrifuge kept at 4°C. After centrifugation, the supernatant fluid is discarded, leaving a red, gelatinous pellet. The loose, pinkish, layer of mitochondria sometimes seen on top of the pellet should be removed as far as possible. The pellets are then resuspended in a total of 30 ml of 0·25 M sucrose, 5 mM $NaHCO_3$ by 5 strokes of a hand-held Teflon–glass homogeniser (Jencons Ltd). After removal of a sample, the suspension is loaded on to an AXII zonal rotor containing a gradient of form AII (*Table 5*), followed by a 50 ml overlay of 0·08 M sucrose. The gradient solutions should all contain 5 mM $NaHCO_3$. It is important that, in all cases, the $NaHCO_3$ should be added at the last minute, from a freshly-prepared solution of AnalaR grade chemical, otherwise the pH will be unacceptably high.

After centrifugation for 1 h at 3700 rev/min, a zone of plasma-membrane fragments will be clearly visible on the centrifugal side of the main mitochondrial peak (see *Figure 4*). The rotor is then decelerated, and the gradient displaced. The tubes containing the plasma-membrane fragments can be identified by inspection of the absorbance profile (at 650 nm) recorded during displacement of the gradient. These are pooled, adjusted to a density of 1·19 g/cm³ with 2 M sucrose, and rehomogenised with 8 strokes of a Potter–Elvehjem

Table 5. GRADIENTS USED IN THE ZONAL ROTORS

	Volume (ml)	*Sucrose concentration* (moles/l)	*Sucrose density at 0°C* (g/cm³)
Gradient AII			
The mixing vessel initially contained	250	0·30	1·039
To this were added sequentially (i)	600	1·25	1·164
(ii)	500	1·66	1·215
Cushion (by-passing mixing vessel)	to fill rotor	2·00	1·265
Gradient HSII			
The mixing vessel initially contained	150	0·35	1·045
To this were added sequentially (i)	300	1·065	1·140
(ii)	230	1·854	1·240
Cushion (by-passing mixing vessel)	to fill rotor	2·00	1·265
Gradient BI			
The mixing vessel initially contained	78	0·30	1·039
To this were added sequentially (i)	139	0·84	1·115
(ii)	332	1·42	1·196
Cushion (by-passing mixing vessel)	to fill rotor	2·00	1·265

All gradients were prepared in apparatus similar to that described by Birnie and Harvey (1968); they were formed at room temperature and cooled to 4°C immediately before entering the rotor. Gradient solutions were either buffered with 5 mM tris-HCl, pH 7·4 (BI), or their pH was adjusted to 7·5 with 5 mM $NaHCO_3$ (AII and HSII); no other salts were added.

homogeniser with the pestle rotated at 2900 rev/min. The pestle should have the same clearance as that used in the initial homogenisation. Samples of this suspension are placed in polycarbonate tubes of an MSE 8 × 50 fixed-angle rotor, overlaid with 0·25 M sucrose containing 5 mM tris-HCl, pH 7·4, and centrifuged for 90 min at 40 000 rev/min. The purified plasma membranes collect as white 'sheets' on the interface between the 0·25 M sucrose and the sample, and can be recovered with a bent pipette.

If unperfused liver is used as the starting material, the same procedure may be used provided that 0·08 M sucrose is used in place of 0·25 M sucrose for the homogenisation and for the resuspension of the crude nuclear fraction. Prolonged lysis, as used in the experiments described earlier, does not seem to be necessary (T. D. Prospero, unpublished experiments).

PREPARATION OF SMALL SHEETS OF PLASMA MEMBRANE FROM HEPATOMA

After killing the animal by cervical fracture, the tumour is extracted intact and put into ice-cold 0·25 M sucrose. After blotting and weighing, the tumour is carefully minced with scissors, and large lumps of the capsule taken out with forceps and put on one side. Homogenisation in 0·25 M sucrose containing 5 mM $NaHCO_3$ and 2 mM $CaCl_2$ is carried out by use of a Potter–Elvehjem homogeniser as described in the previous section. It is essential to use a homogeniser with a fairly wide clearance between pestle and vessel, and to keep the tube axis parallel to the shaft of the pestle during homogenisation to avoid the risk of the pestle jamming. It is not possible to break up the connective tissue of the capsule, which will be left as whitish sheets at the bottom of the homogenisation vessel and can be removed by filtration through two nylon sieves. The connective tissue left on the sieve, and that removed before the start of homogenisation, should be weighed, and this weight subtracted from the total weight of the tumour to give the weight of hepatoma tissue in the homogenate. The volume of the latter should then be adjusted to give a final concentration of at least 10 ml per g of tumour.

Aggregates, nuclei and red blood cells should be removed from the homogenate by centrifugation for 5 min at 500*g*. Mitochondria and membrane fragments are then collected by centrifugation for 10 min at 12 000*g* in the 8 × 50 fixed-angle rotor of an MSE High-speed 18 centrifuge. The surface of the pellet is rinsed gently with the homogenisation medium to remove the pink fluffy layer of microsomes, and the main part of the pellet is resuspended in 25 to 30 ml of the homogenisation medium by 6 slow strokes of a hand-operated Teflon–glass homogeniser. A sample of the suspension is removed, and the remainder is loaded on to an HS rotor containing a gradient of form HSII (*Table 5*). The sample is displaced from the core of the rotor with 40 ml of 0·08 M sucrose and the rotor is accelerated to its operating speed of 9000 rev/min. After centrifugation for 90 min the rotor is decelerated to 1500 rev/min and the gradient is displaced with 2 M sucrose. The extinction at 650 nm and the refractive index of the gradient are monitored, and 20 ml fractions are collected. The plasma membranes are concentrated on the light side of the peak of extinction at 650 nm, which begins at a density of 1·14 g/cm^3 (*Figure 5*). In case of doubt, this region should be monitored for a plasma-membrane enzyme, such as alkaline *p*-nitrophenolphosphatase, which is simple to assay. The tubes containing the plasma-membrane fragments are pooled, the density is adjusted to 1·175 g/cm^3 with 2 M sucrose, and the fragments are rehomogenised with 4 strokes of a Potter–Elvehjem homogeniser with a pestle rotating at 2400 rev/min. The suspension is then loaded into the tubes of an 8 × 50 fixed-angle rotor (not more than 30 ml per tube), overlayed with 0·25 M sucrose (to fill the tube), and centrifuged for 90 min at 40 000 rev/min in an MSE Super-speed 65 centrifuge. The plasma membrane collects as flakes, tending to orange in colour, on the interface between the two layers of sucrose.

PURIFICATION OF PLASMA-MEMBRANE VESICLES FROM THE LIVER MICROSOMAL FRACTION

After homogenisation in a magnesium-free medium as described in connection with the large sheets of membrane, and removal of large particles by centrifugation for 10 min at 10 000*g*, a microsomal pellet is prepared by centrifugation for 1 h at 40 000 rev/min in the 8 × 50 fixed-angle rotor of an MSE Super-speed 65 centrifuge. Microsomes from 5 g of liver are resuspended in 25 ml of 0·25 M sucrose containing 5 mM tris-HCl, pH 7·2, with at least 8 strokes of the Potter–Elvehjem homogeniser, and sonicated in a cooled vessel for 15 s at 150 W (nominal) with a 'Soniprobe' (Dawe Instruments Ltd.). No detectable rise in the temperature of the bulk of the suspension is caused by the sonication. A solution of $Pb(NO_3)_2$ in $\beta\beta$-dimethylglutarate buffer (pH 6·7) is added to give final concentrations of 0·5 mM and 0·016 M for the Pb^{2+} and the buffer, respectively, in a total volume of 60 ml. After incubation for 5 min at 0°C the suspension is sonicated for 30 s and, after removal of a sample for assay, it is loaded into a BXIV rotor containing a sucrose gradient of form BI (*Table 5*). All solutions contain 5 mM tris-HCl, pH 7·4. Finally, an overlay of 50 ml of 0·08 M sucrose is added and the rotor is accelerated to the operating speed of 47 000 rev/min. After centrifugation for 150 min, the rotor is decelerated and the contents are displaced, monitored and collected as described in Reid (1971). The plasma membrane-rich region can be identified from the absorbance profile at 280 nm (see *Figure 7*). The appropriate fractions are then pooled and, after dilution with an equal volume of water, membrane fractions are collected by centrifugation (MSE 8 × 50 rotor at 40 000 rev/min for 90 min).

SEPARATION OF PLASMA-MEMBRANE SHEETS FROM LIVER OR HEPATOMA BY CENTRIFUGATION IN CONVENTIONAL ROTORS

The classical method for isolating plasma-membrane fragments is the modification by Emmelot *et al.* (1964) of the method of Neville (1960). The description of the method given here is based on that given by Emmelot and Bos (1970), with details added from references cited in that article. (It is, however, difficult to track down any detailed description of some of the operations.) The well-minced liver or hepatoma tissue is homogenised in 1 mM $NaHCO_3$, pH 7·5, with the addition, in the case of the hepatoma homogenates, of 2 mM $CaCl_2$, to give a homogenate containing about 0·25 g of tissue per ml. Homogenisation is achieved by 4 to 6 strokes of a Potter–Elvehjem homogeniser with a loosely-fitting pestle rotated at about 1000 rev/min. The homogenate is diluted 5-fold with 1 mM $NaHCO_3$ and, after being stirred for 2 min, it is filtered through a fine cloth to remove connective-tissue fragments. Large particles are then collected by centrifugation for 10 min at 1500*g* (2000*g* for hepatoma). The pellet is resuspended by use of a hand-operated Dounce or Teflon–glass homogeniser in 1 mM $NaHCO_3$ to give a suspension of about 6 ml per g of liver, and the plasma membranes are sedimented by centrifugation for 10 min at 1000*g*. Lipid droplets floating on top of the tube are removed before the rest of the supernatant fluid is sucked off. The pellets are resuspended by stirring with a glass rod. Neville (1960) mentions that the pellet may have a bottom layer of glass fragments and adherent cytoplasm, and that this should not be resuspended. Up to 5 of these 'washing' steps (7 when Ca^{2+} is present in the homogenisation medium) may be required to free the membrane preparation from mitochondria. Neville (1968) and Ray (1970) replace the multiple washing by a single step, though they dilute the crude nuclear pellet to the same volume as the original homogenate. The washed pellet is then resuspended in 1·7 M sucrose (density, 1·22 g/cm³) with a Dounce homogeniser, and a sample is taken and overlaid with sucrose solutions of density 1·20, 1·18 and 1·16 g/cm³. When hepatoma membranes are being prepared a layer of density 1·14 g/cm³ is added. After centrifugation for about 90 min at 100 000*g*, in the case of liver plasma membranes collect at the interface between the sucrose layers of density 1·16 and 1·18 g/cm³. Hepatoma membranes, on the other hand, collect at the 1·14–1·16 g/cm³ interface. Either a swing-out or a fixed-angle rotor may be used (Coleman and Finean, 1966). The membranes are collected with a bent pipette and, after dilution with several volumes of 1 mM $NaHCO_3$, pelleted by centrifugation for 10 min at 3000*g*. The resulting pellet is washed twice by resuspension in 1 mM $NaHCO_3$ and centrifugation for 10 min at 3000*g*. The resulting pellet is the purified plasma-membrane fraction.

SEPARATION OF PLASMA-MEMBRANE FRAGMENTS FROM OTHER TISSUES

Published methods for the separation of plasma-membrane fragments from solid tissues other than liver and hepatoma are summarised in *Table 6*. No mention is made in *Table 6* of methods for the isolation of the intestinal brush border, a specialised region of the plasma membrane, as these are discussed by Porteous (this volume, *Chapter* 7). For reasons which will be discussed below, methods for the preparation of muscle 'cell membranes' have also been omitted. No indication is given of the purity of the membrane preparations, as the absence of well-characterised marker enzymes for extra-hepatic tissue often makes them difficult to assess.

Methods used for the isolation of large sheets of membrane from non-hepatic tissues are clearly similar to those discussed earlier for the isolation of liver membranes. It is not clear whether there are problems with red blood cells similar to those arising with liver preparations. If so, the EDTA which is present in several of the media (when homogenisation is carried out in isotonic sucrose) would probably help to prevent aggregation, but at the cost of causing considerable damage to the cell membranes due to the extraction of Ca^{2+} ions. It will be noted that in the methods described by Keenan *et al.* (1970) for the isolation of the plasma membrane of 'milk gland' and by Benabdelzlil *et al.* (1967) for the isolation of thyroid membranes, homogenisers of the whirling-blade type are used. However, these are best avoided unless it is impossible to get good cell breakage with a Dounce or Potter–Elvehjem type of homogeniser, because if the conditions are not very carefully controlled then the plasma membrane will be totally fragmented.

As has already been mentioned in connection with the isolation of hepatoma-membrane fragments, it is very difficult to break 'free' cells to yield large fragments of the plasma membrane. As can be seen from *Table* 7, three approaches have been adopted; (i) breakage by very 'mild' techniques such as osmotic lysis or a Dounce homogeniser; (ii) toughening of the membrane by some form of pre-treatment; and (iii) deliberate fragmentation of the membrane to vesicles. The first approach is similar to that used with solid tissues and has been applied with notable success by Bosmann *et al.* (1968) to the isolation of plasma membranes from HeLa cells. Marker enzyme measurements showed that a 120-fold purification had been achieved. The second approach, to toughen the membrane by some form of pre-treatment, was developed by Warren *et al.* (1967). While undoubtedly advantageous in yielding large sheets of membrane, techniques based on this approach are open to several objections, notably that enzymes are destroyed, so precluding any quantitative assessment of contaminants. This is especially serious in that electron micrographs show a layer of amorphous material associated with the cytoplasmic scale of the membranes. If this is adherent cytoplasm rather than part of the functional membrane, it will interfere seriously with measurements of the chemical composition of the membranes. The third approach to membrane isolation, the prior conversion of the membrane to vesicles, is exemplified by the work of Kamat and Wallach (1965) on Ehrlich ascites-tumour cells. The principles of this separation have been discussed above in connection with the isolation of membrane fragments from the hepatic microsomal fraction.

So far there have been few systematic studies of the preparation of membrane fragments from 'hard' tissues such as muscle. Methods have been developed which claim to yield the membranes of skeletal muscle cells. Kono and Colowick (1961) employed a series of salt extractions to remove the contents of the muscle cells, followed by a density equilibrium centrifugation to remove collagen and 'granules'. McCollester (1962) found that simple extraction with water at room temperature sufficed to remove the contents of these cells, provided that the tissue had previously been autolysed by a short incubation at 37°C, or a longer incubation at 0°C. Later, McCollester and Semente (1964) found that autolysis could be replaced by treatment with Ca^{2+} at 0°C. Light microscopy was the only means used to assess the purity of these preparations. Both procedures are rather violent, one entailing use of strong salt solutions, the other very hypotonic solutions at a rather high temperature (20 to 24°C). In the absence of any electron micrographs it is impossible to assess the morphological origin of the material, but the presence of whole capilliaries attached to the preparations of McCollester and the collagen-like composition of the preparations of Kono and Colowick make it unlikely that cell-surface membrane constitutes more than a small proportion of either preparation. It is likely that, in the future, surface-membrane fragments will be obtainable by procedures entailing vigorous dispersion of the tissue followed by subfractionation of the resulting microsomes in a zonal rotor (Headon and Duggan, 1971).

Table 6. METHODS FOR PREPARING PLASMA-MEMBRANE FRAGMENTS FROM SOLID TISSUES OTHER THAN LIVER

Tissue	*Pretreatment of tissue*	*Type of homogeniser*	*Homogenisation medium*	*Summary of method*	*Appearance of fraction*	*Means of identification★*	*Method described by*
Calf adrenal			'As Emmelot *et al.* (1964) for liver'				Turkington (1962)
Toad bladder epithelium	None	Dounce	1 mM $NaHCO_3$	Filter through glass beads; low-speed pellet, then isopycnic flotation	Membrane sheets, some desmosomes	lm; em; enz	Hays and Barland (1966)
Rat bladder epithelium	Saturated FMA† for 20 min	Ten Broeck tube	0·02 M tris-HCl, pH 8·0	Low-speed pellet; sedimentation through step gradient	Large sheets and vesicles, luminal surface membrane	em	Hicks and Ketterer (1970)
Rat epithelial fat pad	Prepare free cells; swell in hypotonic medium	Gentle agitation	Complex medium with Mg^{2+}, Ca^{2+} ATP, NAD and NADP	Washed low-speed pellet	'Ghosts' containing part of the original cell contents	lm	Rodbell (1967)
Rat epithelial fat pad	Prepare free cells	Potter–Elvehjem	0·25 M sucrose, 1 mM EDTA, 10 M tris-HCl, pH 7·4	16 000*g* for 15 min pellet; remove aggregate; isopycnic sedimentation in Ficoll gradient	Large vesicles with some junctional complexes	em; enz	McKeel and Jarrett (1970)
Rat kidney	None		'As Emmelot *et al.* (1964) for liver'		Large sheets and vesicles; some desmosomes and brush border vesicles	em; enz	Coleman and Finean (1966)
Rat kidney	None	Dounce and Potter–Elvehjem	0·5 M sucrose	Differential centrifugation	Brush borders with apical cell junctions	em; lm	Thuneberg and Rostgaard (1968)

Rat kidney	None	Potter–Elvehjem	0·25 M sucrose, 1 mM EDTA	Low-speed pellet; isopycnic flotation; dilute, spin down, take upper portion of pellet; wash	Small vesicles	enz	Fitzpatrick *et al.* (1969)
Rat kidney	None	Dounce	20 mM $NaHCO_3$	Isopycnic flotation; rate sedimentation	Brush borders	em; enz	Wilfong and Neville (1970)
Cow milk gland	None	'Whirling Blade' Polyton	1 mM $KHCO_3$	Remove large lumps; low-speed pellet, take upper portion	Large sheets of membrane with junctional complexes	em	Keenan *et al.* (1970)
Rat synaptosome	None	Teflon–glass	0·32 M sucrose	Separate synaptosomes, lyse in distilled water; isopycnic centrifugation	Membrane sheets, some with synaptic thickening and vesicles	em; enz	Rodriguez de Lopes Arnaz *et al.* (1967)
Rat synaptosome	None		10% (w/w) sucrose, pH 7·0	Separate synaptosomes, lyse in distilled water; isopycnic centrifugation in BXV zonal rotor	Membrane sheets, some with synaptic thickening and vesicles	em; enz	Cotman *et al.* (1968)
Calf thyroid	None		'As Emmelot *et al.* (1964) for liver'				Turkington (1962)
Sheep thyroid	None	'Whirling Blade' Servall Omnimixer	Earle's medium	Low-speed pellet, treat with DNAase; layer over 2 M sucrose; spin, take interface	Apical pole fraction microvilli, terminal web; contaminating cytoplasm	em	Benabdelzlil *et al.* (1967)

* as *Table 1*.
† FMA, fluorescein mercuric acetate.

Table 7. METHODS FOR PREPARING PLASMA-MEMBRANE FRAGMENTS FROM CELL SUSPENSIONS

Tissue	*Pretreatment of tissue*	*Type of homogeniser*	*Homogenisation medium*	*Summary of method*	*Appearance of fraction*	*Means of identification**	*Method described by*
Ehrlich ascites	None	Pressure homogeniser (Nitrogen bomb)	0·25 M sucrose, 5 mM tris-HCl, pH 7·4, 0·2 mM $MgSO_4$	Microsomal fraction; lyse with dilute tris-HCl; isopycnic centrifugation in Ficoll gradient	Vesicles	enz	Kamat and Wallach (1965)
Chick fibroblasts	0·9 mM $ZnCl_2$	Potter–Elvehjem	0·5% (v/v) Tween-20 followed by addition of 0·5 vol of 0·01 M $ZnCl_2$	High-speed pellet; remove aggregate; treat with EDTA; isopycnic flotation	Large sheets	em	Perdue and Sneider (1970)
Chick fibroblasts	None	Potter–Elvehjem	0·16 M NaCl	High-speed pellet; remove aggregate; treat with EDTA; isopycnic flotation	Small sheets and vesicles	em; enz	Perdue and Sneider (1970)
HeLa	None	Dounce	0·01 M EDTA, 0·02 M tris-HCl, pH 7·4	4000*g* for 10 min supernatant fluid; isopycnic centrifugation	Very large sheets	em; lm	Bosmann *et al.* (1968)
HeLa	None	Dounce	0·01 M tris-HCl, pH 7·0, 1 mM $MgCl_2$	Low-speed pellet; isopycnic flotation; sonication, gradient centrifugation	Cell ghosts before sonication, vesicles after	lm; em; imm	Boone *et al.* (1969)

Krebs ascites	None	Potter–Elvehjem	1 mM tris-HCl, pH 7·0, +8 mg/l NaCl and 0·4 mg/l KCl	Low-speed pellet; layer over 35% (w/w) sucrose; take interface and recycle	Flat sheets	em	Stonehill and Huppert (1968)
L cells	Saturated FMA†	Dounce	Saturated (approx. 2 mM) FMA†, 0·02 M tris-HCl, pH 8·0	Flotation from 60% (w/w) sucrose; sedimentation through 35% (w/w) sucrose; isopycnic centrifugation	Large sheets with attached amorphous material	em	Warren *et al.* (1968)
L cells	1 mM $ZnCl_2$	Dounce	0·02% (v/v) Tween-20 followed by 0·05 vol of 0·01 M $ZnCl_2$	Filter through glass beads; low-speed pellet; gradient flotation; gradient sedimentation (or as above)	Large sheets with attached amorphous material	em	Warren *et al.* (1968)
Rabbit leucocytes	None	Potter–Elvehjem	11·6% (w/w) sucrose	Gradient sedimentation (steps); gradient sedimentation (continuous gradient)	Vesicles	leucocidin elimination	Woodin and Wiencke (1966)
Pig lymphocytes	None	Stirring	0·15 M KCl, 0·01 M tris-HCl, pH 7·4	Heavy microsome fraction; isopycnic centrifugation	Smooth vesicles	em; enz	Allan and Crumpton (1970)
Pig thymocytes				As for pig lymphocytes			Allan and Crumpton (1970)

* as *Table 2*.
† FMA, fluorescein mercuric acetate.

PURITY OF PLASMA-MEMBRANE PREPARATIONS

Contamination of plasma-membrane preparations may occur in three ways. First, particles such as mitochondria may co-sediment with the fraction. Contaminants of this type can be removed by repeated centrifugation. Second, in the case of large sheets of plasma membrane, particles may be trapped inside the membrane as in an empty sac and will be released only if the membrane is fragmented. Third, proteins or small vesicles may adhere to the membrane, probably due to an electrostatic interaction with the acidic mucopolysaccharide coat possessed by the plasma membranes of most cells (Rambourg and Leblond, 1967). Contaminants of the third type can often be removed by extraction with salt solutions (Benedetti and Emmelot, 1968).

The purity of plasma-membrane preparations may be assessed in three ways; (i) by morphological examination, with either the light or the electron microscope; (ii) by measurement of 'marker' enzymes for other cell organelles; and, (iii) by immunological techniques. Until techniques developed for the cytochemical study of isolated fractions (El-Aaser, 1971) are fully tested, little information is supplied by electron-microscopic examination of plasma-membrane preparations. One membrane vesicle looks very like another, and large organelles, such as mitochondria, can be recognised only with difficulty if the original homogenisation was in a hypotonic medium. With the plasma membrane prepared from a crude nuclear fraction, examination under the phase-contrast microscope is an extremely useful method for detecting large-sized contaminants such as pieces of connective tissue, and even for detecting mitochondria and determining whether they are trapped inside the sheets of membrane or are present free in the medium (see *Figure 1d*). Mitochondria are, however, the smallest particles which can be detected with the light microscope; contaminants smaller in size are most conveniently detected by measuring their 'marker' enzymes.

ENZYMOLOGICAL ASSESSMENT OF CONTAMINANTS OF THE PLASMA-MEMBRANE FRACTION

Most authors who describe a method for the purification of plasma-membrane fragments give specific activity values for the plasma-membrane marker 5′-nucleotidase, for the mitochondrial marker succinate dehydrogenase, and for the endoplasmic reticulum marker glucose-6-phosphatase*. However, the absence, in many cases, of these values for the whole homogenate, and, in some cases, the absence of any data for other fractions, makes interpretation of the results very difficult. Even when comparative values are given, only too often they are activities in a microsomal fraction or some other subcellular fraction whose composition will vary with the experimenter's technique, rather than values for the whole homogenate.

Such results as are available are collected in *Table 8*. In spite of the variations, some pattern emerges. In particular, it will be noticed that there is no correlation between the decrease in the amount of enzymes (such as glucose-6-phosphatase) which indicate contaminating elements, and an increase in the relative specific activity of 5′-nucleotidase. In other words, if the reasonable assumption is made that the variations in the enrichment with respect to 5′-nucleotidase are not due to random error in the assays, contamination of the plasma-membrane fraction must be due mainly to some material which is not indicated by any of the marker enzymes. This is particularly clear if the enzyme patterns of the preparations of plasma membrane from perfused and unperfused liver, prepared by zonal centrifugation, are compared (*Table 8*). The purification of the 5′-nucleotidase-containing fragments is twice as great in the preparation from perfused liver as in that from unperfused liver, yet the level of the contaminating enzymes is lower in the latter.

It is clear from the results in *Table 8* that, for all plasma-membrane preparations, the amount of mitochondrial and lysosomal contamination is very small. Contamination by endoplasmic reticulum, as indicated by glucose-6-phosphatase activity, is more important, and indicates that endoplasmic reticulum vesicles make up 7 to 15% of the protein of the plasma-membrane fraction. It has been proposed that the glucose-6-phosphatase activity of the plasma-membrane fraction may be due not to contamination by endoplasmic reticulum fragments as suggested here, but to the presence of glucose-6-phosphatase activity in the plasma membrane (Benedetti and Emmelot, 1968). However, cytochemical staining of isolated fractions (Fitzsimons, 1969; El-Aaser, 1971), which shows the activity to be located in small vesicles lying alongside the plasma-membrane sheets, makes this hypothesis seem very unlikely. Emmelot and Benedetti (1967) have suggested that the rough vesicles which can be seen associated with the plasma-membrane sheets derived from hepatoma may indicate a real

* Assay methods for these enzymes are described in *Chapter 5* of this volume.

Table 8. YIELD AND PURITY OF RAT-LIVER PLASMA-MEMBRANE FRAGMENTS SEPARATED BY VARIOUS TECHNIQUES (GIVEN IN THE ORDER USED IN *Tables 1, 2* AND *4*

Method used	*Yield* (μg protein/ g wet weight of tissue)	*Percentage of homogenate 5′-nucleotidase*	*Purification relative to homogenate of markers for*				*Reported by*
			Plasma membranes (5′-nucleotidase)	*Microsomes* (glucose-6-phosphatase)	*Mitochondria* (succinate dehydrogenase)	*Lysosomes* (acid phosphatase)	
Pfleger *et al.* (1968)	880	—	—	—	—	—	Pfleger *et al.* (1968)
Weaver and Boyle (1969)	—	5	15	—	—	—	Weaver and Boyle (1969)
Hinton *et al.* (1970)							
Unperfused liver	1060	6·9	15·7	0·87	0·21	0·11	Hinton *et al.* (1971)
Perfused liver	336	6·0	32·3	0·49	0·12	0·34	Hinton *et al.* (1971)
Evans (1970)	300–500	4·2	19·5–30·5	—	—	—	Evans (1970)
Newkirk and Waite (1971)	—	8	35	0·95[a]	1·0[b]	0·35	Newkirk and Waite (1971)
Neville (1960)	—	—	—	—	0·264[b]	—	Vassilitz *et al.* (1967)
Neville (1960)	200	1·4	11·6	—	—	—	Song and Bodansky (1967)
Neville (1960)	100	—	26	—	—	—	Simon *et al.* (1970)
Neville (1968)	480	—	—	0·38	0·31	—	Trams *et al.* (1968)
Neville (1968)	—	—	10–17	—	—	—	Kawasaki and Yamashima (1971)
Emmelot *et al.* (1964)	410	—	12–18[c]	0·6[d]	0	0	Emmelot *et al.* (1964)
Emmelot *et al.* (1964)	2200	—	>12·8	>0·15	0·1	—	Graham *et al.* (1968)
Emmelot *et al.* (1964)	710	3·9	24	0·38	0·07[b]	0·71	Wattiaux-de-Coninck and Wattiaux (1969)
Emmelot *et al.* (1964)	—	—	24·6	1·07	0·01	—	Dod and Gray (1968)
Song *et al.* (1969)	—	3	—	—	—	—	Song *et al.* (1969)
Ray (1970)	1300	32[e]	30[e]	0·15[d]	—	—	Ray (1970)
Coleman *et al.* (1967)	1300	14·4	17·9	0·95	0·06	0·13	Coleman *et al.* (1967)
Coleman *et al.* (1967)	396	—	8	0·26	0·40	—	House and Weidemann (1970)
Stein *et al.* (1968)	1010	—	15	0·6	0·12[b]	—	Stein *et al.* (1968)
Berman *et al.* (1968)	440	—	11·7	0·32	—	—	Berman *et al.* (1969)
House and Weidemann (1970)	2610	—	22	0·11	0	—	House and Weidemann (1970)
Norris *et al.* (1972)	—	8·3	15·1	0·34	0	0	Hinton *et al.* (1971)

a. NADH cytochrome *c* reductase.
b. cytochrome oxidase
c. assuming that 5′-nucleotidase is purified 2-fold in a microsomal fraction
d. assuming that glucose-6-phosphatase is purified 3-fold in a microsomal fraction
e. alkaline *p*-nitrophenylphosphatase

morphological continuity between the endoplasmic reticulum and the plasma membrane in this tissue. The electron micrographs which they present to justify this hypothesis are not totally convincing. The alternative hypothesis that the glucose-6-phosphatase activity is simply due to microsomal contamination is strongly supported by the presence of a number of other microsomal enzymes with similar purification values to that found for glucose-6-phosphatase (*Table 9*). Benedetti and Emmelot (1968) lay great stress on the absence of a number of drug-metabolising enzymes from their preparations. In the absence of any data on the detection limits or recoveries of these rather delicate and feebly-active enzymes, it is uncertain how much weight to attach to their observations.

It is clear that contamination by other cell organelles can account for only a small proportion of the material in the plasma-membrane fraction, and cannot explain the differences between the various preparations. However, at least some of the difference can be explained by the presence of serum proteins, or proteins from the soluble fraction absorbed on to the acidic groups of the surface of the plasma membrane. The presence of haemoglobin in preparations of plasma membranes from unperfused liver is easily discerned by the colour of the packed membrane preparation, and the presence of serum and soluble cytoplasmic proteins has been demonstrated immunologically (Benedetti and Emmelot, 1968). These serum proteins must be bound very firmly for serum protein can be detected immunologically even in preparations made from perfused liver (Allen, 1969). Emmelot *et al.* (1964) believe that these contaminating proteins are removed specifically by washing with 0·15 M NaCl, and it is very clear that this is true for some absorbed enzymes, notably triose phosphate dehydrogenase (Emmelot and Benedetti, 1967) and ATP pyrophosphohydrolase (Lansing *et al.*, 1967). Only 15% of the protein of the plasma membrane is removed by this treatment. Thus even if this protein is exclusively from the adsorbed fraction—and there is some evidence that plasma-membrane enzymes such as 5′-nucleotidase are also partly removed (Emmelot *et al.*, 1964)—one cannot explain the difference between the purification of 5′-nucleotidase in plasma-membrane fractions prepared from unperfused liver either by Emmelot's technique or by use of an AXII zonal rotor and the preparations made from perfused liver.

The great similarity in the amounts of 'contaminating' enzymes in our preparations made from perfused and unperfused liver (*Table 8*) suggests that adsorbed serum proteins may be responsible for a very large proportion of the protein of the fraction. Emmelot

Table 9. COMPARISON OF THE ACTIVITY OF GLUCOSE-6-PHOSPHATASE IN LIVER PLASMA-MEMBRANE FRACTIONS WITH THE AMOUNT OF OTHER PRESUMED ENDOPLASMIC RETICULUM COMPONENTS

Purification with respect to					
Activity in homogenate		*Activity in microsomes*		*Method used to prepare plasma membranes*	*Reported by*
Glucose-6-phosphatase	Other component	Glucose-6-phosphatase	Other component		
0·38	0·24[a]	—	—	Emmelot *et al.* (1964)	Wattiaux-de-Coninck and Wattiaux (1969)
—	—	0·2	0·2[a]	Emmelot *et al.* (1964)	Emmelot *et al.* (1964)
0·95	2·46[a]	—	—	Coleman *et al.* (1967)	Coleman *et al.* (1967)
—	—	0·09	0·09[a]	Berman *et al.* (1969)	Berman *et al.* (1969)
—	—	0·2	0·25[b]	Emmelot *et al.* (1964)	Emmelot *et al.* (1964)
—	—	0·2	0·11[c]	Emmelot *et al.* (1964)	Emmelot *et al.* (1964)
—	—	0·088	0·09[d]	Neville (1960)	Vassilitz *et al.* (1967)
—	—	0·063[f]	0·01[d,f]	Neville (1968)	Fleischer *et al.* (1971)
—	—	0·088	0[e]	Neville (1960)	Vassilitz *et al.* (1967)
—	—	0·063[f]	0[e,f]	Neville (1968)	Fleischer *et al.* (1971)

a. NADH cytochrome *c* reductase
b. Esterase (α-napthyl laurate or caprylate substrate)
c. NADH nucleosidase
d. Cytochrome b_5
e. Cytochrome P_{450}
f. Bovine liver; other entries are for rat liver.

and Benedetti (1967) describe experiments in which a labelled soluble fraction from liver (which would also contain labelled serum proteins, as it was prepared from unperfused liver) was mixed with unlabelled plasma membranes. These were then separated by a further flotation and their radioactivity measured to determine the degree of contamination by soluble cytoplasmic proteins. It was found to be only 5% of the total protein of the re-isolated plasma membranes. This may well be a considerable underestimate of the proportion of adsorbed soluble protein, however, since only freely-exchangeable adsorbed proteins will be estimated by this method. There may well be proteins, bound firmly by the membranes immediately after cell breakage, which will exchange only very slowly with their labelled homologues, and will therefore not be estimated this way.

GENERAL COMMENTS

It is difficult to draw any firm conclusions from the wide field covered in this article. A vital point which must be emphasised is the necessity of monitoring a preparation of membrane fragments at all stages using a variety of criteria. As has already been mentioned, the light microscope is an extremely useful tool in following the purification of membrane sheets, but it must be complemented by enzymological studies to detect small fragments. It is hardly necessary to stress the importance of checking that the enzymes used as markers are really located in the organelles they are supposed to indicate in the tissue under study. Cytochemical studies may be necessary, but normally a literature search will suffice. While marker enzymes are the most versatile of tools for detecting contamination, they may give false indications, either due to low activity in the plasma membrane itself, or because the plasma membrane is contaminated by fragments of organelles which lack the marker employed. The most notable example of the latter is the contamination of microsomes by fragments of outer membrane from mitochondria. The solution is to find a 'marker' for the fragment. In the example quoted it would be monoamine oxidase.

Another type of contaminant is adsorbed soluble protein. This is extremely difficult to detect and measure. Immunological techniques may be used, but there are enormous problems of specificity. One elegant approach to the problem is to add labelled soluble proteins to the membranes (Benedetti and Emmelot, 1968), but it is probably best to add the labelled proteins to the homogenisation medium rather than, as was done by these authors, at a later stage.

A final problem encountered with any method is the risk of getting an unrepresentative preparation of membrane fragments. Such subfractionation has now been achieved experimentally (Evans, 1970), but it is still too early to discuss the possibility that the various membrane fractions may differ in morphological origin. The author will, therefore, thankfully leave all discussion of possible plasma-membrane subfractionations to some future writer.

Acknowledgements. I should like to thank my past and present colleagues at the Wolfson Bioanalytical Centre, especially Mr M. Dobrota, Dr J. T. R. Fitzsimons and Mr K. A. Norris, the results now reported being theirs as much as mine. In particular I thank Mr T. D. Prospero for allowing me to describe his unpublished method, Dr E. Reid for his advice both during the work and in the preparation of this article and Mrs S. Wreyford for her skill in typing a difficult manuscript. Financial support was given by the Cancer Research Campaign and the Science Research Council.

REFERENCES

ALLAN, D. and CRUMPTON, M. J. (1970). *Biochem. J.*, **120**, 133

ALLEN, M. R. (1969). *B.Sc. Project Report*, University of Surrey, Guildford

AMAR-COSTESEC, A., BEAUFAY, H., FEYTMANS, E., THINES-SEMPOUX, D. and BERTHET, J. (1969). In *Microsomes and Drug Oxidations*, (ed., Gillette, J. R., Conney, A. H., Cosmides, G. J., Estabrook, R. W., Fonts, J. R. and Mannering, G. J.), p. 41, New York and London (Academic Press)

ANDERSON, N. G., LANSING, A. I., LIEBERMAN, I., RANKIN, C. T. and ELROD, H. (1968). *Wistar Inst. Monog.*, **8**, 23

BEAUFAY, H., JACQUES, P., BAUDHUIN, P., SELLINGER, O. Z., BERTHET, J. and DE DUVE, C. (1964). *Biochem. J.*, **92**, 184

BENABDELZLIL, C., MICHEL-BECHEL, M. and LISSITZKY, S. (1967). *Biochem. biophys. Res. Comm.*, **27**, 74

BENEDETTI, E. L. and EMMELOT, P. (1968). In *The Membranes*, (ed. Dalton, A. J. and Hagenau, F.), p. 33, New York and London (Academic Press)

BERMAN, H. M., GRAM, W. and SPIRTES, M. A. (1969). *Biochim. biophys. Acta*, **183**, 10

BIRNIE, G. D. and HARVEY, D. R. (1968). *Analyt. Biochem.*, **22**, 171

BOONE, C. W., FORD, L. E., BOND, H. E., STUART, D. C. and LORENZ, D. (1969). *J. Cell Biol.*, **41**, 378

BOSMANN, H. B., HAGOPIAN, A. and EYLAR, E. H. (1968). *Arch. Biochem. Biophys.*, **128**, 51

COLEMAN, R. and FINEAN, J. B. (1966). *Biochim. biophys. Acta*, **125**, 197

COLEMAN, R. MICHELL, R. H., FINEAN, J. B. and HAWTHORNE, J. N. (1967). *Biochim. biophys. Acta*, **135**, 573

COTMAN, C., MAHLER, H. R. and ANDERSON, N. G. (1968). *Biochim. biophys. Acta*, **163**, 272

DAVYDOVA, S. YA. (1968). *Biokhimiya*, **33**, 685

DOD, B. J. and GRAY, G. M. (1968). *Biochim. biophys. Acta*, **150**, 397

DODGE, J. T., MITCHELL, C. and HANAHAN, D. (1962). *Arch. Biochem. Biophys.*, **100**, 119

EL-AASER, A. A. (1971). In *Separations with Zonal Rotors*, (ed. Reid, E.), p. B-5, Guildford (Wolfson Bioanalytical Centre, University of Surrey)

EL-AASER, A. A. and REID, E. (1969). *Histochem. J.*, **1**, 417

EL-AASER, A. A., REID, E., KLUCIS, E., ALEXANDER, P., LETT, J. T. and SMITH, J. (1966). *Natl. Cancer Inst. Monog.*, **21**, 323

EMANUEL, C. F. and CHAIKOFF, I. L. (1957). *Biochim. biophys. Acta*, **24**, 254

EMMELOT, P. and BENEDETTI, E. L. (1967). *Protides of the Biological Fluids*, **15**, 315

EMMELOT, P. and BOS, C. J. (1962). *Biochim. biophys. Acta*, **58**, 374

EMMELOT, P. and BOS, C. J. (1970). *Internat. J. Cancer*, **4**, 605

EMMELOT, P., BOS, C. J., BENEDETTI, E. L. and RUMKE, PH. (1964). *Biochim. biophys. Acta*, **90**, 126

EVANS, W. H. (1970). *Biochem. J.*, **116**, 833

EVANS, W. H. (1971). In *Separations with Zonal Rotors*, (ed. Reid, E.), p. S-3, Guildford (Wolfson Bioanalytical Centre, University of Surrey)

FITZPATRICK, D. F., DAVENPORT, R., FORTE, L. and LANDON, E. J. (1969). *J. biol. Chem.*, **244**, 3561

FITZSIMONS, J. T. R. (1969). *Ph.D. Thesis*, University of Surrey, Guildford

FLEISCHER, B. and FLEISCHER, S. (1969). *Biochim. biophys. Acta*, **183**, 265

FLEISCHER, B. and FLEISCHER, S. (1970). *Biochim. biophys. Acta*, **219**, 301

FLEISCHER, S., FLEISCHER, B., AZZI, A. and CHANCE, B. (1971). *Biochim. biophys. Acta*, **225**, 194

GRAHAM, J. M., HIGGINS, J. A. and GREEN, C. (1968). *Biochim. biophys. Acta*, **150**, 303

HARTMAN, G. C. and HINTON, R. H. (1971). In *Separations with Zonal Rotors*, (ed. Reid, E.), p. S-4, Guildford (Wolfson Bioanalytical Centre, University of Surrey)

HAYS, R. M. and BARLAND, P. (1966). *J. Cell Biol.*, **31**, 209

HEADON, D. R. and DUGGAN, P. F. (1971). In *Separations with Zonal Rotors*, (ed. Reid, E.), p. V-2, Guildford (Wolfson Bioanalytical Centre, University of Surrey)

HERZENBURG, L. A. and HERZENBURG, L. A. (1961). *Proc. Natn. Acad. Sci. U.S.*, **47**, 762

HICKS, R. M. and KETTERER, B. (1970). *J. Cell Biol.*, **45**, 542

HINTON, R. H., DOBROTA, M., FITZSIMONS, J. T. R. and REID, E. (1970). *Europ. J. Biochem.*, **12**, 349

HINTON, R. H., NORRIS, K. A. and REID, E. (1971). In *Separations with Zonal Rotors*, (ed. Reid, E.), p. S-2, Guildford (Wolfson Bioanalytical Centre, University of Surrey)

HOUSE, P. D. R. and WEIDEMANN, M. J. (1970). *Biochim. biophys. Res. Comm.*, **41**, 541

KAMAT, V. and WALLACH, D. F. H. (1965). *Science, N.Y.*, **148**, 1343

KAWASAKI, T. and YAMASHIRA, I. (1971). *Biochim. biophys. Acta*, **225**, 234

KEENAN, T., MORRE, D. J., OLSON, D. E., YUNGHANS, W. N. and PATTON, S. (1970). *J. Cell Biol.*, **44**, 80

KONO, T. and COLOWICK, P. (1961). *Arch. Biochem. Biophys.*, **93**, 520

LANSING, A. I., BELKHODE, M. L., LYNCH, W. E. and LIEBERMAN, I. (1967). *J. biol. Chem.*, **242**, 1772

MARINETTI, G. V. and GRAY, G. M. (1967). *Biochim. biophys. Acta*, **135**, 573

MCCOLLESTER, D. L. (1962). *Biochim. biophys. Acta*, **57**, 427

MCCOLLESTER, D. L. and SEMENTE, G. (1964). *Biochim. biophys. Acta*, **90**, 146

MCKEEL, D. W. and JARRETT, L. (1970). *J. Cell Biol.*, **44**, 417

NEVILLE, D. M. (1960). *J. biophys. biochem. Cytol.*, **8**, 413

NEVILLE, D. M. (1968). *Biochim. biophys. Acta*, **154**, 540

NEWKIRK, J. D. and WAITE, M. (1971). *Biochim. biophys. Acta*, **225**, 224

NORRIS, K. A., EL-AASER, A. A. and REID, E. (1972). In preparation

PERDUE, J. F. and SNEIDER, J. (1970). *Biochim. biophys. Acta*, **196**, 125

PFLEGER, R. C., ANDERSON, N. G. and SYNDER, F. (1968). *Biochemistry*, **7**, 2826

PROSPERO, T. D. and HINTON, R. H. (1972). In press

RAMBOURG, A. and LEBLOND, C. P. (1967). *J. Cell Biol.*, **32**, 27

RAY, T. K. (1970). *Biochim. biophys. Acta*, **196**, 1

REID, E. (1966). In *Enzyme Cytology*, (ed. Roodyn, D. B.), p. 321, London and New York (Academic Press)

REID, E. (1971). Ed., *Separations with Zonal Rotors*, Guildford (Wolfson Bioanalytical Centre, University of Surrey)

RODBELL, M. (1967). *J. biol. Chem.*, **242**, 5744

RODRIGUEZ DE LOPES ARNAZ, G., ALBERICI, M. and DE ROBERTIS, E. (1967). *J. Neurochem.*, **14**, 215

ROSENBURG, M. D. (1969). *Biochim. biophys. Acta*, **173**, 11

SIMON, E. R., BLUMENFELD, O. O. and ARIAS, I. M. (1970). *Biochim. biophys. Acta*, **219**, 349

SONG, C. S. and BODANSKY, O. (1967). *J. biol. Chem.*, **242**, 694

SONG, C. S., RUBIN, W., RIFKIND, A. B. and KAPPAS, A. (1969). *J. Cell Biol.*, **41**, 124

STEIN, Y., WIDNELL, C. and STEIN, O. (1968). *J. Cell Biol.*, **39**, 185

STONEHILL, E. H. and HUPPERT, J. (1968). *Biochim. biophys. Acta*, **155**, 353

TAKEUCHI, M. and TERAYAMA, H. (1965). *Expl Cell Res.*, **40**, 32

THUNEBERG, L. and ROSTGAARD, J. (1968). *Expl Cell Res.*, **51**, 123

TRAMS, E. G., STAHL, W. L. and ROBINSON, J. (1968). *Biochim. biophys. Acta*, **163**, 472

TURKINGTON, R. W. (1962). *Biochim. biophys. Acta*, **65**, 386

VASSILITZ, I. M., DERKATCHEV, E. F. and NEIFAKH, S. A. (1967). *Expl Cell Res.*, **46**, 419

WALLACH, D. F. H. (1967). In *The Specificity of Cell Surfaces*, (ed. Davis, B. D. and Warren, L.), p. 129, Englewood Cliff, New Jersey (Prentice Hall)

WARREN, L., GLICK, M. C. and NASS, M. K. (1967). *J. Cell Physiol.*, **68**, 269

WATTIAUX-DE-CONINCK, S. and WATTIAUX, R. (1969). *Biochim. biophys. Acta*, **183**, 118

WEAVER, R. A. and BOYLE, W. (1969). *Biochim. biophys. Acta*, **173**, 377

WILFONG, R. F. and NEVILLE, D. M. JR. (1970). *J. biol. Chem.*, **245**, 6106

WOODIN, A. M. and WIENEKE, A. A. (1966). *Biochem. J.*, **99**, 479

7

THE ISOLATION OF BRUSH BORDERS (MICROVILLI) FROM THE EPITHELIAL CELLS OF MAMMALIAN INTESTINE

J. W. Porteous

Brush borders form a specialised part of the outermost surface of the epithelial cells of mammalian small intestine. The first part of this article summarises current knowledge and hypotheses which underline the probable importance of the brush borders (that is, the microvillus membrane and its immediately adjacent substructure) in transintestinal transport. For this purpose sugar transport is taken as but one example of several transport phenomena which occur in the intact small intestine *in vivo* and *in vitro*. Methods which are at present in use or have been proposed for the isolation of intestinal brush borders are collected in the second section of this article; the third contains the results of attempts to improve the purity of brush borders isolated by one of these methods, while the last section is devoted to discussion of criteria of purity of isolated intestinal epithelial cell brush borders.

THE ROLE OF THE BRUSH BORDER IN TRANSINTESTINAL TRANSPORT

The small intestine (*Figure 1*) consists of four main substructures: the innermost mucosa (facing the lumen of the intestine), the submucosa, the circular muscle coat and the longitudinal muscle coat. A fifth structure, the peritoneum, covers most of the intestine and all of its supporting mesenteries, which in turn carry the arterial supply, the venous and lymph drainage and the nerve network serving the small intestine. The mucosa itself is complex and consists of (i) approximately cylindrical or tongue-shaped villi which arise as extrusions of the *lamina propria* overlying a thin muscle, the *lamina muscularis mucosae*, which marks the division between mucosa and submucosa; and (ii) a unicellular layer of four cell types which completely covers the villi—simple columnar epithelial cells interspersed with a variable number of mucin-secreting goblet cells cover the lateral walls and the tips of the villi, while a much smaller number of Paneth cells and argentaffine (chromaffine) cells occur in the crypts between the villi. The nerves, lymph ducts and blood vessels traverse all four main layers of intestinal tissue but do not enter the unicellular layer of cells lining the villi.

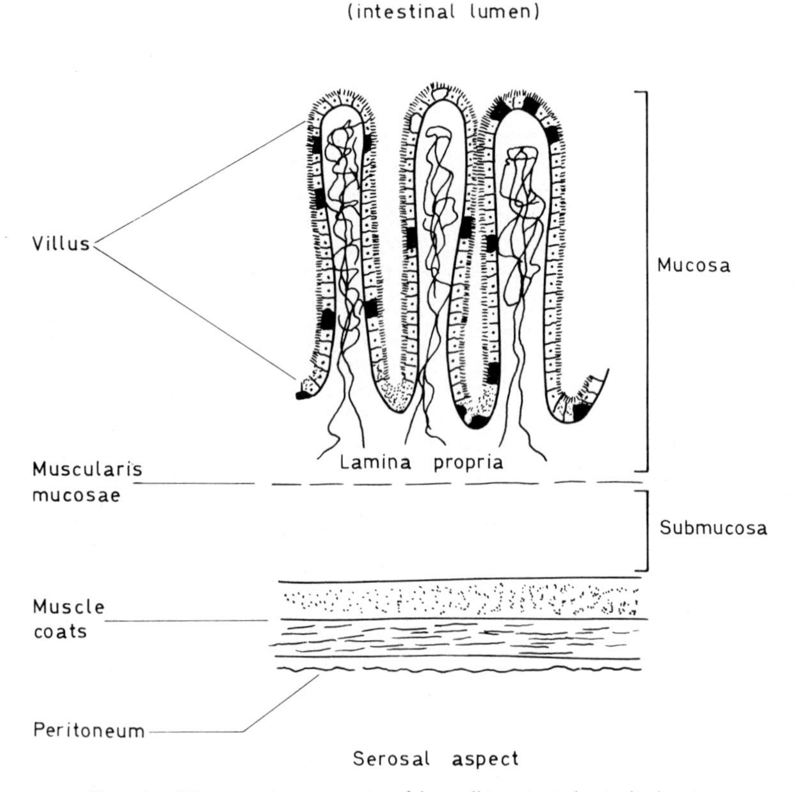

Figure 1. Diagrammatic representation of the small intestine in longitudinal section

It follows that *in vivo* the basic physiological unit for intestinal absorption is the villus (*Figure 2*), but from the biochemical point of view there is now good evidence (Barry *et al.*, 1965; Crane, 1966*a,b*; Eichholz, 1967; Rhodes *et al.*, 1967; Smyth, 1967*a,b*; Iemhoff *et al.*, 1970; Kimmich, 1970; Evans *et al.*, 1971) that the basic unit of intestinal transport activity is the individual columnar epithelial cell. Since these cells concentrate certain specific sugars initially present on the mucosal aspect of *in vitro* preparations of

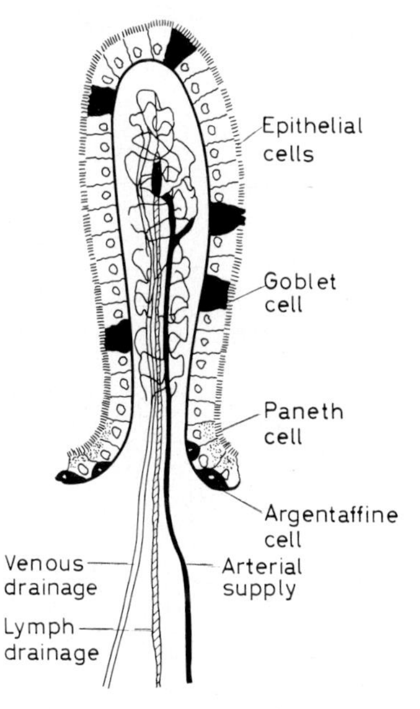

Figure 2. Diagrammatic representation of a villus

small intestine, and since the concentration of sugar in the underlying tissue is lower than in the epithelial cells themselves, McDougal *et al.* (1960), Bihler and Crane (1962) and Bihler *et al.* (1962) postulate that a sugar-specific carrier is present in or near the lumenal pole of the highly polarised epithelial cells, that is, in or near the so-called brush-border or microvillus structure of these cells (*Figure 3*). The results of autoradiographic studies by Kinter and Wilson (1965) using everted sacs of intestine are consistent with these postulates are also the results of Porteous and Herford (1972).

Two other developments in the field of intestinal sugar transport are relevant. First, Riklis and Quastel (1958) showed that active transintestinal transport of glucose is Na^+-dependent. This observation has been confirmed for many species (Robinson, 1967, 1970). Crane (1966*a,b*) postulates that this Na^+-dependence arises from the need to form a ternary complex between sugar, Na^+ and a specific sugar carrier located in the brush-border membrane in order to facilitate entry of sugar across the brush-border membrane into the cell. A separate outwardly-directed and energy-dependent Na^+ 'pump' would then ensure a high concentration of Na^+ outside the cell and a low concentration inside; in Crane's view, such a Na^+ gradient maintained by a functioning Na^+ 'pump' would convert *facilitated entry* into *active transport* of glucose into the cell. There is uncertainty about the location of the appropriate Na^+ 'pump'.

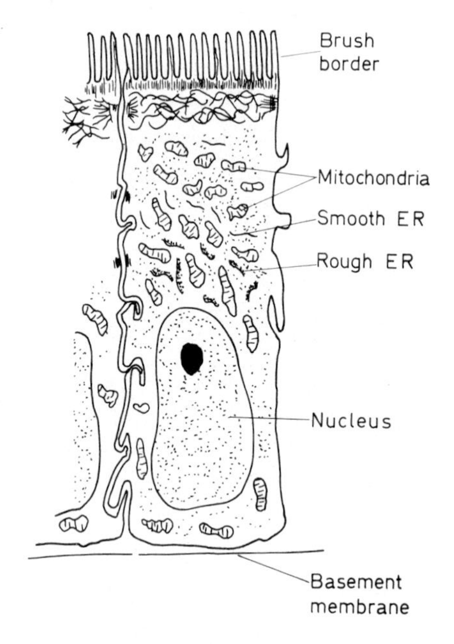

Figure 3. Intestinal epithelial cell structure

The second relevant development was the isolation of a crude brush-border fraction from golden hamster intestine by Miller and Crane (1961) who showed that the fraction contained about 75% of the invertase and maltase activities of the original cell homogenate. Carnie and Porteous (1962) showed that rabbit intestinal invertase was Na^+-activated. This last observation has been greatly extended by Semenza and colleagues (Semenza *et al.*, 1964; Semenza, 1967; Gitzelmann *et al.*, 1970).

Thus the available evidence points to the brush border of the epithelial cells as the site of the terminal stages of intestinal digestion of sugars, and also as the site of a sugar-specific transport mechanism. Hydrolysis of sucrose (and maltose, trehalose, lactose), the binding of specific sugars to the postulated sugar carrier and the mechanisms responsible for 'uphill transport' of sugar into the epithelial cell are each Na^+-dependent. Other evidence (Porteous and Clark, 1965; Rhodes *et al.*, 1967) indicates that the brush

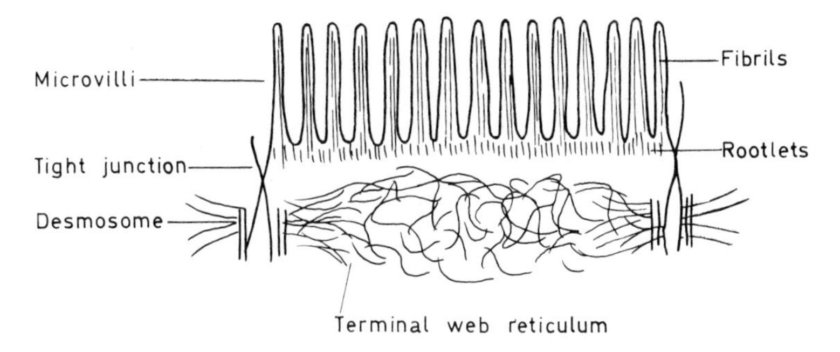

Figure 4. Brush border structure

border may also be the site for the terminal stages of digestion of protein. The brush border is an interesting subcellular structure from the morphologist's point of view—it may well be the most interesting subcellular structure in respect of intestinal cell function *in vivo*.

Substantiation and extension of current concepts of the role of the brush border in intestinal function clearly depend upon the availability of efficient methods for the isolation of this subcellular structure in substantial quantities and essentially free from other cell components. Features that should be present in clean intact brush borders are illustrated in *Figure 4*.

METHODS AVAILABLE FOR THE ISOLATION OF BRUSH BORDERS

Miller and Crane (1961) pioneered the technique of isolating intestinal epithelial cell brush borders. Most authors have used this original method, or slight modifications of it. Miller and Crane (1961) suspended the mucosa of golden hamster intestine in a hypotonic solution, 5 mM EDTA, pH 7·4, and accelerated the disruption of the epithelial cells by vigorous treatment in a Waring Blendor. Low-speed centrifugation and repeated washing of the sediment in 5 mM EDTA yielded brush borders apparently free of nuclei and other cell components, and containing about 75% of the intestinal invertase and maltase activities. In the absence of EDTA the brush borders disintegrated.

Porteous and Clark (1965) published the first account of a procedure which preserved and isolated all the readily identifiable subcellular components of intestinal epithelial cells, viz. the brush borders, nuclei, mitochondria, endoplasmic reticulum and cell sap. The technique involved homogenisation of rabbit mucosal tissue in 0·3 M sucrose, 5 mM EDTA, pH 7·4 under carefully defined conditions in a Teflon–glass motor-driven homogeniser, filtration through nylon cloth to remove extraneous tissue components, and differential centrifugation of the filtrate. A shortcoming of the method was that brush borders were isolated together with nuclei. Despite extensive efforts, Porteous and Clark (1965) and Porteous and Paterson (1965) failed to isolate either pure brush borders or pure nuclei from rat or rabbit intestine after homogenisation of the mucosa in sucrose–EDTA. These authors also failed to isolate clean brush borders from hamster, rat or rabbit intestine using Miller and Crane's (1961) technique of homogenisaton in hypotonic EDTA. In a few experiments with hamsters, and throughout a large number of experiments with rats and rabbits, invariably a mixture of nuclei and brush borders was isolated. Hübscher *et al.* (1965) modified and extended the technique of Porteous and Clark (1965) in several respects but again failed to separate rabbit and guinea-pig intestinal nuclei from brush borders.

Gallo and Treadwell (1963) and Bailey and Pentchev (1964) on the other hand modified the homogenisation conditions used by Miller and Crane (1961) and apparently succeeded in isolating rat intestinal brush borders. Ruttloff *et al.* (1964) and Noack and Schenk (1965) also isolated rat intestinal brush borders. Unfortunately, none of these authors published any quantitative information which would determine the freedom of their preparations from contamination by other subcellular components, nor did Miller and Crane (1961).

Harrison and Webster (1964) devised a rather different approach to the isolation of intestinal brush borders; they everted the complete small intestine from a rat onto the lower end of a glass spiral, immersed this in 2·5 mM EDTA, pH 7·4 and attached the upper end of the spiral to an electromagnetic vibrator. Vibration under critically-defined conditions preferentially released brush borders from the epithelial cells still attached to the intestine, thus minimising contamination of brush borders with nuclei and other subcellular components. Any nuclei released were subsequently preferentially adsorbed onto glass-fibre tissue leaving a suspension of pure brush borders. These authors comment on the Miller and Crane (1961) method as applied to rat intestine:- 'the isolation of rat brush borders is made difficult by the co-sedimentation of nuclear aggregates' (cf. Porteous and Clark, 1965).

Millington *et al.* (1966), following the work of Millington and Finean (1963), investigated the effect of pH upon the isolation of rat intestinal brush borders after homogenisation of mucosa in tris-buffered EDTA. They found that rat brush borders were always contaminated with cytoplasmic components if unbuffered EDTA was used as in the original Miller and Crane (1961) technique. Unfortunately these authors, like Harrison and Webster (1964), did not carry out any quantitative chemical analyses or enzyme assays on their preparations.

Apart from work by Crane and associates, and apart from a few attempts (Porteous and Clark, 1965) to isolate clean brush borders from hamster intestine by the Miller and Crane (1961) method, all the other work reviewed here has been carried out on rat, rabbit, guinea-pig or cat intestine. At least so far as the rat and rabbit tissue is concerned there seems to be consistent qualitative evidence from several authors that the Miller and Crane technique involving homogenisation of mucosa in hypotonic EDTA does not yield pure brush borders but, as in the original sucrose–EDTA method (Porteous and Clark, 1965), a mixture of nuclei and brush borders. The results of Eichholz (1967) suggest that this may also be true of preparations from hamster intestine.

There is an obvious need for an investigation of the precise conditions required to isolate clean brush borders from a variety of common species of small intestine, and a clear need for a quantitative assessment of the purity of the preparations obtained. The work presented in the following section of this article represents a start in this direction.

THE ISOLATION OF PURIFIED BRUSH BORDERS FROM RAT-INTESTINE EPITHELIAL CELLS

The method of isolation eventually devised is based on that of Miller and Crane (1961) with the addition of a step which selectively destroys the integrity of nuclei and then leads to the flocculation of nuclear debris. As a result 90 to 99% of the DNA and the RNA present in the original cell homogenate is removed. Substantially purified brush borders can then be isolated. The use of solutions of phosphate containing citrate and chloride to achieve this selective removal of nuclei from brush-border preparations was suggested by an observation (Porteous and Paterson, 1965) that epithelial cells, segregated from intact intestine by incubation with citrate (Stern and Jensen, 1966) and subsequently added to phosphate-buffered isotonic NaCl, gave a flocculent sediment leaving a turbid supernatant fraction containing a high proportion of brush borders. The flocculent sediment contained disintegrated and aggregated nuclei together with other cellular debris.

A male Wistar rat (approx. 200 g) is anaesthetised with ether, the abdomen opened with a midline incision and the upper and lower limits of the small intestine located and cut. The intestine is washed out *in vivo* with cold 0·9% (w/v) NaCl (100 ml) under slight pressure, then removed to a beaker of fresh 0·9% (w/v) NaCl. The intestine is next washed inside and out with cold 2·5 mM EDTA, pH 7·0, drained, and layed on a glass plate. Excess lumenal fluid is expelled and discarded by clamping one end of the intestine and gently stroking the length of intestine with the edge of a microscope slide. Mucosa is then expressed by a similar, but vigorous, treatment with a microscope slide. All subsequent steps in the procedure, certain details of which are given in *Chart 1*, are carried out at 0 to 5°C.

Mucosa is homogenised under carefully controlled conditions in one of two ways: (*a*) 2–3 g of mucosa is suspended in 60 ml of 2·5 mM EDTA, pH 7·0, using a hand operated Teflon–glass homogeniser (5 strokes; dimensions of homogeniser as stated in *Chart 1*). This suspension is then homogenised in the same motor-driven Teflon–glass homogeniser and the homogenate is diluted to 120 ml with 2·5 mM EDTA, pH 7·0. Alternatively (*b*) 2–3 g of mucosa is homogenised in 120 ml of 2·5 mM EDTA, pH 7·0, using the MSE top-drive macerator (*Chart 1*).

The homogenate is filtered through nylon cloth (*step (ii), Chart 1*). On average, 95% of the homogenate volume is recovered in the filtrate, which is then centrifuged. The supernatant fluid is discarded and the sediment is resuspended in half the original volume of 5 mM EDTA, pH 7·0, using the hand-operated Teflon–glass homogeniser in the manner already described. A suspension of nuclei and brush borders is isolated by repeated centrifugation and sedimentation as described in *Chart 1*. The suspension of nuclei and brush borders is diluted with an equal volume of 0·1 M potassium phosphate, 0·05 M potassium citrate, 0·154 M potassium chloride, pH 7·0, (PCC) and left to stand 20 to 30 min in a tall glass cylinder. The turbid supernatant suspension is decanted through fine nylon cloth (*Chart 1*) which retains any of the flocculent sediment accidently decanted from the cylinder. The sediment formed by low-speed centrifugation of the filtrate is resuspended in 2·5 mM EDTA, pH 7·0, using a hand-operated homogeniser as

Chart 1. Flow sheet for the preparation of intestinal epithelial cell brush borders and nuclei, and for the selective removal of nuclei

(i) (a) Teflon–glass homogeniser; 60 ml of EDTA, pH 7·0.

(b) MSE top-drive macerator; 120 ml of EDTA, pH 7·0.

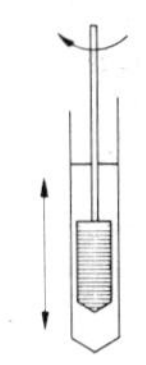

2 to 3 g mucosa homogenised

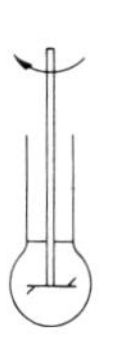

By hand, 5 strokes, then by motor, 2000 rev/min, 60 seconds, 12 strokes.

4000 rev/min, 40 seconds.

(ii) Homogenate diluted (if necessary) to 120 ml with EDTA (2·5 mM, pH 7·0) then filtered through 100 μ mesh nylon cloth.

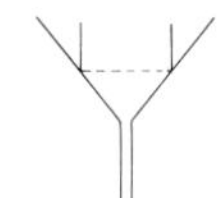

(iii) Filtrate centrifuged at 1000*g* for 10 min

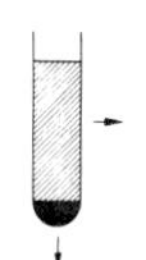

Supernatant fluid discarded.

Sediment resuspended (hand-operated Teflon–glass homogeniser, 5 strokes) in 5 mM EDTA, pH 7·0.

(iv) Step (iii) repeated two or three times until the sediment contained only free nuclei and free brush borders.

(v) Sediment suspended in 5 mM EDTA, pH 7·0 (hand-operated Teflon–glass homogeniser, 5 strokes) and mixed with an equal volume of 100 mM potassium phosphate, 50 mM potassium citrate, 154 mM potassium chloride (PCC), pH 7·0.

(vi) Turbid supernatant fluid decanted from flocculent sediment through 25 μ mesh nylon cloth.

(vii) Brush borders sedimented (1000*g*, 10 min); steps (v) and (vi) repeated once.

(viii) Brush borders sedimented (1000*g*, 10 min). Supernatant fluid discarded; sediment resuspended in 2·5 mM EDTA, pH 7·0.

Note: *the Teflon–glass homogeniser was fashioned from Veridia precision-bore glass tubing, nominal diameter 1½ in, overall length 9 in. The pestle was turned from a cylinder of Teflon 2½ in × 1½ in to give a radial clearance of 0·004 in. The stainless-steel driving shaft was threaded into the Teflon pestle*

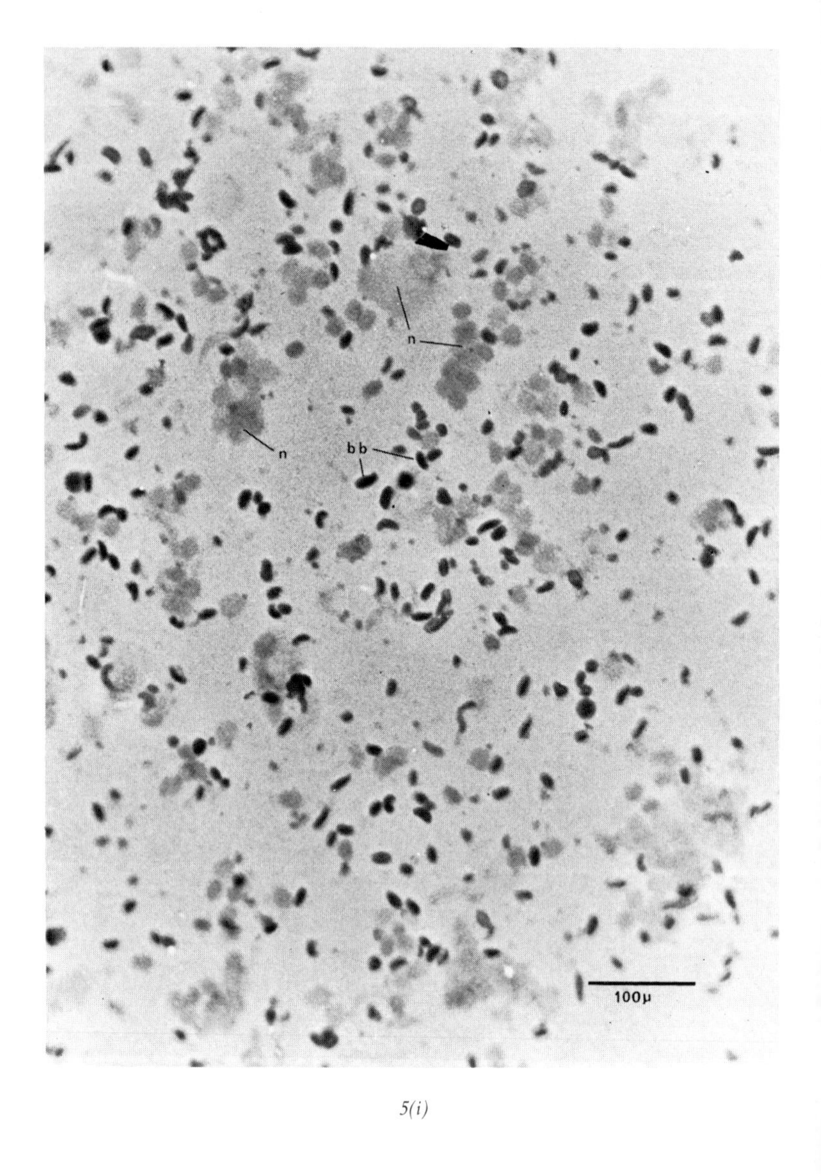

5(i)

Figure 5(i) and (ii). Typical preparations of rat intestinal epithelial cell brush borders (bb) *and nuclei* (n), *at step* (iv) *of* Chart 1. *(Unstained preparations: phase-contrast microscopy)*

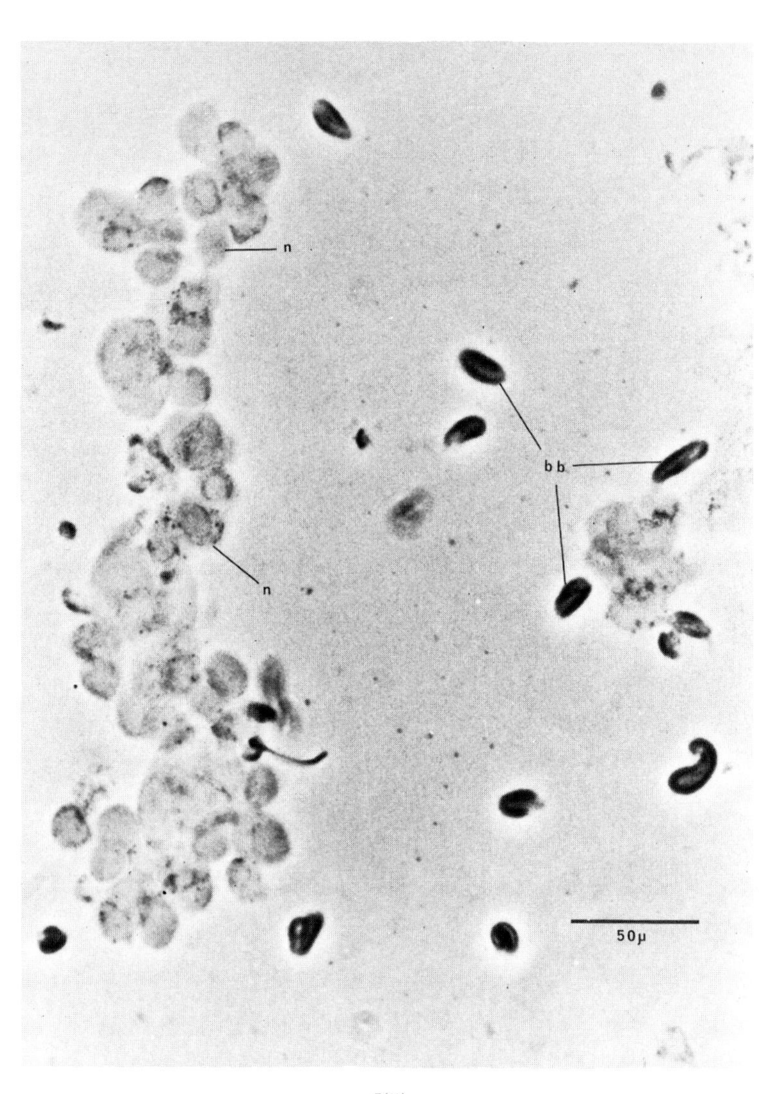

5(ii)

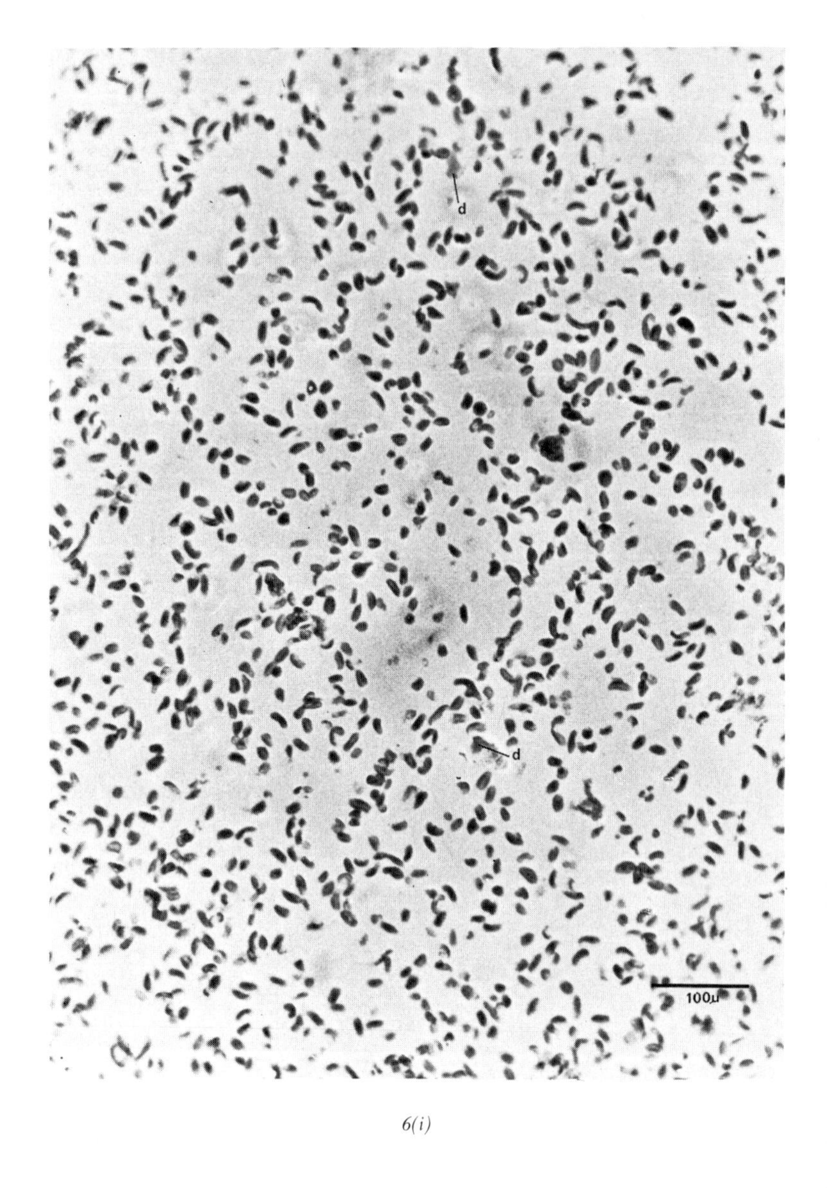

6(i)

Figure 6(i), (ii) and (iii). The same preparation shown in Figure 5 *after removal of nuclei (steps* (v) *to* (viii) *of* Chart 1*). All fields of view contain clean intact brush borders; occasional patches of debris* (d) *are visible and are probably degraded nuclei. Microvilli* (mv) *are seen at high resolution in* Figure 6(iii). *(Unstained preparations: phase-contrast microscopy)*

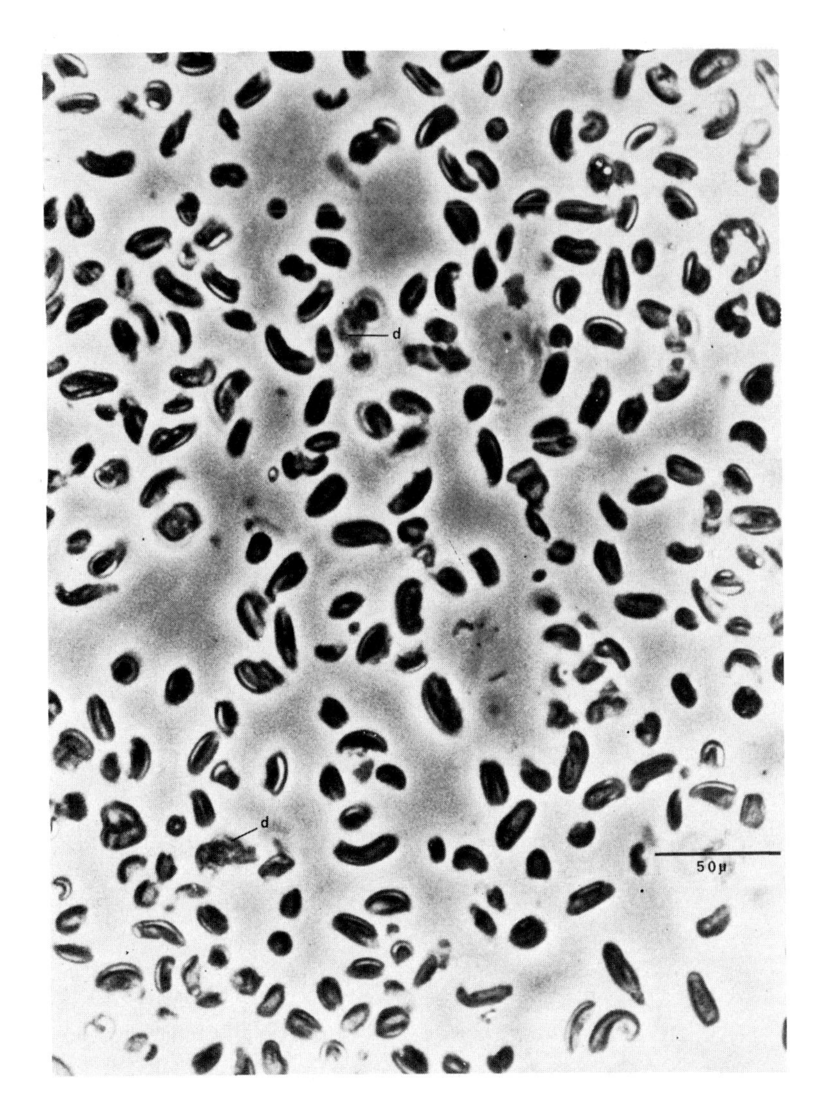

6(ii)

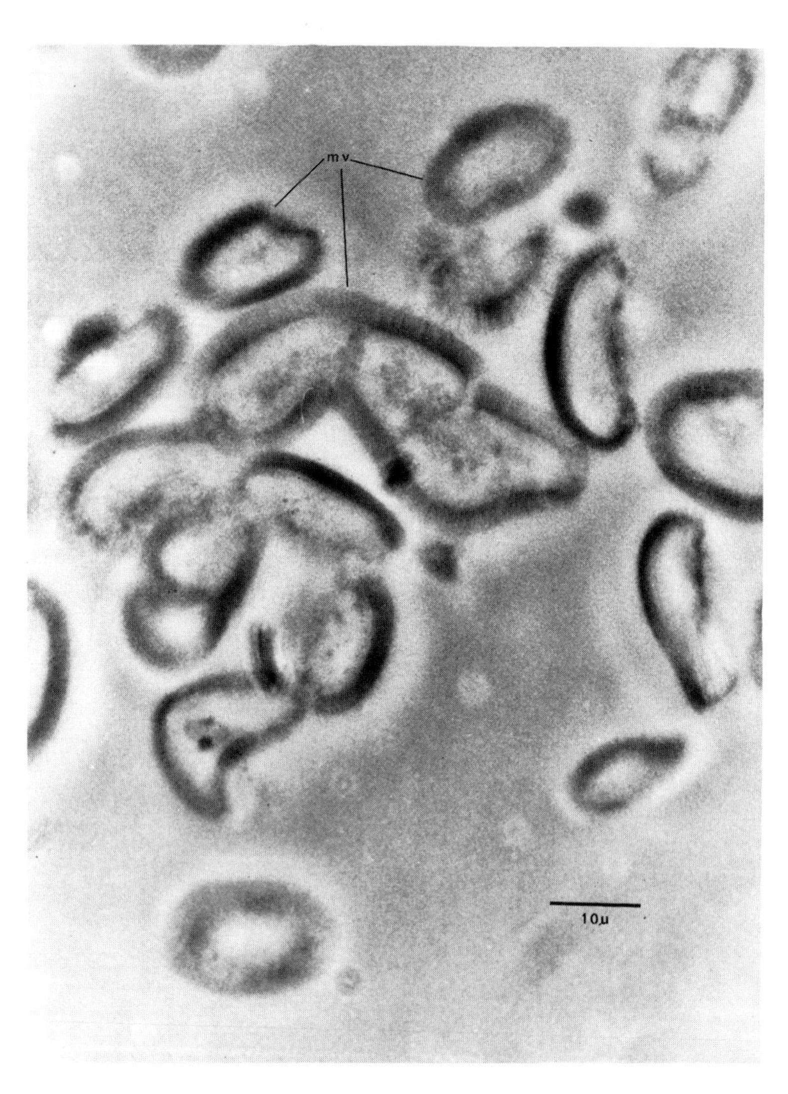

6(iii)

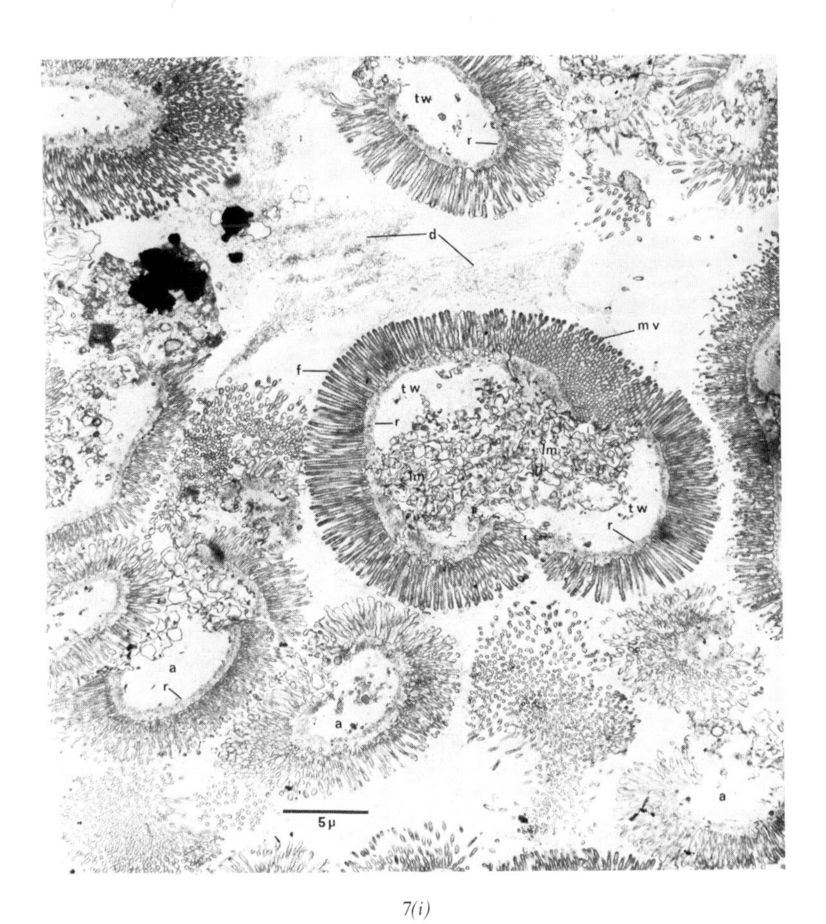

7(i)

Figure 7 (i), (ii), (iii). Electron micrographs prepared from a pellet of the brush borders shown in Figure 6. *The pellet was fixed with glutaraldehyde, post-fixed with* OsO_4, *dehydrated by treatment with graded concentrations of ethanol then with propylene oxide, and embedded in Epon. Microvilli* (mv), *fibrils* (f) *and rootlets* (r) *are visible in each brush border structure; the terminal web* (tw) *is visible in most brush borders but apparently absent* (a) *from some brush borders. Lateral cell membrane* (lm) *is easily identified in* Figure 7 (iii) *but in some brush borders* (Figure 7 (i) and (ii)) *the features marked* (lm) *may also include some vesticular derivatives of the endoplasmic reticulum; this is particularly true when adjacent brush borders are attached by tight junctions and desmosomes (indicated by arrows;* cf. Figure 4*). Debris* (d) *is very probably of nuclear origin*

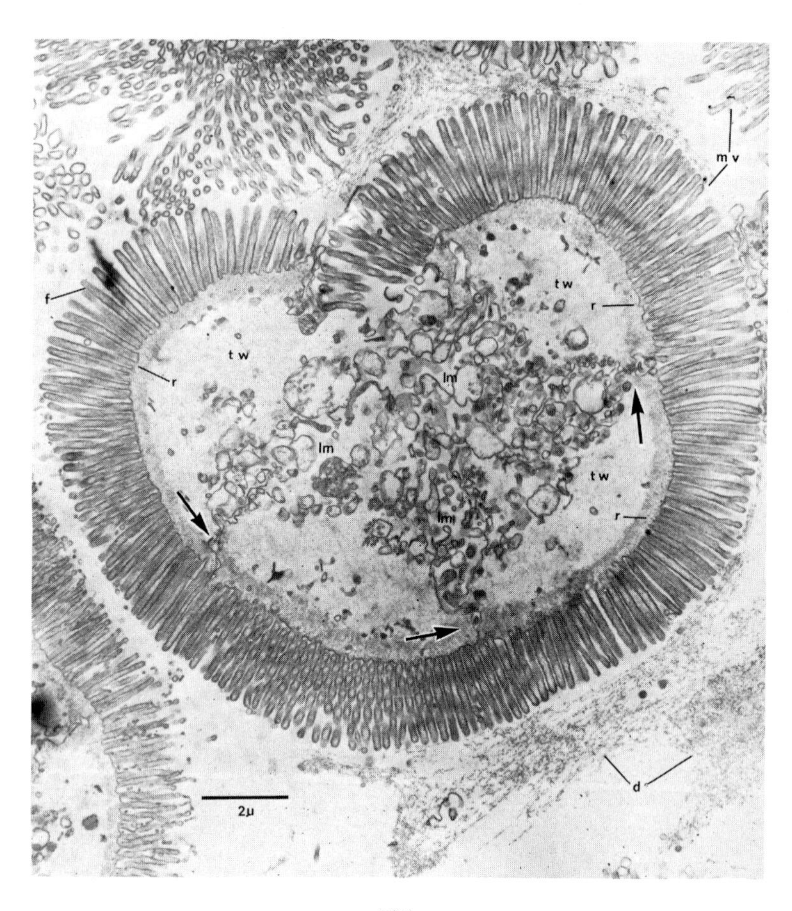

7(ii)

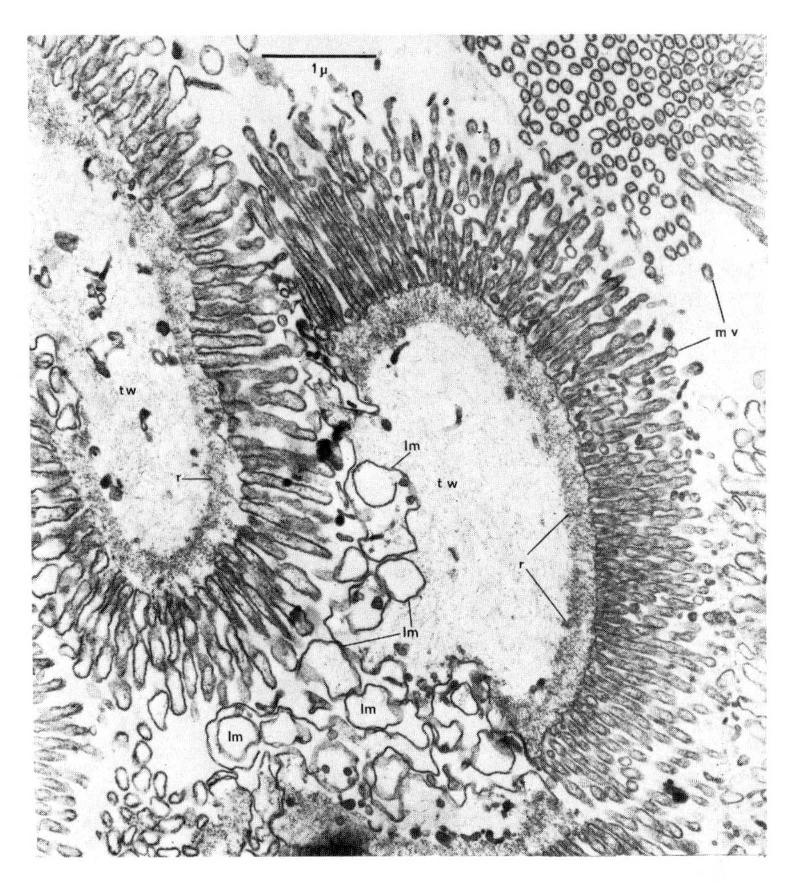

7(iii)

before and the treatment with PCC solution is repeated, followed by filtration and centrifugation as before. The final sediment is resuspended in 2·5 mM EDTA, pH 7·0, and stored at −20°C if analyses are not carried out immediately. *Figure 5* shows a typical preparation of rat intestinal brush borders and nuclei before treatment with PCC solution while *Figures 6* and *7* show the brush borders obtained after removal of the nuclei by PCC treatment. Microscopic observations of this kind are an essential but insufficient criterion of purity of the brush borders.

Table 1 summarises the results of analyses for protein, DNA, RNA, invertase activity, succinate dehydrogenase activity and glucose-6-phosphatase activity at various stages in the isolation procedure described. *Table 2* shows the results of similar analyses in experiments designed to compare the efficiency of 2·5 mM EDTA, 77 mM NaCl solution, pH 7·0, and of 2·5 mM EDTA, 50 mM KH_2PO_4, 25 mM citrate, 77 mM KCl, pH 7·0, in removing nuclei from a suspension of nuclei and brush borders.

Table 1. PURIFICATION OF BRUSH BORDERS FROM RAT INTESTINAL EPITHELIAL CELLS BY TREATMENT WITH PHOSPHATE–CITRATE–CHLORIDE (PCC), PH 7·0. (THE VALUES SHOWN ARE THE MEANS FROM TWO TYPICAL DETERMINATIONS)

Preparation (Chart 1)	*Protein*	DNA	RNA	*Invertase*	SDH*	G6P-ase†
(1) Filtered cell homogenate	100	100	100	100	100	?
(2) Brush borders + nuclei (before PCC treatment)	16	100	23	64	—	—
(3) Brush borders (after PCC treatment)	5	8	6	42	0	0
(4) PCC sediment	10	80	10	5	—	—
(5) Combined supernatants	75	4	75	60	—	—
Per cent recoveries	90	92	91	107	—	—

Table 2. PURIFICATION OF BRUSH BORDERS FROM RAT INTESTINAL EPITHELIAL CELLS BY TREATMENT WITH SODIUM CHLORIDE OR WITH PHOSPHATE–CITRATE–CHLORIDE (PCC), PH 7·0. (THE VALUES SHOWN ARE THE MEANS FROM FOUR SEPARATE DETERMINATIONS)

Preparation (Chart 1)	*Invertase* / *Protein* (units/mg)	*Invertase* / DNA (units/mg)	*Invertase* / RNA (units/mg)	*Per cent invertase*
(1) Filtered cell homogenate	5·5	120	66	100
(2) Brush borders (after NaCl treatment)	70	800	1000	35
(3) Brush borders (after PCC treatment)	70	4900	2400	40

* SDH = succinate dehydrogenase
† G6P-ase = glucose-6-phosphatase
? See text.

QUALITATIVE AND QUANTITATIVE CRITERIA OF THE PURITY OF ISOLATED BRUSH BORDERS

MICROSCOPIC OBSERVATIONS

Routine light-microscopic observation is an essential technique in subcellular fractionation studies involving the isolation of mitochondria and larger subcellular organelles (Simkin and Porteous, 1972). The use of stained preparations is to be avoided if possible, partly because their preparation takes up valuable time but, more important, because staining nearly always leads to flocculation of some organelles. On the other hand, the nuclei shown in *Figure 5* were scarcely visible when viewed with some phase-contrast microscopes, particularly those fitted with a fixed phase ring optical system. Porteous and Clark (1965) resorted to the use of methylene blue staining to detect nuclei. A Leitz Ortholux microscope fitted with positive phase-contrast objectives and a Heine substage condenser was used throughout the present work and enabled nuclei to be detected with ease provided certain precautions were taken. The Heine substage condenser is continuously adjustable to give, with any phase-contrast objective, dark field, phase-contrast or bright-field illumination without any colour fringes in the background field of view. Furthermore, it is possible with this optical system to stop down the substage iris diaphragm so as to increase resolution with high-power objectives. The lack of colour, the flat field of view and the continuously adjustable but even lighting conditions afforded by this instrument make it ideal for rapid but certain searching of fractions at each stage of subcellular fractionation. As to precautions to be observed, these relate to the properties of nuclei rather than to the quality of the microscope used. Nuclei tend to adhere to glass surfaces, other organelles tend to float freely in the suspension fluid on a microscope slide. It is thus quite easy, even with a high-quality microscope, to miss the poor-contrast nuclei immobilised on the glass surfaces and to focus only on the high-contrast, crisply outlined organelles floating and moving in the suspension fluid. In *Figure 5* the brush borders have been put deliberately out of focus in order to bring the immobilised nuclei almost into focus. It will also be noted that the nuclei tend, even when apparently intact, to adhere one to another under the conditions used in this work.

The light-microscope micrographs shown here were obtained by observation of subcellular fractions suspended in 2·5 mM EDTA as described in the preceding section. They thus represent survey views of relatively large samples of the subcellular fraction so that occasional or trace contaminants might not be readily detected. Concentration of such suspensions by centrifugal sedimentation, followed by fixation, embedding and sectioning for electron microscopy affords the opportunity not only of examining the detailed structure of the principle subcellular organelle present but also of detecting and examining trace contaminants which have now likewise been concentrated into a small volume. Thus *Figure 6* shows brush borders, to most of which are attached some apical cytoplasm, but it is not possible to see whether this cytoplasm includes mitochondria (see Porteous and Clark, 1965) or endoplasmic reticulum, or whether lateral cell membranes are still attached to the brush borders. It is quite impossible to see whether the terminal web (*Figure 4*) is intact or missing. In *Figure 6* occasional patches of ill-defined material are visible and are presumed to be nuclear debris. Electron microscopic examination (*Figure 7*) shows quite clearly that some nuclear debris is in fact present in these purified brush-border preparations and, further, that the brush borders themselves may still have attached to them considerable amounts of lateral cell membrane and/or endoplasmic reticulum but are completely free of mitochondria (Porteous and Clark, 1965). Most of the brush borders are complete with the terminal web but some appear to have lost this structure while retaining the fibrils and rootlets of the microvilli themselves (cf. *Figure 4*). Occasionally several adjacent brush borders remain attached at the tight junction and adjacent desmosome (*Figures 4* and *7*) and, in these instances, lateral membrane and possibly endoplasmic reticulum is most prominent.

QUANTITATIVE ANALYSES

In *Table 1*, the recoveries of components and activities are the sums of the appropriate values for fractions 3, 4 and 5. As these recoveries are satisfactory, it is legitimate to compare the analytical values for fractions 2 and 3. Fraction 2 is the result of the application of the Miller and Crane (1961) technique to rat intestinal mucosa; this fraction contains 16% of the protein of the homogenate, 100% of the DNA, 23% of the RNA and 64% of the invertase. If it is assumed that invertase activity is associated solely with the brush borders (Miller and Crane, 1961; Carnie and Porteous, 1962*b*; Gallo and Treadwell, 1963; Bailey and Pentchev, 1964; Ruttloff *et al.*, 1964; Porteous and Clark, 1965; Hübscher *et al.*, 1965; Eichholz, 1967; Johnston, 1967; Semenza, 1967; Forstner *et al.*, 1968; Porteous, 1968) and that the enzyme is not inactivated during the isolation of the brush borders, and if it is further assumed that DNA is predominantly associated with nuclei but not associated with brush borders, then the analytical figures quoted for fraction 2 (*Table 1*) suggest that the Miller and Crane (1961) technique, when applied to rat intestinal mucosa, preserves nuclei rather better than it does brush borders. It is clear that treatment of fraction 2 with PCC solution leads to the loss of about one-third of the invertase activity but simultaneously affords a considerable purification of the brush borders in respect of protein and, more particularly, in respect of DNA and RNA. Thus the relative specific activity for invertase (per cent invertase activity/per cent protein content) for fraction 1 is 1, for fraction 2 is 4 and for fraction 3 is 8. Fraction 3 (*Table 1*) contains no detectable succinate dehydrogenase activity, confirming the electron microscopic observation that mitochondria are absent from this preparation. Glucose-6-phosphatase activity is also not detectable in fraction 3 but there is some doubt about the interpretation of this finding. Fraction 1 (*Table 1*) contains lower and more variable glucose-6-phosphatase activity than does an analogous rabbit preparation (Porteous and Clark, 1965), in agreement with the findings of Clark and Sherratt (1967) for the microsome fraction of rat intestinal epithelial cells. Salomon *et al.* (1964) found low glucose-6-phosphatase activity and high phospholipase A activity in rat intestine and surmised that the latter enzyme inactivated the membrane-bound phosphatase. If a similar situation applied to fraction 3 (*Table 1*) the result shown would not reflect the true absence of glucose-6-phosphatase from the brush borders. The assay was nevertheless included in an attempt to resolve doubts about the intracellular location of intestinal epithelial cell glucose-6-phosphatase (Porteous and Clark, 1965).

Results shown in *Table 1* are taken from early experiments and represent the minimum purification of rat intestinal epithelial cell brush borders obtainable by the technique elaborated. In later experiments, incorporating detailed refinements in technique (*Chart 1*), the absolute amount of DNA and of RNA present in the purified brush borders was decreased from 8% and 6%, respectively, (*Table 1*) to less than 1% of that present in the homogenate. The protein content was decreased to less than 1% of that present in the homogenate without further loss of invertase activity. Results which are typical of these later experiments are presented in *Table 2* together with results which provide a direct comparison of the ability of NaCl and of PCC solution to remove DNA and RNA from the brush border plus nuclei preparation. It is clear that the latter treatment is the more effective in eliminating DNA and RNA from the final brush-border preparation. The quantitative information contained in *Table 2* is consistent with the microscopic observation that some nuclear debris is still present in the purified brush-border preparation.

It may be noted that the three specific activities listed in *Table 2* provide a useful index of the purification of a brush-border preparation. On the basis of assumptions already made concerning the intracellular localisation of invertase activity and of DNA, the ratio invertase to protein should increase and reach a high plateau value as purification of brush borders proceeds, whereas the ratio of invertase to DNA and of invertase to RNA should simultaneously increase towards infinity. That the first ratio has increased twelve-fold between fraction 1 and fraction 3 (*Table 2*) whereas the other two ratios have increased only forty-fold suggests that further purification should be possible and for some purposes may be necessary. Nevertheless the analytical values shown in *Tables 1* and *2* provide substantial quantitative support for the claim that invertase activity is a specific attribute of the intestinal epithelial cell brush border. Previous work has provided either insufficient quantitative information to eliminate other possibilities or has provided no quantitative information at all. One exception, already noted, is provided by Eichholz (1967) who found a ratio of protein to DNA of unity for hamster brush borders isolated by a slight modification of the Miller and Crane (1961) technique. Porteous and Clark (1965) found a ratio of 7 for their subcellular fraction containing both nuclei and brush borders. Exact significance cannot be attached to a comparison of such ratios determined in different ways but it seems very probable that hamster mucosa (Eichholz, 1967), like rat and rabbit mucosa (Porteous and Clark, 1965) yields a mixture of brush borders and nuclei when subjected to the Miller and Crane (1961) procedure. Eichholz (1967) used the elegant techniques elaborated by Eichholz and Crane (1965) and by Overton *et al.* (1965) to separate the brush borders into a microvillus membrane fraction, a fibrillar and two unidentified fractions. The first fraction showed the highest specific activities for alkaline phosphatase, ATPase and several disaccharidase activities. The quantitative results of Eichholz (1967) and those presented in *Tables 1* and *2* are complementary. Both strongly suggest that the brush border is the unique location of intestinal epithelial cell invertase activity. Eichholz's results go further and show that this enzyme is confined to the microvillus membrane of the brush borders. An attempt to demonstrate the expected Na^+-dependent binding of D-glucose to such membrane preparations (Eichholz *et al.*, 1969) gave disappointing results; future work will doubtless reveal the reasons for these results. A further exception to the general lack of quantitative information on intestinal brush-border preparations was provided by Forstner *et al.* (1968) who isolated purified rat intestinal brush borders by techniques essentially similar to those reported here and by Porteous (1968). These authors also succeeded in isolating a membrane subfraction of the purified brush borders.

Table 3 summarises the qualitative and quantitative information currently available on isolated intestinal brush borders. Several points concerning the quantitative results are of interest in so far as they bear on the purity of the isolated brush borders and subfractions thereof. First, the increase in the value of the ratios in columns (2) and (4) (*Table 3*) reflects a steady purification of the brush borders through stages* (*a*), (*b*) and (*c*) of the present work; purification in respect to RNA greatly exceeds that in respect of protein. The value of the ratios in columns (1) and (3) show a marked decrease between stages (*a*) and (*b*), reflecting the relative conservation of nuclei and destruction of brush borders at this stage of purification. Between stages (*b*) and (*c*) the values of these two ratios show sharp increases, that for invertase to DNA being much the greater. Between stages (*a*) and (*c*) there is a fourteen-fold purification of brush borders with respect to protein and a forty-fold purification with respect to DNA and RNA. These results suggest that further purification from adventitious DNA and RNA should be possible or that these compounds are an integral part of the brush border of intestinal epithelial cells. The former view seems preferable until substantial evidence to the contrary is produced. Second, the ratios calculated from the results of Forstner *et al.* (1968) are remarkably similar in magnitude to those quoted from the present work with but one exception, viz. the value of the ratios in columns (1) and (3) do not decrease sharply between stages (*a*) and (*b*) because these authors found only 5% (instead of 100%) of the original DNA in their crude brush-border fraction. This may be accounted for partly by the higher pH of their homogenisation medium and partly by the more vigorous homogenisation conditions used by them, a supposition which would also account (Carnie and Porteous, 1962*a*,*b*) for the lower yield of invertase activity obtained by Forstner *et al.* (1968) (40% instead of 64%) at stage (*b*). Third, at stage (*c*) Forstner *et al.* (1968) achieved the same order of magnitude of purification of the brush borders in respect of protein (column 2) but somewhat better purification in respect of DNA and RNA (columns 3 and 4) at the expense of some overall loss of invertase activity (25% recovery instead of

* See footnote to *Table 3* for definitions of stages *a*, *b*, *c* and *d*.

Table 3. SUMMARY OF PUBLICATIONS CONCERNING THE ISOLATION OF INTACT BRUSH BORDERS (OR BRUSH BORDERS PLUS NUCLEI) FROM INTESTINAL EPITHELIAL CELLS

Brush borders or brush borders + nuclei preparation		*Animal*					*Per cent of total activity in homogenate*				
Paper	*Authors*		(1) *protein*/DNA (mg/mg)	(2) *Invertase*/*protein* (units/mg)	(3) *Invertase*/DNA (units/mg)	(4) *Invertase*/RNA (units/mg)	(5) *Invertase*	(6) *Maltase*	(7) *Alkaline phosphatase*	(8) *Succinate dehydrogenase*	(9) *Glucose-6-phosphatase*
1	Miller and Crane (1961)	hamster	⋆	100 ▲	⋆	—	80	70	—	—	—
2	Gallow and Treadwell (1963)	rat	—	—	—	—	96	—	—	—	—
3	Eichholz and Crane (1963)	hamster and guinea pig	≠	—	≠	—	—	—	—	—	—
4	Bailey and Pentchev (1964)	rat	—	—	—	—	76	—	—	—	—
5	Harrison and Webster (1964)	rat	—	—	—	—	—	—	—	—	—
6	Ruttloff *et al.* (1964)	rat	—	—	—	—	55–77	55–66	—	—	—
7	Porteous and Clark (1965)	rabbit (*a*)	20	2	40	65	100	—	100	100	100
		(*b*)	7	4	30	230	70		57	30	41
8	Hübscher *et al.* (1965)	guinea pig (*a*)	18 ∅	1	60 ∅	—	100	—	100	100	100
		(*b*)	3 ∅	4	49 ∅		75		65	7	9
		rabbit (*a*)	33 ∅	1	125 ∅	—	100	—	—	100	100
		(*b*)	5 ∅	5	84 ∅		60			5	12

9	Eichholz and Crane (1965)	hamster		—	—	★	—	—	—	—	—	—
10	Overton *et al.* (1965)	hamster		—	—	★	—	—	—	—	—	—
11	Noack and Schenk (1965)	rat		—	—	—	—	—	—	—	—	—
12	Millington *et al.* (1966)	rat		—	—	—	—	—	—	—	—	—
13	Eichholz (1967)	hamster	(*b*)	1	75+	75■	—	—	—	—	—	—
			(*d*)	—	280+							
14	Forstner *et al.* (1968)	rat	(*a*)	30	3	90	90	100	—	100	—	—
		(Sprague-	(*b*)	25	30	750	1400	38	—	40	—	—
		Dawley	(*c*)	60	60	7500	4300	24	—	25	—	—
			(*d*)	220	90	20 000	9000	20	—	—	—	—
15	Porteous (1968) and present paper, *Table 2*	rat	(*a*)	24	5	120	66	100	—	—	100	100
		(Wistar)	(*b*)	4	20	80	200	64	—	—	—	—
			(*c*)	70	70	4900	2400	40	—	—	0	0

a. homogenate
b. crude brush-border fraction
c. purified brush-border fraction
d. brush border membrane fraction
— no analyses reported.
★ no quantitative analyses for DNA; nuclei reported absent or extremely infrequent.
▲ recalculated as μmoles of glucose/mg of protein per h.
≠ noted presence of DNA as artefact of isolation procedure.
ϕ DNA determined as DNA-phosphorus; recalculated assuming DNA/P = 12 (Zamenhof, 1958).
+ values given by Eichholz (1967) have been multiplied by 60 to make them comparable with others quoted in the same column.
■ calculated from Eichholz's (1967) values in two preceding columns.

Some of these papers give additional quantitative information as follows:-
(2) about 25% of homogenate cholesterol esterase activity found in brush borders; about 50% total cholesterol found in brush borders.
(4) 6% of homogenate mutarotase activity found in brush borders.
(7) determined distribution of acid phosphatase, aminopeptidase and cholesterol amongst subcellular fractions.
(8) determined distribution of wide range of enzyme activities amongst subcellular fractions; purified guinea-pig brush borders had specific invertase activity of 13 μmole of glucose/mg protein per h.
(13) determined specific activities of alkaline phosphatase, ATPase, and five disaccharidases; and content of carbohydrate, cholesterol, phospholipid, esterified fatty acid in brush borders and subfractions thereof.
(14) determined content of phospholipid, cholesterol, ATPase, aminopeptidase and several disaccharidase in purified brush borders and subfractions.

40% in the present work, column 5). Fourth, Forstner *et al.* (1968) prepared a membrane component of the purified brush borders thereby achieving further purification of the invertase activity relative to protein content (column 2) but a markedly greater purification relative to DNA and RNA content (columns 3 and 4). Fifth, the less extensive information available from Eichholz (1967) does not accord as closely as might be expected with the corresponding values quoted from the present work and from Forstner *et al.* (1968). This may reflect differences between the hamster and the rat, or differences arising from the use of a quite different technique for isolating brush-border membranes directly from what was probably a mixture of brush borders and nuclei. Eichholz's (1967) ratio of invertase to protein for the membrane fraction is considerably greater than that obtained by Forstner *et al.* (1968). Unfortunately, Eichholz (1967) does not give analytical results which would allow calculation of values for columns 1, 3, 4 and 5. Sixth, the results quoted from the present work and from Forstner *et al.* (1968) illustrate the usefulness of the three ratios (invertase to protein, invertase to DNA, invertase to RNA) in assessing the purification of intestinal brush borders from other organelles on the basis of the assumptions made earlier in this article. On the information available, these two brush-border preparations and the membrane preparation of Forstner *et al.* (1968) are probably the purest and certainly the best characterised preparations available to date. There is a pressing need for more critical quantitative analyses of isolated organelles and membrane preparations. It would be of considerable interest to compare results obtained in one laboratory with one animal species using all three methods of isolation of brush borders and brush-border membranes (Eichholz, 1967; Forstner *et al.*, 1968; Porteous, 1968).

Acknowledgements. The author thanks Mrs Jeff Thain, Mrs Barbara Stroud and Keith Patrick for technical assistance at various times during this work, and David Sim for processing photographic material. Mrs Christa Jeffery undertook all the electron microscopic investigations; some of the micrographs are reproduced here. The Medical Research Council provided generous financial assistance.

REFERENCES

BARRY, R. J. C., SMYTH, D. H. and WRIGHT, E. M. (1965). *J. Physiol.*, **181**, 410

BAILEY, J. M. and PENTCHEV, P. (1964). *Proc. Soc. exp. Biol.*, **115**, 796

BIHLER, I. and CRANE, R. K. (1962). *Biochim. biophys. Acta*, **59**, 78

BIHLER, I., HAWKINS, K. A. and CRANE, R. K. (1962). *Biochim. biophys. Acta*, **59**, 74

CARNIE, J. A. and PORTEOUS, J. W. (1962*a*). *Biochem. J.*, **85**, 450

CARNIE, J. A. and PORTEOUS, J. W. (1962*b*). *Biochem. J.*, **85**, 620

CLARK, B. and SHERRATT, H. S. A. (1967). *Comp. Biochem. Physiol.*, **20**, 223

CRANE, R. K. (1966*a*). *Gastroenterology*, **50**, 254

CRANE, R. K. (1966*b*). In *Intracellular Transport, Symp. Int. Soc. Cell Biol.*, (ed. Warren, K. B.), vol. 5, p. 71, New York and London (Academic Press)

EICHHOLZ, A. and CRANE, R. K. (1965). *J. Cell Biol.*, **26**, 637

EICHHOLZ, A. (1967). *Biochim. biophys. Acta*, **135**, 475

EICHHOLZ, A., HOWELL, K. E. and CRANE, R. K. (1969). *Biochim. biophys. Acta*, **193**, 179

EVANS, E. M., WRIGGLESWORTH, J. M., BURDETT, K. and POVER, W. F. R. (1971). *J. Cell Biol.*, **51**, 452

FORSTNER, G. G., SABESIN, S. M. and ISSELBACHER, K. J. (1968). *Biochem. J.*, **106**, 331

GALLO, L. L. and TREADWELL, C. R. (1963). *Proc. Soc. exp. Biol.*, **114**, 69

GITZELMANN, R., BÄCHI, T., BINZ, H., LINDENMANN, J. and SEMENZA, G. (1970). *Biochim. biophys. Acta*, **196**, 20

HARRISON, D. D. and WEBSTER, H. L. (1964). *Biochim. biophys. Acta*, **93**, 662

HÜBSCHER, G., WEST, G. R. and BRINDLEY, D. N. (1965). *Biochem. J.*, **97**, 629

IEMHOFF, W. G. J., VAN DEN BERGH, J. W. O., DE PJIPER, A. M. and HÜLSMANN, W. C. (1970). *Biochim. biophys. Acta*, **215**, 229

JOHNSTON, C. F. (1967). *Science, N.Y.*, **155**, 1670

KIMMICH, G. A. (1970). *Biochemistry*, **9**, 3659

KINTER, W. B. and WILSON, T. H. (1965). *J. Cell Biol.*, **25**, 19

MCDOUGAL, D. B., LITTLE, K. D. and CRANE, R. K. (1960). *Biochim. biophys. Acta*, **45**, 483

MILLER, D. and CRANE, R. K. (1961). *Biochim. biophys. Acta*, **52**, 293

MILLINGTON, P. F. and FINEAN, J. B. (1963). In *Biochemical Problems of Lipids*, (ed. Frazer, A. C.), p. 116, Amsterdam (Elsevier)

MILLINGTON, P. F., CRITCHLEY, D. R. and TOVELL, P. W. A. (1966). *J. Cell Sci.*, **1**, 415

NOACK, R. and SCHENK, G. (1965). *Ernährungsforschung*, **10**, 77

OVERTON, J., EICHHOLZ, A. and CRANE, R. K. (1965). *J. Cell Biol.*, **26**, 693

PORTEOUS, J. W. and CLARK, B. (1965). *Biochem. J*, **96**, 159

PORTEOUS, J. W. and PATERSON, A. C. (1965). (unpublished observations)

PORTEOUS, J. W. (1968). *FEBS Letters*, **1**, 46

PORTEOUS, J. W. and HERFORD, P. J. (1972). *Biochem. J.*, **126** (in press)

RIKLIS, E. and QUASTEL, J. H. (1958). *Canad. J. Biochem. Physiol.*, **36**, 347

RUTTLOFF, H., NOACK, R., FRIESE, R. and SCHENK, G. (1964). *Biochem. Z.*, **341**, 15

RHODES, J. B., EICHHOLZ, A. and CRANE, R. K. (1967). *Biochim. biophys. Acta*, **135**, 959

SALOMON, L. L., JAMES, J. and DAGINAWALA, H. F. (1964). *Proc. 6th Int. Congr. Biochem.*, **32**, 526

SEMENZA, G., TOSI, R., VALLOTON-DELACHAUX, M. C. and MÜLHAUPT, E. (1964). *Biochim. biophys. Acta*, **89**, 109

SEMENZA, G. (1967). *Protoplasma*, **63**, 70

SIMKIN, J. L. and PORTEOUS, J. W. (1972). In *The Glycoproteins*, 2nd edition, (ed. Gottschalk, A.), Ch. 10, Amsterdam (Elsevier)

SMYTH, D. H. (1967*a*). *Proc. Roy. Soc. Med.*, **60**, 321

SMYTH, D. H. (1967*b*). *Brit. Med. Bull.*, **23**

STERN, B. K. and JENSEN, W. E. (1966). *Nature, Lond.*, **209**, 789

ZAMENHOF, Z. (1958). In *Biochemical Preparations*, vol. 6, p. 11, New York (Wiley)

8

PREPARATION AND PROPERTIES OF MICROSOMAL AND SUBMICROSOMAL FRACTIONS FROM SECRETORY AND NON-SECRETORY TISSUES

J. R. Tata

Unlike nuclei and mitochondria, the term 'microsomes' does not refer to a discrete cytoplasmic organelle. It is an operational term which denotes the particulate fraction that sediments upon centrifugation at speeds exceeding 10 000*g* (de Duve, 1964; Reid, 1967). Different types of microsomal fractions can, therefore, vary a great deal in both structure and biochemical properties, not only according to intrinsic features like species, age and type of tissue but also to extraneous conditions such as diet and hormonal status. The term microsomes, however, often refers to a mixture of membraneous elements and ribosomes (or polysomes) present either as separate entities or as complexes. These are derived from the smooth, or agranular, and the rough, or granular, endoplasmic reticulum. Like all cellular membranes, microsomes are not stable structures and turn over at a rapid rate relative to the life-time of the cell (Omura *et al.*, 1967; Siekevitz *et al.*, 1967; Arias *et al.*, 1969).

MICROSOMAL ACTIVITIES AND FUNCTIONS

A wide variety of enzymic activities are associated with the membranes of the granular and agranular endoplasmic reticulum, whereas the ribonucleoprotein particles or ribosomes are mainly concerned with protein biosynthesis. Glucose-6-phosphatase, NADPH-cytochrome *c* reductase and the haem P_{450} are exclusively associated with microsomal membranes and are often used as markers for microsomal preparations. *Table 1* lists a few of the membrane-linked enzymes which participate in such processes as extra-mitochondrial electron transport, ATPase and ion movements, relaxation and contraction for muscle, drug detoxication and carbohydrate metabolism. Of course, the type of activity that would predominate in a given microsomal preparation depends

Table 1. SOME ACTIVITIES AND FUNCTIONS ASSOCIATED WITH THE TWO PRINCIPAL MICROSOMAL CONSTITUENTS: MEMBRANES AND RIBOSOMES

Fraction	*Activity*	*Function*
Membrane	Glucose-6-phosphatase	Carbohydrate metabolism and gluconeogenesis
	Cytochrome b_5, P_{450}, hydroxylases	Drug detoxication
	NADPH-cytochrome *c* reductase	Extramitochondrial electron transport
	Na^+/K^+- and Mg^{2+}-ATPase	Ion movement, muscular contraction and relaxation
	Steroid reduction	Steroid biogenesis
Ribosomes (polysomes)	Enzymes and processes concerned with amino acid incorporation into protein	Protein synthesis
	Ribonuclease	RNA metabolism

For detailed references see: Ernster *et al*., 1962; Dallner, 1963; Orrenius *et al*., 1965; Imai *et al*., 1966; Reid, 1967; Elson, 1967.

on the nature and age of the tissue from which it is obtained. Thus, liver microsomes exhibit a strong drug-detoxication activity, muscle and brain preparations an active ATPase, and adrenal preparations are active in steroidogenesis. A further indication of the complexity of the enzymic make-up of the microsomal membranes is evident from the different rates of turnover of different proteins and phospholipid components (Omura *et al*., 1967; Arias *et al*., 1969). Studies on the biogenesis of microsomes have shown that the different membrane-associated enzymes are not formed simultaneously (Dallner *et al*., 1966*a*, *b*; Pollak and Ward, 1967; Siekevitz *et al.*, 1967). The heterogeneity established during early developmental stages often persists throughout the life-time of the fully differentiated cell, as evidenced by the recent demonstration by Glaumann and Dallner (1970) of the subfractionation of smooth microsomes of adult rat liver into fractions of different enzymic composition. Reid (1967), Dallner and Ernster (1968) and Dallner (1969) have exhaustively reviewed the distribution and function of the various microsomal components.

MAJOR PROCEDURES FOR MICROSOMAL PREPARATIONS

Although the isolation of a microsomal fraction was described by Claude in 1941, the most successful procedure is that of differential centrifugation devised by Palade and Siekevitz (1956). Dallner (1963) and Reid (1967) have written excellent historical accounts of the problem. Even now, most isolation procedures for unfractionated microsomes from many different tissues employ a modification of the original procedure of Palade and Siekevitz (1956). This involves the prior separation of nuclei and mitochondria from the tissue homogenate by low-speed centrifugation, followed by ultracentrifugation of the mitochondria-free supernatant fraction, usually at 105 000*g* for 1 to 2 h. The resulting pellet contains about half of the cellular RNA and a large part of the membrane-associated constituents shown in *Table 1*. Numerous procedures used currently for preparation of unfractionated microsomes from a wide variety of tissues are based on this procedure (Zomzely *et al.*, 1964; Imai *et al.*, 1966; Cunningham and Crane, 1966; Teng and Hamilton, 1967; Morais and Goldberg, 1967; Lim and Adams, 1967; Chen and Young, 1967; Bullock *et al.*, 1968; Earl and Morgan, 1968; Dunn, 1970; Andrews and Tata, 1968, 1971).

With the rapidly-growing interest in protein synthesis, the procedure of Palade and Siekevitz (1956) was adopted for the preparation of ribosomes or polysomes by treating the mitochondria-free supernatant fraction or the microsomal pellet with an excess of a detergent like sodium deoxycholate to free the particles from the membranes (Siekevitz, 1956; Siekevitz and Palade, 1960; Wettstein *et al.*, 1964; see Elson, 1967). It has been shown, however, that the addition of sodium deoxycholate to isolated microsomes leads to poor incorporation of amino acids due to nuclease attack. This is much attenuated if the detergent is added to the mitochondria-free supernatant fluid before the separation of ribosomes since the cell sap, especially of rat liver, contains a powerful ribonuclease inhibitor (Bloemendal *et al.*, 1967; Blobel and Potter, 1967). Hunter and Korner (1966) found that the non-ionic detergent, Triton X-100, was more suitable than deoxycholate for deriving ribosomes from rat liver for studies of amino acid incorporation. No explanation was then given, but recently Olsnes (1970) has suggested that deoxycholate, but not Triton X-100, may be removing one of the three major proteins associated with the cytoplasmic 'informosome' (messenger ribonucleoprotein) particle thus making the messenger RNA more accessible to attack by nuclease.

Many attempts have also been made to sub-fractionate microsomes into smooth and rough membranes to correspond to the granular and agranular endoplasmic reticulum (Fawcett, 1964). Usually the microsomal pellet can be seen to be composed of a 'heavy' and a 'light' layer which can be partially separated by careful rinsing and suspension. Ernster *et al.* (1962) described a method for rat liver in which microsomes were first isolated in isotonic sucrose and then treated with small amounts (0·26%) of sodium deoxycholate to yield, upon further ultracentrifugation, a tight pellet consisting of rough membranes and free ribosomes, on top of which was a loose 'fluffy' layer of smooth or agranular membranes. The use of detergent to bring about this separation necessarily meant a loss of part of the membranes and perhaps even a disturbance of the normal distribution of microsomal components.

A method developed by Moulé and her colleagues (Moulé *et al.*, 1960; Chauveau *et al.*, 1962) exploits the differences in density of particles derived from the agranular and granular endoplasmic reticulum. These workers achieved a separation of the smooth and rough microsomal membranes by dispersion of microsomes in hypertonic sucrose followed by prolonged ultracentrifugation (22–24 h). This separation is quite effective, with little loss of enzymic activity associated with membranes, but it does not achieve a separation of free and membrane-bound ribosomes. Hallinan and Munro (1965*a*) modified this procedure by treating the rough and smooth membrane fractions with iso-octane, a method previously used by Hawtrey and Schirren (1962) to subfractionate microsomes. The use of iso-octane speeded up the separation of free ribosomes from membrane-attached ribosomes, and good amino acid incorporation activity by these particles was reported (Hallinan and Munro, 1965*b*). However, the use of an organic solvent would be a distinct disadvantage in studying the relationship between membrane structure and functions as well as the attachment of ribosomes to membranes.

In 1963, Dallner described a detergent-free procedure which

allowed a quantitative separation of smooth and rough membranes. It has proved very useful in studies of the biogenesis of microsomes (Dallner *et al.*, 1966*a*, *b*; Pollak and Ward, 1967; Arias *et al.*, 1969; Tata, 1970), in which good recovery of membranes was essential. Dallner's procedure for submicrosomal fractions involves ultracentrifugation of the mitochondria-free supernatant fraction through a discontinuous sucrose density gradient of 0·25 M and 1·3 M sucrose. Smooth membranes accumulate at the interface, whereas rough membranes and free polysomes sediment through the bottom dense layer. Inclusion of 0·15 M CsCl, which causes aggregation of membranes (Dallner and Nilsson, 1966), was found to facilitate the separation. This procedure is well suited for studying membrane-associated enzymes, but the composition of the medium (especially the presence of CsCl) is detrimental to ribosomes and precludes studies of protein synthesis. Bloemendal and his colleagues (Bloemendal *et al.*, 1964, 1967), as well as other workers (Blobel and Potter, 1967; Tata and Williams-Ashman, 1967; Andrews and Tata, 1971), have described detergent-free methods in which centrifugation through a discontinuous sucrose density gradient allows the separation of free polyribosomes which are active in incorporating amino acids into protein.

FACTORS IMPORTANT IN ISOLATION PROCEDURES

It is now becoming increasingly clear that, although free (non-membrane-bound) polyribosomes can effectively incorporate amino acids into protein *in vitro*, in intact cells the population of polysomes attached to membranes of the endoplasmic reticulum is as actively, or more actively, engaged in protein synthesis (Henshaw *et al.*, 1963; Hendler, 1964; Redman *et al.*, 1966; Campbell *et al.*, 1967; Campbell, 1970; Tata, 1967*a*, *b*, 1970; Andrews and Tata, 1968, 1971). This is particularly evident in those tissues that secrete a large amount of protein, such as the liver (Maganiello and Phillips, 1965; Sabatini *et al.*, 1966) and thyroid (Morais and Goldberg, 1967). Attachment of ribosomes to membranes which are not normally recognisable as endoplasmic reticulum has also been suggested for non-protein-secreting cells (Hendler, 1968; Andrews and Tata, 1968, 1971). A co-ordination has been observed for the rates at which ribosomes and microsomal membranes proliferate under the influence of growth and developmental hormones (Tata, 1967*a*, *b*, 1970). For these reasons, an ideal method for preparing microsomes and sub-microsomal fractions should satisfy the following requirements:

(a) quantitative recovery of membranes with minimal loss of structure and enzymic activity;
(b) quantitative recovery of polysomes and free ribosomes showing high activity in incorporation of amino acids into protein *in vitro*;
(c) minimal disturbance of the attachment of polysomes to membranes and the localisation of nascent protein synthesised *in vivo*.

RECOMMENDED METHODS FOR THE ISOLATION OF MICROSOMES AND SUBMICROSOMAL FRACTIONS

The relative amount of smooth and rough endoplasmic reticulum varies a great deal from one type of tissue to another. It is very high in predominantly protein-secreting tissues, such as liver, pancreas and thyroid and low in predominantly non-secretory tissues such as kidney, brain and muscle. The optimal media and methods for fractionating microsomes and submicrosomal fractions are necessarily different for different tissues. Below are described methods used in the author's laboratory over the last 4–5 years for (a) liver, (b) cerebral cortex and (c) skeletal muscle. All three requirements mentioned are satisfied by the method for liver, but not by those for brain and muscle.

LIVER

From morphometric and biochemical studies Weibel *et al.* (1969) concluded that the surface area of endoplasmic reticulum in 1 cm^3 of rat-liver tissue measures 11 m^2 of which about 7 m^2 is rough endoplasmic reticulum containing approx. 2×10^{13} ribosomes. Thus one is dealing with an extremely large amount of material and the method described below is basically a combination of Dallner's (1963) procedure for the separation of smooth from rough membranes and that of Bloemendal *et al.* (1964, 1967) and Blobel and Potter (1967) for the preparation of free polysomes. Brief descriptions of the method have been published for rat (Tata and Williams-Ashman, 1967) and tadpole liver (Tata, 1967*b*).

It is common practice to starve animals used for preparation of microsomes, especially from liver, in order to reduce contamination with glycogen. Starvation is not recommended for any physiological or functional studies, since a reduction in total food intake or any major food component is known to reduce protein synthesis, cause detachment of ribosomes attached to membranes, affect enzymic activity associated with these structures and accelerate their breakdown and that of polysomes (Tata *et al.*, 1963; Munro *et al.*, 1964; Weber *et al.*, 1964; Quirin-Stricker and Mandel, 1967; see Reid, 1967). In our experience the presence of glycogen does not substantially affect amino acid incorporation by microsomes or submicrosomal fractions or affect the distribution of nascent protein synthesised *in vivo*.

Rats fed *ad libitum* are killed by a blow on the head or by cervical dislocation and the livers are rapidly removed and chilled in ice-cold homogenisation *medium A*. All further procedures are carried out at about 2°C. The homogenisation medium A contains 0·35 M sucrose, 0·025 M KCl, 0·01 M $MgCl_2$* and 0·05 M tris-HCl buffer, pH 7·6. The optimal homogenate concentration for studies on amino acid incorporation *in vitro* by microsomes and submicrosomal particles is 2·5 ml of homogenisation medium per g liver (Tata and Williams-Ashman, 1967). However, more dilute homogenates (10 to 20%) are suitable for analytical studies. The composition of the homogenisation medium needs to be varied to suit the use to which the preparations are to be put and the type of tissue, as was done for tadpole-liver microsomes (Tata, 1967*b*). The most commonly-used procedure for homogenisation nowadays is to disrupt the tissue in a Teflon-glass Potter–Elvehjem homogeniser. It is important to have a large clearance (0·10–0·15 mm), not so much for preserving microsomal integrity as for preventing contamination with fragments of nuclei and mitochondria, since even small amounts of nuclear and mitochondrial extracts would seriously vitiate experiments on labelling of microsomal RNA and phospholipids. In such experiments it is advisable to determine the DNA and cytochrome oxidase content of microsomes as a guide to contamination with nuclear and mitochondrial fragments. The speed of homogenisation should vary according to age of the tissue, its toughness and its subcellular particles and to the size of the homogeniser, but it is good practice not to exceed 8 to 12 strokes at 1000 rev/min. For liver it is most convenient to chop up the tissue finely with a pair of scissors before homogenising, but other forms of pre-treatment are necessary for other tissues, especially skeletal

* This concentration of Mg^{2+} is essential for studying amino acid incorporation into protein *in vitro*.

muscle (Breuer *et al.*, 1964; Florini and Breuer, 1966; Chen and Young, 1968; Andrews and Tata, 1971; see below) and uterus (Teng and Hamilton, 1967; Widnell *et al.*, 1966).

The homogenate is then freed of erythrocytes, cell debris, nuclei and mitochondria by centrifuging it at the speed necessary to sediment mitochondria (6000 to 9000*g*). It is always better to centrifuge the initial mitochondria-free supernatant fraction once again, or even twice, in order to remove the small amount of residual 'light' mitochondria. The speed of centrifugation is important, as it should allow the removal of almost all mitochondria without loss of microsomes or submicrosomal fractions. Howell and his colleagues (Howell *et al.*, 1964; Loeb *et al.*, 1967) have centrifuged nuclei-free homogenates at 20 000*g* for 20 min to find that the majority of rough microsomal membranes sediment with mitochondria. This illustrates the danger to the yield of microsomes of overcentrifugation to remove mitochondria. There is relatively less risk of losing microsomes if the homogenate is centrifuged at speeds under 8000*g* to remove mitochondria (Roodyn *et al.*, 1965).

The mitochondria-free supernatant fraction can now be processed in several ways according to the microsomal preparation required. For obtaining unfractionated microsomes the mitochondria-free supernatant fluid is centrifuged at 105 000*g* for a minimum of 1 h (preferably 2 h). The pellet of microsomes can be suspended in the homogenisation medium with a hand-operated all-glass homogeniser and centrifuged again. In order to obtain polyribosomes, the most commonly-used procedure is to free the particles of the membranes with 0·5 to 1·0% (w/v) sodium deoxycholate, although other ionic and non-ionic detergents (Triton X-100 or Lubrol) have also been used. Use of excessive amounts of detergent is detrimental to the structure and function of polyribosomes. In earlier studies, microsomes were first separated from the cell sap and then treated with detergents (Siekevitz and Palade, 1960; see Reid, 1967), but this practice should be avoided because of the presence of a ribonuclease associated with the endoplasmic reticulum (de Lamirande *et al.*, 1966) and the presence of a ribonuclease inhibitor in the cell sap. It is therefore best to treat the mitochondria-free supernatant fluid directly with the detergent before ultracentrifugation. The presence of ribonuclease inhibitors in the cell sap, at least in rat liver, is thought to minimise damage to polyribosomes (Bont *et al.*, 1965; Blobel and Potter, 1966). The amount of detergent necessary to bring about complete separation of polyribosomes from membranes depends on the microsomal phospholipid content and can vary a great deal from one tissue to another or according to the degree of maturation of the tissue (Tata,

1967*b*). The minimum amount necessary should first be determined in every type of preparation (Tata, 1967*b*).

Until recently, there was no single method available for the preparation of the major submicrosomal fractions suitable for yielding preparations active both in enzymes associated with microsomal membranes as well as incorporation of amino acids into protein. With the realisation that the association of polysomes with membranes is important for protein synthesis, and that the rapid rate of proliferation of microsomal membranes is associated with high protein synthetic activity, it became highly desirable to have such a procedure. The one used in our laboratory is as follows.

The mitochondria-free supernatant fraction is layered onto 0·4 to 0·5 vol of 1·3 M sucrose, with the same composition in salts and buffer as the homogenisation medium, in MSE 8 × 10, 8 × 25 or 8 × 50, or Beckman No. 40, fixed-angle rotor tubes. The volume of the dense sucrose solution should be varied according to the type of rotor used so as to obtain the optimal separation between smooth and rough membranes. The material is centrifuged at 105 000*g* for 2·5 h (or at 65 000*g* for 3·5 h for the MSE 8 × 50 rotor). Shorter periods of centrifugation may be used if the rough membranes are to be processed further. After centrifugation, the upper supernatant fluid (cell sap), an opaque layer of smooth membranes at the interface of the two sucrose solutions and the dense sucrose solution over the pellet of rough membranes are each carefully siphoned off with a Pasteur pipette. The suspension of smooth membranes is diluted about 5 times in the homogenisation medium and centrifuged once again over 1·3 M sucrose containing 10 mM $MgCl_2$ as before or pelleted directly by centrifuging at 105 000*g* for 1·5 h. The rough membrane pellet is suspended by mild hand-homogenisation in an all-glass homogeniser either in the homogenisation medium or in the cell sap (especially for amino acid incorporation studies) to yield a suspension of 0·25 to 1·5 g-equiv./ml of liver. The suspension is then layered over 1·3 M sucrose medium and centrifuged again as before to free it of any contaminating smooth membranes, or it may be retained for further processing.

For subfractionation of rough membranes, the suspension is layered over a discontinuous sucrose density gradient in swing-out rotor tubes (5 to 7 ml, 7 to 9 ml or 15 to 20 ml in the Beckman SW 25.1, SW 27 or SW 25.2 rotors, respectively). The density gradient consists of a lower layer of 2 M sucrose and an upper one of 1·5 M sucrose, both containing 10 mM $MgCl_2$ (11 and 12 ml, 12 and 14 ml or 19 and 21 ml in the SW 25.1, SW 27 or SW 25.2 rotors, respectively). A further layer of 3 to 6 ml of 1·15 M sucrose solution is sometimes used to trap any contaminating smooth membranes.

The tubes are then centrifuged at 25 000 rev/min for 14 to 16 h. For small amounts of samples, a more rapid fractionating (in about 4 h) can be achieved in an MSE 3 × 10 or Spinco SW 39L swing-out rotor run at 40 000 rev/min and containing 3·0 and 3·7 ml of the 2·0 M and 1·5 M sucrose solutions, respectively (Tata, 1967*b*). The same separation can be achieved in the MSE 6 × 15 or Beckman SW 27.1 swing-out rotors with proportional increases in the volumes of the sucrose solutions of different densities. After centrifugation, the following major fractions are obtained: (i) an upper layer, termed 'light rough membranes', at the interface of the 1·5 M sucrose solution and the less dense reddish-brown supernatant

Chart 1. Summary of procedure for preparing submicrosomal fractions from rat liver

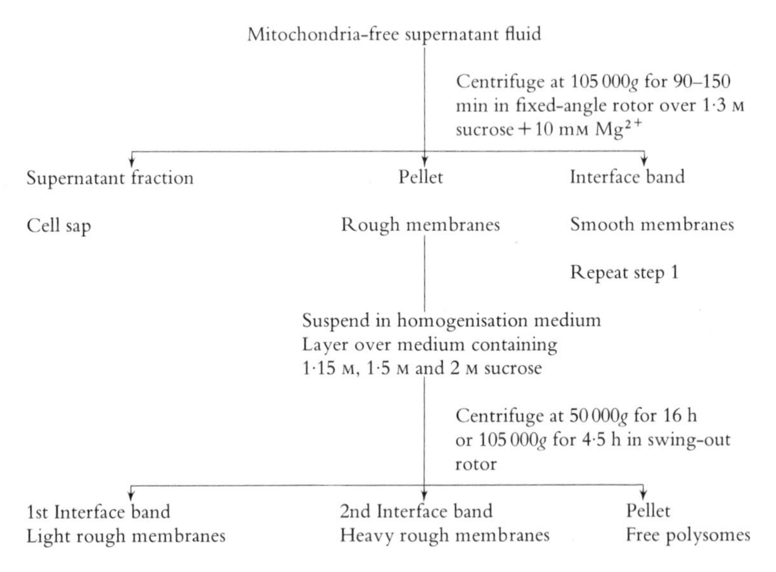

layer; (ii) 'heavy rough membranes' which collect at the interface of the 1·5 M and 2·0 M sucrose solutions; and (iii) a pellet containing 'free' or non-membrane-bound ribosomes and polysomes plus the major part of the glycogen. The light and heavy rough membranes are removed with a Pasteur pipette and gently homogenised by hand in an all-glass apparatus. The pellet at the bottom of the tube is very gently rinsed with the homogenisation medium and then similarly suspended. The composition of suspending medium and the concentration of the particles depend on the use to which the submicrosomal fractions are to be put.

Murty and Hallinan (1969) have reported that free polysomes

obtained by sedimenting through 2 M sucrose can be slightly contaminated with some rough microsomal membranes. This contamination, although rarely observed in our procedure, was not eliminated by prolonged starvation and could be best removed by a repeated centrifugation through 2 M sucrose. Lee *et al.* (1969) have recently used the zonal ultracentrifuge to prepare smooth and rough microsomes from rat liver.

The scheme in *Chart 1* summarises the fractionation procedure for liver.

BRAIN

Because of the greater morphological complexity of brain, both at the whole tissue and subcellular levels, only the cerebral cortex was studied. Several different homogenisation media have been described earlier for brain microsomes (see Dallner and Ernster, 1968) but it is only recently that media, with or without detergent, have been developed specifically for studying protein synthesis by ribosomes (Zomzely *et al.*, 1964; Campbell *et al.*, 1966; Adams and Lim, 1966; Mahler and Brown, 1968; Merits *et al.*, 1969). The procedure described below was developed in our laboratory for studying amino acid incorporation into protein by free and bound ribosomes of rat cerebral cortex (Andrews and Tata, 1968, 1971).

Cortical shells, pooled from 4 to 8 rats, are collected in ice-chilled *medium B* (0·25 M sucrose, 0·1 M KCl, 0·012 M $MgCl_2$ and 0·05 M tris-HCl buffer, pH 7·4). They are finely chopped and washed twice in 20 ml of medium B before homogenisation in 3·5 vol of medium B with 3 strokes of a Teflon-glass homogeniser (clearance 0·15 mm) at 800 rev/min. More than 5 strokes of the pestle reduced the amino acid incorporation activity of the final preparations. A mitochondria-free supernatant fraction is obtained by centrifuging the homogenate at 10 000*g* for 10 min and repeating the centrifugation on the first supernatant fluid. Removal of mitochondria at higher centrifugal forces (Zomzely *et al.*, 1964) carries the risk of losing the membrane-bound ribosomal fraction. Whole microsomes were obtained from the second mitochondria-free supernatant fraction by centrifugation at 105 000*g* for 2 h.

For submicrosomal fractionation, 14·0 ml of mitochondria-free supernatant fraction are layered over 6·5 ml of 0·8 M sucrose containing the other components of medium B, and centrifuged at 105 000*g* for 2 h in an MSE Superspeed 50 ultracentrifuge (8 × 25 fixed-angle rotor). After the interface layer and the supernatant fluid are discarded, the walls of the tubes are carefully wiped with paper tissue and the pellet is gently suspended in 10 ml of medium B. The suspension is layered over a discontinuous sucrose gradient consisting of 9 ml of 2·0 M sucrose and 11 ml of 1·0 M sucrose, with KCl, $MgCl_2$ and tris-HCl as in medium B, in swing-out rotor tubes (SW 25.1). The samples are then centrifuged to equilibrium for 16 h at 60 000*g* at 0°C. Three submicrosomal fractions are thus obtained: a hazy layer just below the 0·25/1·0 M sucrose interface, a dense layer around the 1·0/2·0 M sucrose interface which contains the bulk of membrane-bound RNA, and a pellet of free ribosomes. The interface fractions are aspirated and, along with the pellet, are suspended in medium B to a total volume of 20 ml, pelleted at 105 000*g* for 1·5 h and resuspended in medium B. These three fractions are termed 'light rough membranes', 'heavy rough membranes' and 'free polysomes', respectively, without, however, implying morphological identity with similar fractions obtained from liver.

MUSCLE

Earlier methods (Breuer *et al.*, 1964; Rampersad *et al.*, 1965) based on homogenisation of skeletal muscle in media of low ionic strength, give very low yields of muscle ribosomes because of their co-precipitation with myosin (Heywood *et al.*, 1967). As the larger polysomes (both free and membrane-bound) are preferentially lost with the myofibrilar fraction the final ribosomal fraction is unrepresentative of the general population of muscle polysomes. Chen and Young (1968) reported higher yields of polysomes (300–400 μg of RNA per g muscle) by homogenising muscle in the medium of high ionic strength (0·25 M KCl), described by Heywood *et al.* (1967), and then sedimenting the polysomes by lowering the concentration of KCl to 0·06 M. However, after preliminary trials, the ionic composition of media described by Earl and Morgan (1968) and Bullock *et al.* (1968), for obtaining detergent-treated polysomes from cardiac and skeletal muscle, was found to be more suitable.

Batches of 50 to 100 g of muscle from hind limbs are finely chopped with scissors and homogenised in 5 vol of *Medium C* (0·1 M KCl, 0·01 M $MgCl_2$ and 0·5 M tris-HCl buffer, pH 7·6) with an Ultra-Turrax disintegrator (operated at 80 v for a 240 v rating) for 2 periods of 15 s each. The homogenate is centrifuged at 40 000*g* for 30 min in an MSE 8 × 50 fixed-angle rotor. This results in the loss of a substantial amount of ribosomal material in the sediment, but is a step deemed necessary for the eventual purity of microsomal fractions. A crude microsomal pellet is prepared from the supernatant fraction by centrifugation at 105 000*g* for 1·5 h. A suspension of the pellet in 50 ml of medium C is layered directly over 2·0 M sucrose in medium C and centrifuged at 60 000*g* for 16 h to obtain a single 'membrane-bound' polysomal fraction at the interface and separated from free polysomes. The two fractions are suspended in medium C, centrifuged at 105 000*g* for 1·5 h and the pellets finally suspended in 2 to 4 ml of medium C.

COMPOSITION AND PROPERTIES OF MICROSOMAL AND SUBMICROSOMAL FRACTIONS

ELECTRON MICROSCOPY

In *Figures 1* to *5* are shown some electron micrographs* of typical submicrosomal fractions obtained from rat liver by the method described. For the preparations stained with osmic acid and glutaraldehyde, the membranes and free polysomes were pelleted from their suspensions by centrifugation at 105 000*g* for 90 min and sections cut across different regions of the pellet. The presence of large amounts of glycogen in the free polysome fraction prevented its proper staining, and because of this the rats had to be starved overnight (*Figure 5*). Negative staining was performed on a 100-fold dilution of the suspension of the particles without pelleting; this procedure was suitable for visualising the membrane component but not the ribosomes.

Even after the first centrifugation, extremely few ribosomal profiles were found in either the pellets or suspensions of smooth membranes (*Figures 1* and *2*). These were almost eliminated after a second centrifugation of the suspension layered over 1·3 M sucrose. Size and shape of smooth membrane vesicles are similar to those observed by others both after osmic acid fixation and negative staining (Ernster *et al.*, 1962; Cunningham and Crane, 1966; Bloemendal *et al.*, 1967). Both the light and heavy rough membranes are composed of vesicles studded with ribosomes. The difference between the two preparations was not so much a contamination of the lighter preparation with smooth membranes but a higher density of packing of ribosomes per unit length of the membranes of the heavier preparation (*Figures 3* and *4*). This difference may be related to that of the greater protein synthesising

* I am indebted to Dr J. A. Armstrong and the late Dr R. C. Valentine for electron microscopy.

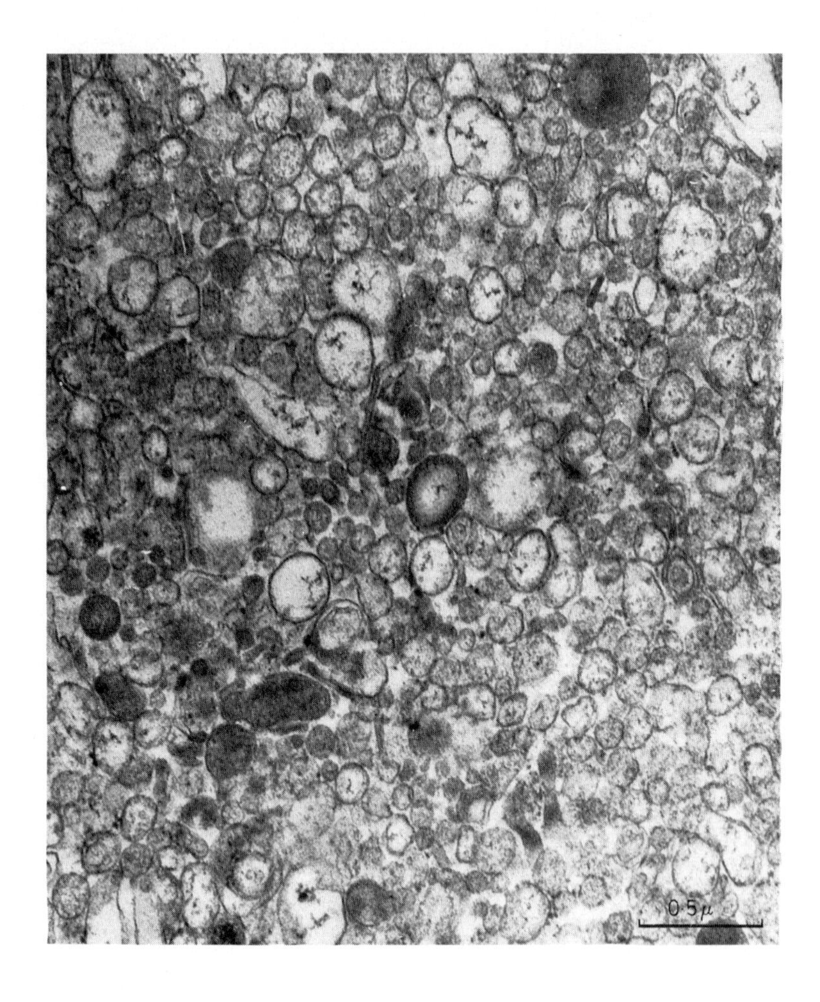

Figure 1. Electron micrograph of a section through a pellet of the smooth membrane fraction from rat liver; fixed with glutaraldehyde and osmic acid

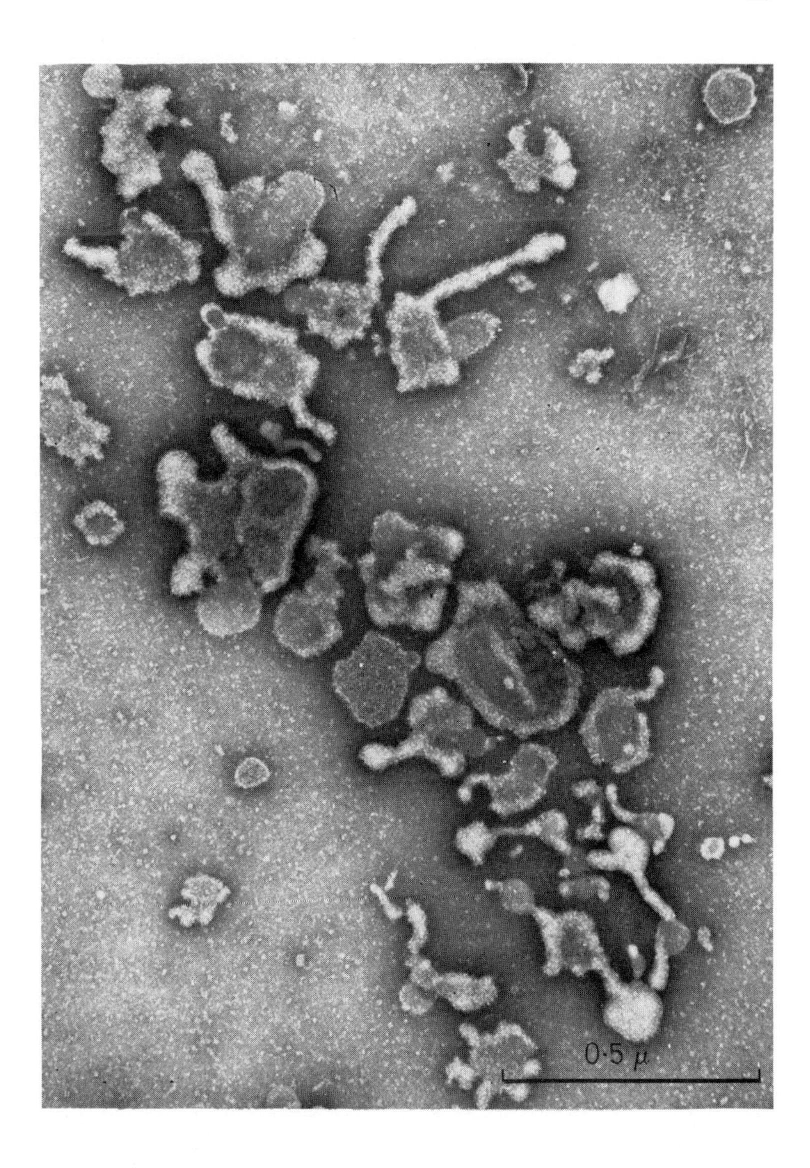

Figure 2. Electron micrograph of smooth membranes from rat liver; negatively stained with sodium tungstosilicate, pH 7·0, in a suspension of 1 : 100 dilution

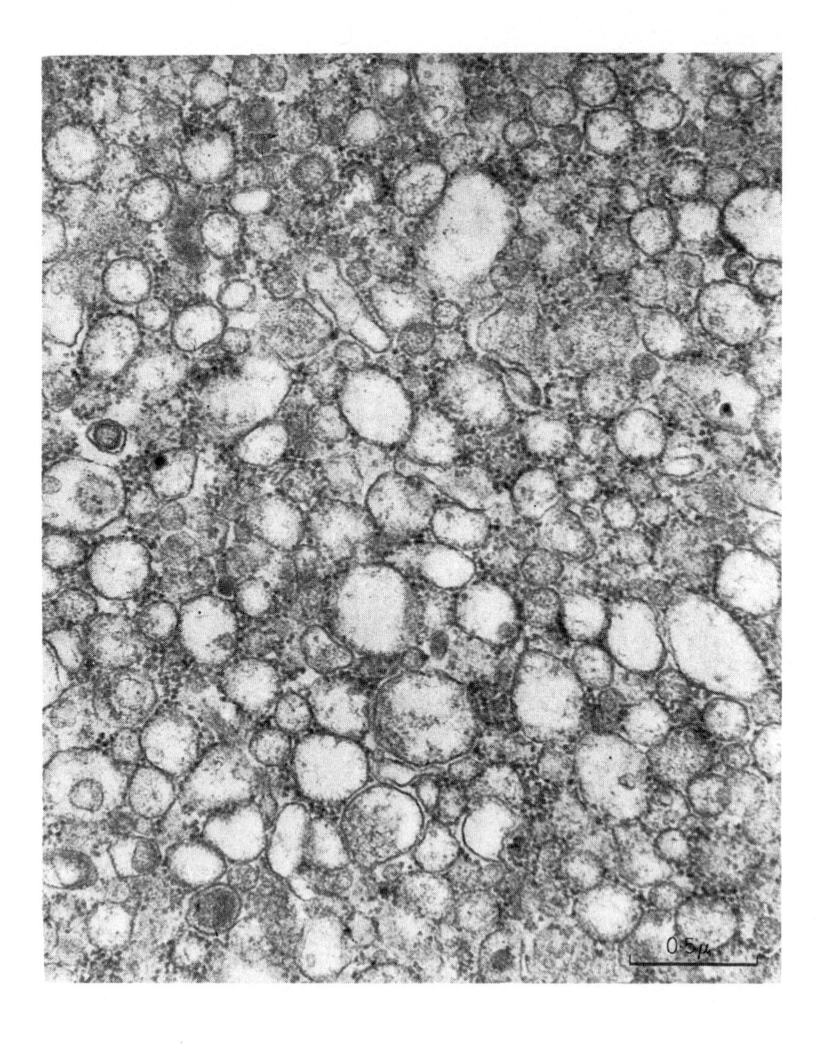

Figure 3. Electron micrograph of a section through a pellet of the light rough membrane fraction from rat liver; fixation as Figure 1

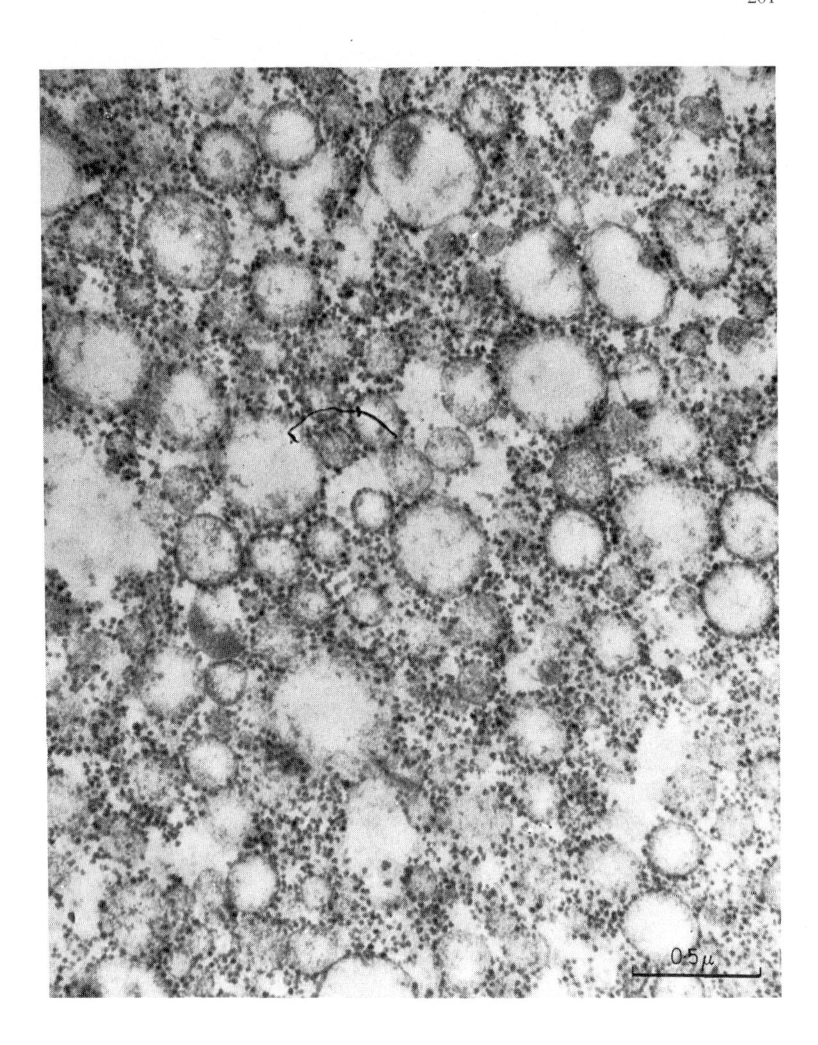

Figure 4. Electron micrograph of a section through a pellet of the heavy rough membrane fraction from rat liver; fixation as Figure 1

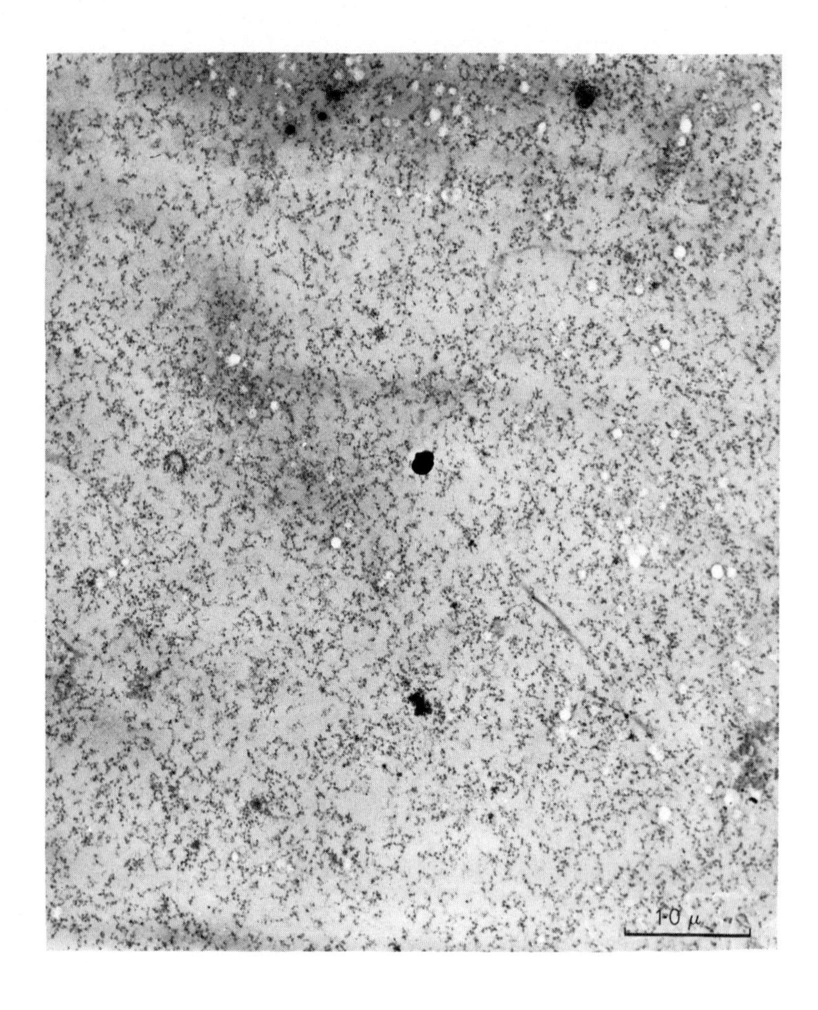

Figure 5. Electron micrograph of a section through a pellet of the free polysome fraction from livers of rats starved overnight; fixation as Figure 1

capacity of the heavy rough membranes (*Tables 4* and *5*). Virtually no membrane-like structures are detected in the polyribosomal material sedimenting through 2 M sucrose (*Figure 5*). Rather similar electron micrographs of free and membrane-attached ribosomes have been presented by Bloemendal *et al.* (1967), except that they did not separate the smooth and rough membranes before fractionating the latter. The same group (Benedetti *et al.*, 1966) had also observed earlier two types of polysomes in the electron microscope —one tightly bound, the other loosely bound to membranes of the endoplasmic reticulum.

Whereas the liver preparations were quite homogeneous in appearance, those from brain and muscle were not. The major contaminants of microsomal membranes (smooth and rough) from the cerebral cortex were myelin fragments with smaller amounts of synaptosomes (see Whittaker, 1970) and possibly 'reticulosomes' (Adams and Lim, 1966). In muscle the major contaminant of microsomes or sarcotubular fragments were small mitochondria. The detection of rough microsomal vesicles by electron microscopy of isolated submicrosomal fraction from skeletal muscle (Andrews and Tata, 1971) is interesting since the presence of such structures had been denied earlier (Wool *et al.*, 1968), although membrane-bound ribosomes have been seen in tissue sections (Gustafsson *et al.*, 1965).

CHEMICAL COMPOSITION

The chemical composition of whole and fractionated microsomes from rat liver (*Table 2*) is compatible with both the structural patterns seen above and their enzymic and protein-synthesising activities (see below, *Tables 4* and *5*). The amount of protein, phospholipid, RNA and glycogen recovered in whole hepatic microsomes and subfractions obtained by this procedure are similar to those described by others in freely-fed adult rats (see Reid, 1967). The composition of smooth and rough endoplasmic reticulum can vary a great deal during the early stages of growth and development (Dallner *et al.*, 1966*a*). Many workers use starved animals, and the effect of starvation is to increase the RNA and protein content in the free polysome fraction and that of protein and phospholipid recovered in the smooth membrane fraction (Bloemendal *et al.*, 1967; Blobel and Potter, 1967). At the same time, starvation, or reduction in protein intake, causes an overall breakdown of all elements of the endoplasmic reticulum (Munro *et al.*, 1964).

Although the amount of RNA in the smooth membranes represents only a very small proportion of total microsomal RNA, it may be a true membrane component. Several workers have also recently found small amounts of it in smooth membranes of rat and tadpole liver (Bergeron-Bouvert and Moulé, 1966; Shapot and Pitot, 1966; Rodionova and Shapot, 1966). The relative distribution of RNA of different sizes in smooth membranes is quite different from that associated with free polysomes in which over 90% of RNA is made up of 28*S* and 18*S* molecules.

Table 2. CHEMICAL COMPOSITION OF WHOLE MICROSOMES AND SUBMICROSOMAL FRACTIONS* OF RAT LIVER

Fraction	*Protein*	*Phospholipid* (mg per g-equiv. of liver)	RNA	*Glycogen*
Whole microsomes	19·4±2·2	7·5±1·3	3·8±0·2	43·8±9·6
Smooth membranes	7·6±1·5	3·5±0·7	0·1±0·06	2·0±0·5
Light rough membranes	2·9±0·8	0·9±0·3	1·0±0·1	4·3±1·2
Heavy rough membranes	4·5±0·5	1·3±0·3	2·1±0·4	8·0±2·0
Free polysomes	0·8±0·2	0·02	0·5±0·1	25·2±6·0

* Fractionation (as *Chart 1*) using Sprague-Dawley rats of 150 to 180 g fed *ad libitum*.
These values do not represent total amount of hepatic microsomes but about 60% of homogenate RNA is recovered as whole microsomes

The significance of membrane-associated RNA in the regulation of protein synthesis is not known, but for Scott-Burden and Hawtrey (1969) it is mostly tRNA. These workers also suggest that the 7*S* RNA seen by others in smooth microsomal membranes may be largely degraded cytoplasmic ribosomal RNA.

The composition of phospholipid is almost identical in both the smooth and rough membranes of the liver (Maganiello and Phillips, 1965; Dallner *et al.*, 1966*a*; Tata, 1967*b*). However, we have recently observed that the kinetics of incorporation of [^{32}P] into phospholipids of the two fractions show some differences. At short time intervals (up to 3 h) after [^{32}P] administration, the pattern of labelling of the major (phosphatidyl choline and phosphatidyl ethanolamine) and minor (sphingomyelin and lysolecithin) components is virtually identical for smooth and rough membranes. With increasing duration of exposure to the isotope (up to 24 h) there is a marked increase in incorporation of [^{32}P] into the minor components of smooth membranes relative to the major ones, but the distribution of label in all phospholipid components of rough membranes remains unchanged with time. The rate of turnover of minor phospholipid components of smooth and rough membranes is, therefore, not the same, and it would be interesting to determine if this difference may be related to the ability of only part of microsomal membranes to bind ribosomes (Sabatini *et al.*, 1966). With regard to the composition of microsomes from rat brain and skeletal muscle, the heavy contamination of smooth microsomal membranes with other subcellular fractions rules out a meaningful comparison. In *Table 3*, however, a comparison is made of the composition of 'heavy rough membranes' and 'free polysomes' from these two predominantly non-protein secreting tissues. The ratio of RNA to phospholipid of rough microsomes reflects the

Table 3. RELATIVE RNA, PROTEIN AND PHOSPHOLIPID CONTENT OF MEMBRANE-BOUND AND FREE RIBOSOMES FROM BRAIN AND MUSCLE

Tissue	*Fraction*	RNA / *Protein*	*P-lipid* / *Protein*	RNA / *P-lipid*
Brain	Membrane-bound	0·13	0·23	0·55
	Free polysomes	0·70	0·10	7·00
Muscle	Membrane-bound	0·35	0·38	0·91
	Free polysomes	0·27	0·10	2·00

Microsomal subfractions were prepared as described in the text. Each value is the mean of duplicate determinations for 8 to 12 preparations of brain and 4 of muscle. The variation was within $\pm 5\%$. Membrane-bound fraction refers only to the 'heavy rough microsomes' (from Andrews and Tata, 1971).

relative density of packing of ribosomes on endoplasmic reticulum membranes so that the 'heavy' rough fraction from brain and muscle resembles the 'light' rather than the 'heavy' fraction from the liver. The lower RNA to phospholipid ratio of brain heavy rough membranes may also be due to contamination with such membranous material as myelin sheath fragments. The low RNA to protein ratio of 0·27 for free polysomes from muscle (compared with 0·7 to 0·75 for brain and liver preparations) is largely due to contamination with myosin fibrils.

ENZYMIC ACTIVITIES

The method described above for liver preserves the enzymic activity associated with the membrane components of microsomes. A few examples are given in *Table 4*. These values agree well with those described by other workers for unfractionated microsomes and for smooth and rough membranes (Ernster *et al.*, 1962; Dallner, 1963; Maganiello and Phillips, 1965; Dallner *et al.*, 1966*b*; see Reid, 1967). Glaumann and Dallner (1970) have recently shown that in rat liver the smooth membranes, in what appears to be a homogeneous fraction, can be separated into subfractions of heterogeneous composition. Thus the values shown in *Table 4* may just represent an average for a mosaic pattern formed from membrane units of different chemical and enzymic composition. Earlier work from the same laboratory (Dallman *et al.*, 1969) on subfractionation of rough microsomes of rat liver had revealed a heterogeneous distribution of membrane enzymes in the lighter and heavier subfractions obtained after sonication. As with the chemical composition of these preparations, the distribution of enzymic activity varies a great deal according to a variety of factors such as age (Dallner *et al.*, 1966*b*; Pollak and Ward, 1967), diet (Munro *et al.*, 1964; Weber *et al.*, 1964), hormonal status (Tata *et al.*, 1963; Suzuki *et al.*, 1967) and drug administration (Orrenius *et al.*, 1965). For example, Dallner *et al.* (1966*b*) have observed that

Table 4. DISTRIBUTION OF SOME MEMBRANE-ASSOCIATED ENZYMES IN WHOLE MICROSOMES AND SUBMICROSOMAL FRACTIONS OF RAT LIVER

Fraction	*Specific activity of enzymes* (per mg of protein)		
	ATPase	*Glucose-6-phosphatase*	*NADPH-cytochrome c reductase*
	(mμmol P_i released/min)		(mμmol NADPH oxidised/min)
Whole microsomes	68±25	95±20	73±32
Smooth membranes	75±21	81±23	104±20
Light rough membranes	83±20	122±35	58±7
Heavy rough membranes	53±17	80±12	70±21

the specific activities of glucose-6-phosphatase and NADPH-cytochrome *c* reductase at birth are markedly higher in the rough than in the smooth membranes of the rat liver, whereas the opposite is true for ATPase. With development, the concentrations of these enzymes change rapidly until they are not very different in the two submicrosomal fractions in adult animals. The hepatic enzymes summarised in *Table 4* could be barely detected in the free polysome fraction, a result which agrees well with the virtual absence of phospholipid (*Table 2*) and membrane-like structures in electron micrographs (*Figure 5*). The enzymic activities of the membranes from microsomal fractions of brain and muscle are not given because of the contamination with other organelles, but it is well known that microsomal fractions from these tissues contain a strong Na^+-K^+-ATPase activity.

PROTEIN SYNTHESIS

Much of the interest in microsomal protein synthesis has been focused on the role of attached ribosomes in the export of proteins in such predominantly secretory tissues like liver and pancreas (Palade, 1966; Campbell, 1970). However, it is now important to consider that (i) membrane-bound ribosomes in secretory tissues may synthesise intracellular proteins (Pitot *et al.*, 1969), (ii) membrane-bound ribosomes not only exist in predominantly non-secretory tissues but are active in protein synthesis (Andrews and Tata, 1968, 1971) and (iii) that membrane-bound and free ribosomes in reticulocytes (Bulova and Burka, 1970) and in bacteria (Chefurka *et al.*, 1970) may carry out the synthesis of different classes of proteins.

Figure 6 shows that in all three tissues (liver, brain and muscle) the heavy membrane-bound fraction, as obtained by the procedures described in this article, are more active than the unattached polysomes. The difference is most marked for liver as is further shown in *Table 5* for the incorporation of three different amino acids. That this difference of 4 to 5-fold between heavy rough membranes, on the one hand, and light rough membranes and free polysomes on the other, is not largely due to differences in endogenous messenger content was shown by experiments in which synthetic messengers coding for phenylalanine (poly U) and isoleucine (poly UA) were used. A 3 to 5-fold higher incorporation activity was still maintained by the heavy particles, thus indicating an inherently higher capacity of these membrane-bound ribosomes to carry out protein synthesis. It is interesting to note the difference in activities of the light and heavy membrane-bound ribosomal fractions, which suggests that the density of packing of ribosomes on membranes may have a bearing on their protein synthetic activity. The results in *Table 5* are at variance with those reported by Bloemendal *et al.* (1964, 1965, 1967) on free and membrane-bound polysomes. This difference is most likely to be due to the prolonged starvation of rats and pre-incubation of the polysomes *in vitro* practised by these workers.

An even greater difference in the activities of membrane-bound and free polysomes was seen when the protein-synthesising capacity was estimated from the distribution of nascent protein synthesised *in vivo* (*Table 6*). The specific activity of nascent protein recovered in heavy rough membrane fraction is over 20 times that associated with free polysomes. A 'balance sheet' for radioactive proteins in all subcellular fractions showed that the low levels of radioactivity in free polysomes could have arisen from a preferential loss during isolation of these particles. Considering that a large amount of microsomal protein is present in rough membranes of the liver (see *Table 2*), the results in *Table 6* demonstrate that the majority of proteins synthesised in the liver cell are made on membrane-

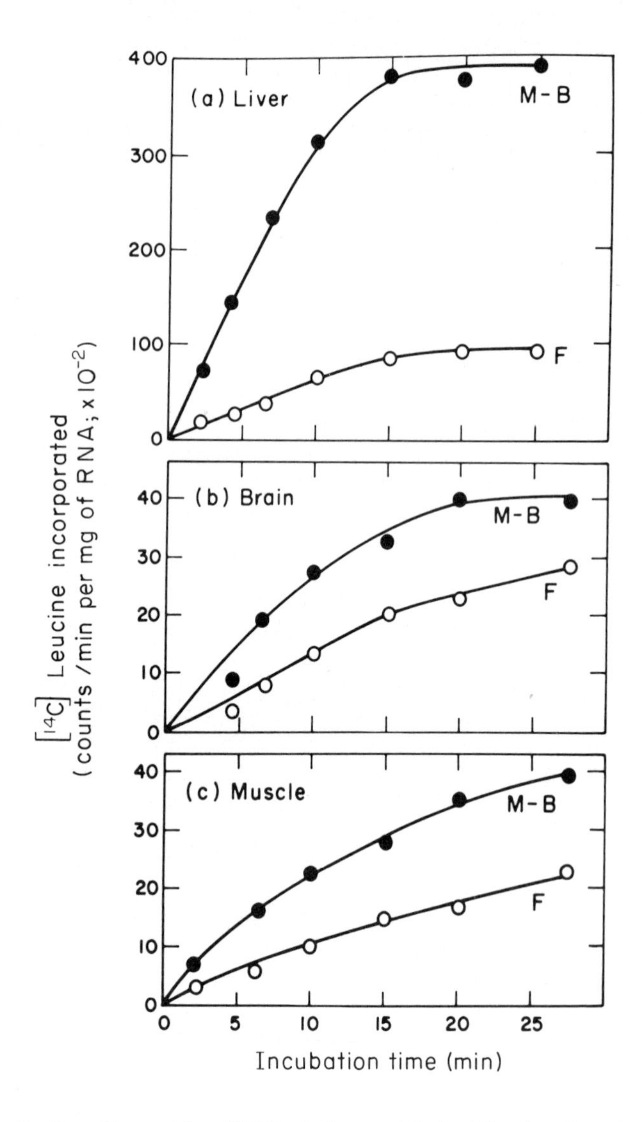

Figure 6. Rate of incorporation of [^{14}C]*leucine into protein by free* (○) *and membrane-bound* (●) *ribosomes from rat* (a), *liver;* (b), *brain; and* (c), *muscle. Incubation conditions were as described by Andrews and Tata (1971), except that only 40 to 50 μg of ribosomal RNA were added per ml of incubation medium. Each point gives the mean value of duplicate samples. The liver membrane-bound fraction as the 'heavy rough membrane' preparation of* Table 2

attached ribosomes, thus confirming the original findings of Henshaw *et al.* (1963). The recovery of relatively large amounts of protein rapidly-labelled *in vivo* in the smooth membrane fraction reflects the movement of newly-synthesised protein from the rough membranes for transport to the outside of the cell (Maganiello and Phillips, 1965; Redman *et al.*, 1966; Siekevitz and Palade, 1966). There was a rapid increase in radioactive protein of the smooth membrane with increasing time after the administration of labelled

Table 5. CAPACITY OF SUBMICROSOMAL FRACTIONS DERIVED FROM ROUGH MEMBRANES OF RAT LIVER TO INCORPORATE AMINO ACIDS INTO PROTEIN *in vitro* IN PRESENCE OF ENDOGENEOUS AND SYNTHETIC MESSENGER RNA (COMPILED FROM DATA OF TATA AND WILLIAMS-ASHMAN, 1967)

Fraction	*Incorporation of* [^{14}C]*amino acid* (counts/min per μg of RNA/10 min)*				
	Phenylalanine	*(+poly U)*	*Isoleucine*	*(+poly UA)*	*Lysine*
Free polysomes	13·6	38·1	5·3	6·4	14·6
Light rough membranes	18·6	41·8	3·6	5·7	12·2
Heavy rough membranes	58·8	128·4	23·8	30·0	59·8

* Amounts of particulate RNA added to each incubation tube in a total volume of 1·05 ml: free polysomes, 60 to 100 μg; light rough membranes, 20 to 30 μg; heavy rough membranes, 50 to 70 μg; 200 μg of polyU (polyuridylic acid or polyUA (copolymer of uridylic and adenylic acids) were added where indicated.

Table 6. DISTRIBUTION OF LABELLED NASCENT PROTEIN IN MICROSOMAL FRACTIONS AND CELL SAP

Fraction	*Specific activity of protein* counts/min per mg
Whole microsomes	4560
Smooth membranes	1100
Light rough membranes	3475
Heavy rough membranes	8400
Free polysomes	370
Cell sap	165

Rats were killed 10 min after intraperitoneal injection of 4·5 μCi of a mixture of [^{14}C]labelled amino acids (*Chlorella* protein hydrolysate); fractionation of livers: see text.

amino acids. The above pattern of vectorial distribution applies only to liver or other tissues like the pancreas and thyroid which synthesise large amounts of proteins for export. It should not be generalised as it is different in non-protein-secreting cells (Andrews and Tata, 1968, 1971). Redman and Sabatini (1966) had shown that much of the newly-formed protein, synthesised *in vivo* or *in vitro*, could be discharged by puromycin from membrane-bound ribosomes into the lumen of the microsomal membrane. A large part of this nascent protein, which is recovered by solubilisation with a detergent following incubation with puromycin, is thought to be destined for secretion. When the same procedure was applied to microsomal fractions from brain and muscle, significantly different results were obtained (Andrews and Tata, 1971), as shown in *Table* 7. Whereas only the free ribosomes of liver released most of their nascent protein directly into the supernatant fraction, both the membrane-bound and free ribosomes of brain and muscle did

Table 7. THE RELEASE OF [^{14}C]PROTEIN LABELLED *in vitro* FROM MEMBRANE-BOUND AND FREE RIBOSOMES FROM LIVER, BRAIN AND MUSCLE BY TREATMENT WITH PUROMYCIN AND DEOXYCHOLATE

Fraction	*Tissue*	*Particulate [^{14}C]protein released (per cent untreated control) by* Deoxycholate (*a*)	Puromycin (*b*)	Puromycin and deoxycholate (*c*)	$c-(a+b)$
Membrane-bound	Liver	24	13	55	+18
	Brain	15	56	70	−1
	Muscle	8	54	60	−2
Free ribosomes	Liver	9	35	44	0
	Brain	7	64	69	−2
	Muscle	4	46	51	+1

One ml of incubation mixture contained the following amount of free and 'heavy' membrane-bound ribosomes: liver; 0·5 to 1·0 mg of RNA; brain; 0·1 to 0·5 mg of RNA; muscle; 0·1 to 0·3 mg of RNA. Where indicated, 10^{-3} M puromycin was added after 8 min of incubation which was continued for another 8 min. At the end of this period half the samples were treated with 0·5% (w/v) sodium deoxycholate and the 105 000*g* particulate fraction separated from the supernatant fraction. The fraction of [^{14}C]protein recovered in the pellet was then measured. Each value is the mean of 4 to 10 experiments with an S.D. of ±5% and the results are expressed as the fraction of [^{14}C]protein recovered in untreated control preparations that is released after treatment with 0·5% sodium deoxycholate (from Andrews and Tata, 1971).

the same. There is little movement of nascent protein synthesised on bound cerebral and muscle ribosomes into the lumen of the membrane to which they are attached, which is in keeping with the non-secretory role of these tissues. Thus the ribosomes bound to membranes of the endoplasmic reticulum in secretory and non-secretory tissues differ in a major way from those in secretory tissues in their modes of discharge of newly-synthesised proteins.

How the ribosome is attached to the membrane is not known except that it is through the large ribosomal subunit (Sabatini *et al.*, 1966). Recently differences in the ribosomal protein of free and membrane-bound polysomes have been reported (Fridlender and Wettstein, 1970) and it is thought that polyamines may somehow be involved in this attachment (Khawaja and Raina, 1970). Whatever the exact mechanism of the attachment, it is reasonable to assume that there are other functions besides secretion of proteins that are determined by the association between ribosomes and membranes of the endoplasmic reticulum.

CONCLUDING REMARKS

There is no 'standard' procedure for the preparation of microsomal or submicrosomal fractions of the cell. As with other subcellular fractions, the fractionation method one adopts depends largely on the type of function that is to be studied. Because of the wide variety of biochemical and physiological functions associated with the microsomal fraction, it is not surprising that a variety of fractionation methods is now available. It is also not difficult to predict that new procedures for the isolation of microsomal and submicrosomal fractions will be introduced in the future as more information is gathered on the complex structures and functions associated with them.

Until now the procedures developed for isolation of microsomal components have been aimed at studying either the enzymes of the membranes of endoplasmic reticulum or the protein-synthesising activity of ribosomes. Very often procedures suitable for one were useless for the other. Because of the importance of the relationship between ribosomes and membranes in the regulation of protein synthesis and the rapid proliferation of all elements of the endoplasmic reticulum during growth and differentiation, it became essential to have one fractionation procedure which yields both the membrane and ribosomal components at optimal activity. Procedures for the preparation of microsomal and submicrosomal fractions of liver, brain and muscle that satisfy this condition and which have been useful for the last five years in the author's laboratory have been described in this article.

REFERENCES

ADAMS, D. H. and LIM, L. (1966). *Biochem. J.*, **99**, 261
ANDREWS, T. M. and TATA, J. R. (1968). *Biochem. biophys. Res. Commun.*, **32**, 1050
ANDREWS, T. M. and TATA, J. R. (1971). *Biochem. J.*, **121**, 683
ARIAS, I. M., DOYLE, D. and SCHIMKE, R. T. (1969). *J. biol. Chem.*, **244**, 3303
BENEDETTI, E. L., BONT, W. S. and BLOEMENDAL, H. (1966). *Lab. Invest.*, **15**, 196
BERGERON-BOUVET, C. and MOULÉ, Y. (1965). *Biochim. biophys. Acta*, **123**, 616
BLOBEL, G. and POTTER, V. R. (1966). *Proc. natn. Acad. Sci. U.S.A.*, **55**, 1283
BLOEMENDAL, H., BONT, W. S. and BENEDETTI, E. L. (1964). *Biochim. biophys. Acta*, **87**, 177
BLOEMENDAL, H., BONT, W. S., DE VRIES, M. and BENEDETTI, E. L. (1967). *Biochem. J.*, **103**, 177
BONT, W. S., REZELMAN, G. and BLOEMENDAL, H. (1965). *Biochem. J.*, **95**, 150
BREUR, C. B., DAVIES, M. C. and FLORINI, J. R. (1964). *Biochemistry*, **3**, 1713

BULLOCK, G., WHITE, A. M. and WORTHINGTON, J. (1968). *Biochem. J.*, **108**, 417
BULOVA, S. I. and BURKA, E. R. (1970). *J. biol. Chem.*, **245**, 4907
CAMPBELL, P. N. (1970). *FEBS Letters*, **7**, 1
CAMPBELL, P. N., LOWE, E. and SERCK-HANSSEN, G. (1967). *Biochem. J.*, **103**, 280
CAMPBELL, M. K., MAHLER, H. R., MOORE, W. J. and TEWARI, S. (1966). *Biochemistry*, **5**, 1174
CHAUVEAU, J., MOULÉ, Y., ROUILLER, C. and SCHNEEBELI, J. (1962). *J. Cell Biol.*, **12**, 17
CHEFURKA, W., YAPO, A. and NISMAN, B. (1970). *Can. J. Biochem.*, **48**, 893
CHEN, S. C. and YOUNG, V. R. (1968). *Biochem. J.*, **106**, 61
CLAUDE, A. (1941). *Cold Spring Har. Symp. quant. Biol.*, **9**, 263
CUNNINGHAM, W. P. and CRANE, F. L. (1966). *Expl Cell Res.*, **44**, 31
DALLMAN, P. R., DALLNER, G., BERGSTRAND, A. and ERNSTER, L. (1969). *J. Cell Biol.*, **41**, 357
DALLNER, G. (1963). *Acta path. microbiol. scand.*, Suppl. 166
DALLNER, G. (1970). *Proc. 4th Internat. Congress Pharmacol. (Basel, Switzerland)*, **4**, 70
DALLNER, G. and ERNSTER, L. (1968). *J. Histochem. Cytochem.*, **16**, 611
DALLNER, G. and NILSSON, R. (1966). *J. Cell Biol.*, **31**, 181
DALLNER, G., SIEKEVITZ, P. and PALADE, G. (1966). *J. Cell Biol.*, **30**, (*a*) 73; (*b*) 97
DE DUVE, C. (1964). *J. theor. Biol.*, **116**, 135
EARL, D. C. N. and MORGAN, H. E. (1968). *Arch. Biochem. Biophys.*, **128**, 460
ELSON, D. (1967). In *Enzyme Cytology*, (ed. Roodyn, D. B.), p. 107, New York (Academic Press)
ERNSTER, L., SIEKEVITZ, P. and PALADE, G. (1962). *J. Cell Biol.*, **15**, 541
FAWCETT, D. W. (1964). In *Intracellular Membraneous Structure*, (ed. Seno, S. and Cowdry, E. V.), p. 15, Okayama (Jap. Soc. Cell Biology)
FLORINI, J. R. and BREUER, C. B. (1966). *Biochemistry*, **5**, 1870
FRIDLENDER, B. R. and WETTSTEIN, F. O. (1970). *Biochem. biophys. Res. Commun.*, **39**, 247
GLAUMANN, H. and DALLNER, G. (1970). *J. Cell Biol.*, **47**, 34
GUSTAFSSON, R., TATA, J. R., LINDBERG, O. and ERNSTER, L. (1965). *J. Cell Biol.*, **26**, 555
HALLINAN, T. and MUNRO, H. N. (1965). (*a*) *J. exp. Physiol.*, **50**, 93; (*b*) *Biochim. biophys. Acta*, **108**, 285
HAWTREY, A. D. and SCHIRREN, V. (1962). *Biochim. biophys. Acta*, **61**, 467
HENDLER, R. W. (1965). *Nature, Lond.*, **207**, 1053
HENDLER, R. W. (1968). *Protein Biosynthesis and Membrane Biochemistry*. New York (John Wiley & Co. Inc.)
HENSHAW, E. C., BOJARSKI, T. B. and HIATT, H. H. (1963). *J. molec. Biol.*, **7**, 122
HEYWOOD, S. M., DOWBEN, R. M. and RICH, A. (1968). *Biochemistry*, **9**, 3289
HOWELL, R. R., LOEB, J. N. and TOMKINS, G. M. (1964). *Proc. natn. Acad. Sci. U.S.A.*, **52**, 1241
HUNTER, A. R. and KORNER, A. (1966). *Biochem. J.*, **100**, 73P
IMAI, K., OMURA, T. and SATO, R. (1966). *J. Biochem. Tokyo*, **60**, 274
KHAWAJA, J. A. and RAINA, A. (1970). *Biochem. biophys. Res. Commun.*, **41**, 512
LAMIRANDE, DE G., BOILEAN, S. and MORAIS, R. (1966). *Can. J. Biochem.*, **44**, 273
LEE, T. -C., SWARTZENDRUBER, D. C. and SNYDER, G. (1969). *Biochem. biophys. Res. Commun.*, **36**, 748
LIM, L. and ADAMS, D. H. (1967). *Biochem. J.*, **104**, 229
LOEB, J. N., HOWELL, R. R. and TOMKINS, G. H. (1967). *J. biol. Chem.*, **242**, 2069
MAGANIELLO, V. and PHILLIPS, A. H. (1965). *J. biol. Chem.*, **240**, 3951
MAHLER, H. R. and BROWN, B. J. (1968). *Arch. Biochem. Biophys.*, **125**, 387
MERITS, I., CAIN, J. C., RDZOK, E. J., and MINARD, F. N. (1969). *Experientia*, **25**, 739
MORAIS, R. and GOLDBERG, I. H. (1967). *Biochemistry*, **6**, 2538
MOULÉ, Y., ROUILLER, C. and CHAUVEAU, J. (1960). *J. biophys. biochem. Cytol.*, **7**, 547
MUNRO, H. N., MCLEAN, E. J. T. and HIRD, H. J. (1964). *J. Nutr.*, **83**, 186
MURTY, C. N. and HALLINAN, T. (1969). *Biochem. J.*, **112**, 269
OLSNES, S. (1970). *Eur. J. Biochem.*, **15**, 464
OMURA, T., SIEKEVITZ, P. and PALADE, G. E. (1967). *J. biol. Chem.*, **242**, 2389

ORRENIUS, S., ERICSSON, J. L. and ERNSTER, L. (1965). *J. Cell Biol.*, **25**, 627
PALADE, G. (1966). *J. Am. Med. Assoc.*, **198**, 815
PALADE, G. E. and SIEKEVITZ, P. (1965). *J. biophys. biochem. Cytol.*, **2**, 171
PITOT, H. C., SLADEK, N., RAGLAND, W., MURRAY, R. K., MOYER, G., SOLING, H. D. and JOST, J. -P. (1969). In *Microsomes and Drug Oxidations*, (Ed. Gillette, J. R., Conney, A. H., Cosmides, G. J., Estabrook, R. W., Fonts, J. R. and Mannering, G. J.), p. 59, New York (Academic Press)
POLLAK, J. K. and WARD, D. B. (1967). *Biochem. J.*, **103**, 730
QUIRIN-STRICKER, C. and MANDEL, P. (1967). *Bull. Soc. Chim. biol.*, **49**, 1517
RAMPERSAD, O. R., ZAK, R., RABINOWITZ, M., WOOL, I. G. and DESALLE, L. (1965). *Biochim. biophys. Acta*, **108**, 95
REDMAN, C. M. and SABATINI, D. D. (1966). *Proc. natn. Acad. Sci. U.S.A.*, **56**, 608
REDMAN, C. M., SIEKEVITZ, P. and PALADE, G. E. (1966). *J. biol. Chem.*, **241**, 1150
REID, E. (1967). In *Enzyme Cytology*, (ed. Roodyn. D.), p. 321, New York (Academic Press)
RODIONOVA, N. P. and SHAPOT, V. S. (1966). *Biochim. biophys. Acta*, **129**, 206
ROODYN, D. B., FREEMAN, K. B. and TATA, J. R. (1965). *Biochem. J.*, **94**, 628
SABATINI, D. D., TASHIRO, Y. and PALADE, G. E. (1966). *J. molec. Biol.*, **19**, 503
SCOTT-BURDEN, T. and HAWTREY, A. D. (1969). *Biochem. J.*, **115**, 1063
SHAPOT, V. and PITOT, H. (1966). *Biochim. biophys. Acta*, **119**, 37
SIEKEVITZ, P. (1965). *Expl Cell Res. Suppl.*, **7**, 90
SIEKEVITZ, P. and PALADE, G. E. (1960). *J. biophys. biochem. Cytol.*, **7**, 619
SIEKEVITZ, P. and PALADE, G. E. (1966). *J. Cell Biol.*, **30**, 519
SIEKEVITZ, P., PALADE, G. E., DALLNER, G., OHAD, I. and OMURA, T. (1967). In *Organizational Biosynthesis* (ed. Vogel, H. J., Lampen, J. O. and Bryson, V.), p. 331, New York (Academic Press)
SUZUKI, M., IMAI, K., ITO, A., OMURZ, T. and SATO, R. (1967). *J. Biochem. Tokyo*, **62**, 447
TATA, J. R. (1967*a*). *Nature, Lond.*, **213**, 566
TATA, J. R. (1967*b*). *Biochem. J.*, **105**, 783; (1967*c*). *ibid*, 47P
TATA, J. R. (1970). *Biochem. J.*, **116**, 617
TATA, J. R., ERNSTER, L., LINDBERG, D., ARRHENIUS, E., PEDERSEN, S. and HEDMAN, R. (1963). *Biochem. J.*, **86**, 408
TATA, J. R. and WILLIAMS-ASHMAN, H. G. (1967). *Eur. J. Biochem.*, **2**, 366
TENG, C. S. and HAMILTON, T. H. (1967). *Biochem. J.*, **105**, 1091
WEBER, G., SINGHAL, R. L., STAMM, N. B., FISHER, E. A. and MENTENDICK, M. A. (1964). *Adv. Enzyme Regul.*, **2**, 1
WEIBEL, E. R., STAUBLI, W., GNÄGI, H. R. and HESS, F. A. (1969). *J. Cell Biol.*, **42**, 68
WETTSTEIN, F. O., STAEHELIN, T. and NOLL, H. (1963). *Nature, Lond.*, **197**, 430
WHITTAKER, V. P. (1970). In *Control Processes in Multicellular Organisms*, (ed. Wolstenholme, G. E. W. and Knight, J.), p. 338, London (J. and A. Churchill)
WIDNELL, C. C., HAMILTON, T. H. and TATA, J. R. (1967). *J. Cell Biol.*, **32**, 766
WOOL, I. G., STIRWALT, W. S., KURIHARA, K., LOW, R. B., BAILEY, P. and OYER, D. (1968). *Recent Progr. in Hormone Res.*, **24**, 139
ZOMZELY, C. E., ROBERTS, S. and RAPPOPORT, D. (1964). *J. Neurochem.*, **11**, 567

9

ISOLATION OF ANIMAL POLYSOMES AND RIBOSOMES

S. A. Bonanou-Tzedaki and H. R. V. Arnstein

Ribosomes are ribonucleoprotein particles which, in animal cells, contain about equal amounts of RNA and protein, have a molecular weight of about 4×10^6 daltons, a diameter of approx. 250 Å and a sedimentation coefficient of approx. 80*S* (for reviews see Arnstein, 1963; Petermann, 1964). They are present mainly in the cytoplasm of the cell, where they occur either as single particles (monoribosomes) or as aggregates of two or more particles, termed polysomes or ergosomes. In cells containing endoplasmic reticulum, that is, most cells other than some embryonic or tumour cells and reticulocytes, which contain little or no endoplasmic reticulum, the majority of the ribosomes are attached to membranes, although some are present free in the cytoplasm (Palade and Siekevitz, 1956; Benedetti *et al.*, 1966). These two classes of ribosomes and polysomes, free and membrane-bound, can be isolated separately (Blobel and Potter, 1967) and there is growing evidence that they are involved in the synthesis of predominantly different proteins (see review by Campbell, 1970). For example, free rat-liver polysomes are reported to synthesise largely ferritin (Redman, 1969; Hicks *et al.*, 1969) but very little albumin (Takagi and Ogata, 1968) or glycoproteins (Hallinan *et al.*, 1968), that is, proteins usually destined for 'export' from the cell.

Ribosomes, or ribonucleoprotein particles, have also been detected inside the nuclei of thymus (Pogo *et al.*, 1962), rat-liver (McCarty *et al.*, 1966; Sadowski and Alcock-Howden, 1967) and HeLa cells (Zimmerman *et al.*, 1969; see also reviews by Perry,

1967; Allfrey, 1970) and inside mitochondria from a variety of sources (Küntzel and Noll, 1967; O'Brien and Kalf, 1967; Rifkin *et al.*, 1967; Attardi and Ojala, 1971; Brega and Vesco, 1971). It appears that ribosomes from the cytoplasm of animal cells are similar in size to cytoplasmic ribosomes from yeasts and higher plants, whereas ribosomes from micro-organisms and chloroplasts are smaller, having sedimentation coefficients of approx. 70*S*. Until recently, it was thought that mitochondrial ribosomes are similar to those present in bacteria, but it has now been established that they are even smaller, with sedimentation coefficients of 55 to 60*S* in the case of *Xenopus laevis* (Swanson and Dawid, 1970) and HeLa cells (Attardi and Ojala, 1971; Brega and Vesco, 1971).

ASPECTS OF POLYSOME STRUCTURE AND FUNCTION

It is now generally accepted that polysomes are complexes of several ribosomes attached to one molecule of messenger RNA (Rich *et al.*, 1963; Noll *et al.*, 1963) and that this is the functional unit in protein biosynthesis, although the existence of animal messenger RNA (mRNA), with the properties originally specified by Jacob and Monod (1961), is still questioned by certain investigators (e.g. Harris, 1970). The polysome model rests largely on two observations, which are (a), the proportionality between polysome size, molecular weight of mRNA and polypeptide synthesised, and (b), the sensitivity of polysomes to ribonuclease. In this section we will consider briefly the evidence for these statements and also mention other factors which govern polysome function and structure and which are, consequently, important in any discussion of the isolation of active polysomes.

The significance of polysomes was first established in rabbit reticulocytes when it was shown that nascent protein, labelled in intact cells which were subsequently lysed and their contents analysed on a sucrose gradient, was not associated with the single 80*S* ribosome peak but with heavier ribosomal aggregates (Warner *et al.*, 1962; Gierer, 1963). These appeared in electron micrographs (Slayter *et al.*, 1963) as clusters of 4 to 6 ribosomes closely held together, with a contour length of about 1500 Å, which at that time was thought to be in good agreement with the expected length of haemoglobin messenger on the assumptions that (i) the polysomes contain messenger for the synthesis of one haemoglobin chain of approximately 150 amino acids, (ii) 3 nucleotides code for each amino acid (Crick *et al.*, 1961) and (iii) the expected 450 nucleotides are stacked with a DNA-like separation of 3·4 Å (Fuller and Hodgson, 1967). At the same time Staehelin *et al.* (1964) established a correlation between polysome size and length of mRNA in an extensive study of rat-liver polysomes. Similarly, lymph-node cells synthesising antibody protein contain two classes of polysomes, a small one (7 to 8 ribosomes) and a large one (16 to 20 ribosomes); these structures are consistent with the sizes expected for the synthesis of light and heavy chains of antibody, respectively (Becker and Rich, 1966). In embryonic chick muscle there is also a range of polysome sizes, but only the heaviest ones, consisting of complexes of about sixty ribosomes, can synthesise myosin (Heywood *et al.*, 1967), in agreement with the size of an mRNA of about 5500 nucleotides required to code for a protein subunit of molecular weight of about 200 000 daltons (Heywood and Nwagwu, 1968). Myoglobin is synthesised by smaller muscle polysomes consisting of 4 to 6 ribosomes (Kagen and Linder, 1969). Another striking example is the re-formation of polysomes in actinomycin-treated HeLa cells upon infection of the cells with polio virus (Penman *et al.*, 1963). The polycistronic viral RNA (molecular weight about 2×10^6 daltons) becomes associated with polysomes sedimenting at about 400*S*, which are more than double the size of the polysomes of uninfected cells and which synthesise virus-specific proteins.

Two exceptions should be mentioned at this stage. First, it has been shown that monosomes (single ribosomes attached to mRNA) in rabbit reticulocytes (Lamfrom and Knopf, 1965) or rat liver (Munro *et al.*, 1964) are capable of synthesising complete new polypeptide chains in a cell-free system. It would appear, therefore, that the polysome model is not obligatory for protein synthesis, although it is usually very advantageous. Second, in certain metabolic conditions, an mRNA may be found associated with much heavier or lighter polysomes than usual. This was shown to occur when intact rabbit reticulocytes were incubated with the isoleucine antagonist *O*-methylthreonine (Hori and Rabinovitz, 1968; Kazazian and Freedman, 1968). Since protein synthesis proceeds from the amino-terminal to the carboxyl-terminal amino acid (Dintzis, 1961) and isoleucine is present near the amino-terminal end of the α chain, but near the carboxyl-terminal end of the β chain, inhibition of isoleucine incorporation into nascent α and β chains resulted in smaller α chain polysomes (aggregates of 2 to 3 ribosomes) and larger β chain polysomes (aggregates of 10 to 14 ribosomes). This treatment therefore differentiated between polysomes containing mRNAs specifying the haemoglobin α or β chains. Some larger polysomes, containing 15 to 20 ribosomes, have been demonstrated in reticulocytes (Rifkind *et al.*, 1964). This number of ribosomes could still be accommodated on a haemoglobin mRNA if, as more recent measurements of stacked and unstacked single-stranded polynucleotides indicate, the interphosphate distance is 6·5 Å (Eisenberg and Felsenfeld, 1967), since this would result in haemoglobin mRNA being as long as 2900 Å. Moreover, there is growing evidence (Labrie, 1969; Gaskill and Kabat, 1971) that haemoglobin mRNA may be considerably larger than 450 nucleotides, as previously assumed.

The sensitivity of polysomes to nucleases or shear forces also supports the model that polysomes are complexes of ribosomes held together by an mRNA thread. For example, excessive handling, or even repetitive pipetting, leads to disruption of mRNA by shearing forces (Oppenheim *et al.*, 1968). Exposure of reticulocyte polysomes to as little as 0·03 μg of ribonuclease per mg of ribosomal RNA converts them into monoribosomes with simultaneous reduction in their capacity to synthesise protein (Arnstein, 1961), and this susceptibility of polysomes to traces of ribonucleases has been employed as a diagnostic tool by many investigators (see Noll, 1969). Nevertheless, in special circumstances, polysomes may not be degraded by nucleases; thus it was observed by Ling and Dixon (1970) that pancreatic ribonuclease could not degrade trout-testis disomes engaged in protein biosynthesis. The greatest hazard during the preparation of polysomes comes from cytoplasmic nucleases which may be released from lysosomes or nuclei during cell fractionation and either act directly or get adsorbed onto ribosomes and become activated later in suitable ionic conditions. There is evidence that tissues with relatively stable mRNAs are low in nucleases (Stavy *et al.*, 1964) and that the ribonuclease activity of a tissue increases with age (Arora and de Lamirande, 1968). A control function in protein synthesis has been suggested for tissue nucleases (Breuer *et al.*, 1969; Kraft and Shortman, 1970). However there is evidence, in bacteria at least, that ribosomes contain their own ribonucleases (Robertson *et al.*, 1968) and one of them, ribonuclease V, has been shown to degrade mRNA specifically in a process coupled to peptide bond formation (Kuwano *et al.*, 1969).

Polysomes from rat lymph nodes (Manner and Gould, 1965), rabbit reticulocytes (Zak *et al.*, 1966), rat liver (Nair *et al.*, 1966) or sea-urchin eggs (Piatigorsky, 1968) may also be disaggregated by proteolytic enzymes (e.g. trypsin, chymotrypsin or papain), an effect which is not apparently due to 'latent' ribonuclease. Although it was shown that, in calf-eye lens (Benedetti *et al.*, 1968) and cultured mammalian fibroblasts (Goldberg and Green, 1967), the sensitivity of polysomes to proteolytic, but not nucleolytic, enzymes was due to nascent peptide chains which establish polyribosomal interactions that lead to giant polysomes, the effect of proteases on polysomes in general may be explained by two considerations. First, studies of the structure of polysomes by electron microscopy (Shelton and Kuff, 1966) suggest that small polysomes may consist of rosettes of 5 or 6 ribosomes closely packed together with the small subunits forming a tight circle and that larger polysomes are arranged as a helix with a turn of about 6 ribosomes, the small subunits again pointing inwards (the helical arrangement of polysomes was also suggested by Pfuderer *et al.* (1965) on the basis of hydrodynamic studies of polysomes). According to this model, the maintenance of polysome structure depends not only on the integrity of the mRNA thread but also on the presence of protein–protein interactions between the small ribosomal subunits. Second, it is now thought that mRNA is not only carried into the cytoplasm as a ribonucleoprotein particle but that it also exists in the polysomes as a messenger–ribonucleoprotein complex (for references see Spohr *et al.*, 1970) from which it may be released by EDTA (Huez *et al.*, 1967; Perry and Kelley, 1968). The exact number, role and relation, if any, of these messenger-associated proteins to the various initiation and other transfer factors of protein synthesis (Miller and Schweet, 1968; Cohen, 1968; Fuhr *et al.*, 1969; Prichard *et al.*, 1970) is far from clear. Nevertheless, it is not difficult to envisage how a destruction of some protein components could undermine the integrity of the polysome structure and reduce its activity.

Polysome breakdown results not only from the random fragmentation of mRNA, but also by the orderly release of single ribosomes from the end of intact messenger strands (Hardesty *et al.*, 1963). This is commonly observed in certain, usually reversible, metabolic conditions in which initiation of new peptide chains relative to chain completion and release is inhibited (for discussion see Noll, 1969). For example, a dramatic breakdown of polysomes occurs upon cooling intact cells, apparently because chain initiation requires a much higher energy of activation than chain extension (Das and Goldstein, 1968; Friedman *et al.*, 1969). Inhibitors of protein synthesis like NaF (Marks *et al.*, 1965; Vesco and Colombo, 1970) or aurintricarboxylic acid (Grollman and Huang, 1970; Lebleu *et al.*, 1970) also convert polysomes into monosomes by interfering with some step in chain initiation. Similarly, inhibitors of energy-generation processes, for example, dinitrophenol, cause breakdown of polysomes, presumably by a differential effect on initiation. The dissociation of run-off ribosomes into subunits, which can subsequently reattach on to the mRNA, is, however, probably also an energy-dependent step (Colombo *et al.*, 1968). Metabolic shifts in intact animals such as from feeding to starvation (Staehelin *et al.*, 1967), or omission of certain nutrients from cells grown in tissue culture (Eliasson *et al.*, 1967), or omission of components of cell-free systems (Baliga *et al.*, 1968) also lead to polysome disaggregation. Disaggregation of polysomes without breakdown of mRNA is also observed as a result of fraudulent chain termination in the presence of puromycin (Villa-Trevino *et al.*, 1964; Lawford, 1969) or when polysomes are exposed to conditions which favour dissociation of ribosomes into subunits (Martin and Wool, 1969). The fate of mRNA in this case is still in dispute. For example, when reticulocyte polysomes are titrated with EDTA an mRNA-protein complex is released (Huez *et al.*, 1967), but when polysomes are treated with pyrophosphate (Holder and Lingrel, 1970) or high concentrations of KCl (Hamada *et al.*, 1968) the mRNA remains attached to the small subunit, as the dissociated preparation can still support endogenous protein synthesis (Bonanou *et al.*, 1968). It is probable that production of ribosomal subunits leads to initial release of the mRNA with subsequent reattachment to one of the small subunits (Pragnell and Arnstein, 1970), provided that, during fractionation, the subunits do not undergo any irreversible conformational changes (as may occur after EDTA treatment), which may interfere with mRNA binding.

METHODS OF ISOLATION

Polysomes and ribosomes have been isolated from a wide variety of animal cells (*Table 1*). Although there is no simple method applicable to the isolation of polysomes from any given source, the most important factors in all cases include suitable conditions of cell lysis and fractionation, rapid handling and maintenance of a low temperature (see also von der Decken, 1967; Noll, 1969).

Some cells, for example, reticulocytes, can be lysed by osmotic shock, but usually it is necessary to disrupt the cells mechanically by homogenising suitably minced tissue in a Dounce tissue grinder or in a Potter–Elvehjem glass homogeniser, often with a motor-driven Teflon pestle, in media containing sucrose (0·15 to 0·45 M). Exposure to hypotonic solutions may facilitate homogenisation; nevertheless it is preferable to disrupt cells in isotonic (0·25 M sucrose) or slightly hypertonic media or, if this is not possible, to adjust the medium to isotonicity, usually with sucrose, immediately after lysis to avoid rupture of mitochondria and lysosymes.

Table 1. KEY REFERENCES FOR THE ISOLATION OF POLYSOMES FROM REPRESENTATIVE ANIMAL TISSUES

Tissue	*Reference*	*Tissue*	*Reference*
Rabbit reticulocytes	Arnstein *et al.* (1964)	Bovine mammary gland	Herrington and Hawtrey (1971)
Rabbit bone marrow	Grau and Favelukes (1968)	Rat uterus	Teng and Hamilton (1967)
Rat lymph nodes	Becker and Rich (1966)	Human placenta	Laga *et al.* (1970)
Mouse plasmacytoma	Williamson and Askonas (1967)	Hen oviduct	Gerlinger *et al.* (1969)
Chick embryonic muscle	Heywood *et al.* (1967)	*Xenopus laevis* ovary	Ford (1966)
Rat skeletal muscle	Florini and Breuer (1966)	Rat testis	Means *et al.* (1969)
Rat-heart muscle	Earl and Morgan (1968)	Rat brain	Zomzely *et al.* (1966)
Rat liver	Wettstein *et al.* (1963)	Calf-eye lens	Benedetti *et al.* (1968)
Dog pancreas	Breillatt and Dickman (1969)	Human fibroblasts	Golberg and Green (1967)
Rat spleen	Talal and Kaltreider (1968)	HeLa cells	Penman *et al.* (1963)
Mouse kidney	Priestley *et al.* (1969)	BHK cells	Ascione and Arlinghaus (1970)
Sheep thyroid	Cartouzou *et al.* (1968)	Sea-urchin embryos	Nemer and Infante (1965)
		Surf-clam embryos	Firtel and Monroy (1970)

For nuclear and mitochondrial ribosomes see references cited on pp. 215–216

With certain tissues, for example, lymph nodes or spleen, good polysomes were obtained by gentle teasing in isotonic medium, but only when no attempt was made to disrupt all cells (Becker and Rich, 1966). With particularly tough tissue, for example, heart muscle, brief treatment with proteinase was found necessary to help break the cellular structure prior to homogenisation (Earl and Korner, 1965). Sonication in a Vir-Tis '45' homogeniser was used for the isolation of skeletal muscle polysomes (Florini and Breuer, 1966), but care is necessary because sonication also disrupts polysomes (Williamson *et al.*, 1969; Zylber and Penman, 1970). The use of detergents has facilitated the disruption of certain cells, for example, deoxycholate and BRIJ-58 for HeLa cells (Huang and Baltimore, 1969) and Nonidet P-40 in high salt for baby hamster-kidney cells (Ascione and Arlinghaus, 1970).

Contamination of liver polysomes with nuclear ribonucleoprotein, apparently caused by leakage during cell disruption and fractionation, has also been observed (Olsnes, 1970*a*).

As a rule, the optimal concentration of monovalent cations (K^+ alone or in combination with Na^+ or NH_4^+) is about 10 times that of Mg^{2+} (range employed varies from 2·5 to 20 times) and the Mg^{2+} concentration is usually 5 mM (range 1 to 13 mM). Much lower Mg^{2+} concentrations lead to polysome disaggregation whereas much higher Mg^{2+} concentrations lead to the formation of non-specific aggregates which do not respond to traces of pancreatic ribonucleases (for a discussion of the effective bivalent cation concentration in incubation media see Manchester, 1970). Occasionally, however, special media have to be used in order to prevent precipitation of cell protein during homogenisation, since this can lead to loss of polysomes by coprecipitation. In the case of polysomes from chick-embryo muscle, homogenisation was carried out in high KCl (0·25 M) to avoid precipitation of myosin (Heywood *et al.*, 1967). There is still no agreement on the relative merits of cations other than K^+ employed in the isolation media although the presence of NH_4^+ in the incubation buffer is reported to improve incorporation. To date, tris-HCl, or sometimes bicarbonate or phosphate, buffers have been used almost exclusively in the isolation media; however, the superiority of other buffer systems has been demonstrated by Good *et al.* (1966) and Huston *et al.* (1970). Although polysomes are stable over the entire pH range of 6 to 8·8, it is customary to isolate them at pH 7·6 (range employed 7·2 to 8·2) as at that pH they show optimal activity in protein synthesis. Nevertheless, given the considerable temperature dependence of tris buffers, the value of these measurements is questionable, since buffers are usually adjusted at room temperature, while polysomes are isolated at 0°C and incorporation activity is assayed at 37°C. The inclusion of glutathione, or other sulphydryl agents, in the extraction medium is also advocated by some workers as it enhances the subsequent rate of amino acid incorporation by polysomes.

Ribonuclease is, of course, the greatest hazard during polysome isolation. Its action can be diminished by shortening the time of fractionation, working in the cold, avoiding rupture of subcellular components and adding various ribonuclease inhibitors to the homogenate. Bentonite (Murthy and Rappoport, 1965), especially when prepared according to exact specifications (Watts and Mathias, 1967), macaloid (Williamson and Askonas, 1967), dextran sulphate (Ascione and Arlinghaus, 1970), heparin (Rowley and Morris, 1966), even RNA (La Via *et al.*, 1967) have proved beneficial for the isolation of polysomes. Ribonuclease inhibitors prepared from rat-liver cell sap (Roth, 1958; Blobel and Potter, 1966; Lawford *et al.*, 1966; Sugano *et al.*, 1967) or cerebral cortex cell sap (Takahashi *et al.*, 1966) have been added prior to homogenisation for the isolation of undegraded polysomes from spleen (Northrup *et al.*, 1967), bone marrow (Grau and Favelukes, 1968) or isolated rat-liver nuclei (Lawford *et al.*, 1967). Similarly, Scharff and Uhr (1965) reported that breakdown of polysomes by nucleases during disruption of lymphocytes could be prevented by the addition of deoxycholate together with cytoplasm from a large (5-fold) excess of HeLa cells to the lymph node cells before homogenisation.

A major problem which complicates the isolation of the majority of animal polysomes is their attachment to membranes; the nature of this interaction is not completely understood. Detergents, either ionic like deoxycholate (DOC) or sodium dodecyl sulphate, or non-ionic like Triton X-100, Lubrol W or iso-octane, have been found necessary to solubilise the membranes, leaving intact ribosomal particles which may be sedimented in the ultracentrifuge. The method originally employed was to add 0·5% (w/v) DOC to the microsome fraction of, for example, rat liver (Littlefield *et al.*, 1955) and a good recovery of ribonucleoprotein particles, with a high RNA to protein ratio and activity for protein synthesis, was obtained. Essentially analogous procedures were used similarly by many workers (e.g. Korner, 1961; Rendi and Hultin, 1960, who also used concentrated salt solutions in conjunction with detergents) but with the discovery of polysomes it was found that these conditions frequently yielded only monomers and dimers (80*S* and 110*S* particles) due to the release of ribonuclease by DOC (Sugano *et al.*, 1967). When, however, DOC was added to rat-liver microsomes together with the post-mitochondrial supernatant fraction (Lawford *et al.*, 1966), which contains a ribonuclease inhibitor, breakdown of polysomes could be prevented. Similarly, undegraded polysomes were obtained when DOC was added to kidney membrane-bound polysomes in the presence of mouse-liver cell sap (Priestley *et al.*, 1969) and to mammary-gland microsomes in the presence of homologous cell sap (Gaye and Denamur, 1970).

Simpler, and much more widely used now, is the method of adding DOC, usually to a final concentration of 1 to 2% (w/v), directly to the post-mitochondrial supernatant fraction. When the temperature during all manipulations was rigorously kept near 0°C, and the post-mitochondrial supernatant fraction, made 1·3% with respect to DOC, was sedimented through two layers of 0·5 M and 2·0 M buffered sucrose, a final rat-liver polysome pellet was obtained which consisted of over 90% polysomes (Wettstein *et al.*, 1963). Bloemendal *et al.* (1964) developed a method employing a discontinuous sucrose gradient for obtaining polysomes free from membranes without using any detergent. With rat liver, 1% Triton X-100 was reported to yield a more active polyribosome preparation than 1% DOC (Hunter and Korner, 1966), but Huston *et al.* (1970) claimed that 1·67% DOC was more effective than Triton X-100. There is some evidence that DOC or high salt removes certain proteins which are usually associated with mRNA in polysomes, whereas Triton X-100 does not (Olsnes, 1970*b*). This perhaps explains why DOC degraded purified polysomes from mouse plasma-cell tumours, whereas Triton X-100 had no effect on the size distribution of polysomes (Kuff and Roberts, 1967).

Chromatography with DEAE-cellulose has been applied to the isolation of pancreatic ribosomes (Dickman and Bruenger, 1969) and ECTHAM-cellulose has been similarly used for the isolation of microsomes and ribosomes from rat liver, plasma-cell tumours and reticulocytes, with retention of activity for cell-free protein synthesis (Peterson and Kuff, 1969; Kedes *et al.*, 1969).

FRACTIONATION AND CHARACTERISATION OF POLYSOMES

The tissue homogenate is freed from nuclei, mitochondria, cell debris, etc. by differential centrifugation at about 10 000*g* for 10 to 20 min. The post-mitochondrial supernatant fraction may then be further centrifuged in a fixed-angle rotor at 100 000*g* for 1 h to sediment microsomes, or it may be layered directly, with or without detergent, onto a discontinuous sucrose gradient and centrifuged at, for example, 100 000*g* for 3 h or more for the sedimentation of purified polysomes. Resuspended liver ribosomal pellets have been purified by precipitation with 0·05 M $MgCl_2$ and dialysis (Takanami, 1960). In cells like the reticulocyte, in which over 90% of the ribosomes are free, the particles can be isolated by acid precipitation at pH 6·0. After a low-speed centrifugation, the pellet is redissolved in buffer and may be further fractionated by rate-zonal or by differential centrifugation (Arnstein *et al.*, 1965). Alternatively, the post-mitochondrial supernatant fluid or the resuspended polysome pellet may be subjected to sucrose-gradient fractionation, and the gradient fractions containing the required polysomes or monosomes pooled, diluted with buffer and pelleted by differential centrifugation or by Mg^{2+} and ethanol precipitation (Falvey and Staehehin, 1970).

Sucrose-gradient centrifugation separates macromolecules according to size, density and shape. The sample to be analysed is layered in a narrow zone over a tube full of a buffered sucrose solution, introduced as a gradient to stabilise the liquid in the tube against convection, and spun at high speed in a swing-out rotor. The solute molecules move outwards and separate into bands according to their individual sedimentation rates. The simplest device for producing linear gradients is that first described by Britten and Roberts (1960) (for a comprehensive discussion of gradient separations see Noll, 1969). Once formed, the sedimenting zones are remarkably stable (at least up to 48 h at 4°C). Sucrose gradients are often employed analytically for determining the sedimentation coefficients of different polysomes in calibrated gradients, as the distance travelled by a band from the top of the tube is proportional to the sedimentation coefficient of the molecules contained in that band (Martin and Ames, 1961). It is important not to overload the gradient as this leads to decreased resolution owing to the diffusion of one band into another. At the end of centrifugation the gradients can be fractionated automatically by pumping out the tube contents continuously, monitoring their extinction in a recording spectrophotometer and collecting appropriate volumes in a fraction collector. If the material on the gradients is labelled, radioactivity may be assayed either by using aqueous samples in a suitable scintillation solvent or by acid precipitation of the nucleoprotein and filtration on suitable disks, the radioactivity of which can be measured by either scintillation or gas-flow counting (Campbell and Sargent, 1967).

When no preparative separation of polysomes is required, a convenient estimation of the relative amounts and *S* values of the different species can be obtained by centrifugation of samples in the analytical ultracentrifuge (see Petermann, 1964). Ribosomes have also been separated from polysomes analytically by polyacrylamide gel electrophoresis (Dahlberg *et al.*, 1969).

Polysomes are usually stored frozen as pellets under a layer of 0·25 M sucrose, either in water or buffered salts. Alternatively, concentrated suspensions (10 mg/ml, or more) may be frozen. There is little or no loss of activity after storage for a few months at −20°C or for over six months in liquid nitrogen. Recently it has been reported that reticulocyte polysomes may be lyophilised and stored at room temperature for several months without loss of structure or activity when tested with polyU for synthesis of polyphenylalanine (Christman and Goldstein, 1971).

Polysome pellets are resuspended in cold buffer by gentle stirring with a glass rod or with a hand-operated glass homogeniser. Alternatively, agitation of the buffer by repeated suction with a Pasteur pipette may be used. Care must be exercised, however, to avoid degradation of the polysomes by shear forces.

Cytoplasmic proteins are often adsorbed onto polysome pellets derived from the post-mitochondrial supernatant fraction and may be removed by resuspending and washing the ribosomes with buffer or strong salt solutions (Moldave and Skogerson, 1967). Alternatively, polysomes may be purified by centrifugation through a layer of concentrated sucrose or through a sucrose gradient containing appropriate salts and buffer.

Clean preparations of polysomes are almost colourless in the pellet, but opalescent in solution. The purity of polysomes may be conveniently assessed by measuring the extinction of an appropriate sample at 240, 260 and 280 nm. For good polysomes, the ratio of extinction at 260 nm to that at 280 nm should be 1·7 or higher, and the ratio of extinctions at 240 nm and 280 nm about 1·0. Contaminants can be estimated from the extinction of the solution at different wavelengths, for example at 415 nm for haemoglobin. Ferritin is particularly troublesome in liver fractionation but may be estimated from the ratios of extinctions at 260, 280 and 320 nm (Wilson and Hoagland, 1965). The polysome yield varies among tissues, but it is usually in the range of 0·3 to 3 mg of ribosomes per g of tissue. The relationship

$$E_{260} \text{ of 1 mg of ribosomes/ml} = 10$$

may be used as a rough guide for calculating the yield, but the value increases with polysome purification and only RNA and protein estimations give reliable results (Campbell and Sargent, 1967).

The ultimate proof of the isolation of good polysomes is their capacity to support protein synthesis in a cell-free system. Unfortunately, both the extent and rate of protein synthesis decreases with cell fractionation in the following order: whole tissue, tissue slices, homogenate, microsomes, crude polysomes and purified polysomes. This is probably due to the disruption of the micro-environment in which polysomes operate in the cell, for example, the exact ionic strength, pH, spatial orientation of ribosomes and association with membranes, as well as to the loss or inactivation of certain components intimately associated with polysomes. Thus, in a cell-free system, the presence of adequate amounts of enzymes,

Table 2. INCORPORATION OF [^{14}C]PHENYLALANINE INTO ACID-INSOLUBLE PROTEIN BY RETICULOCYTE RIBOSOMES

A mitochondria-free lysate from rabbit reticulocytes was fractionated on a sucrose gradient (cf. *Figure 1*) and samples from the pentamer and monomer regions of the gradient were assayed in a cell-free system, essentially according to Arnstein *et al.* (1964). Each tube contained 0·1 mg of ribosomes, 0·5 mg of pH 5 enzymes and, where indicated, 15 μg of polyU. The incorporations, performed at a final Mg^{2+} concentration of 5 mM, were corrected for no energy and ribosome blanks.

Ribosome fractions	*Specific activity* (pmoles of Phe/mg of ribosomes)	
	−*polyU*	+*polyU*
Polysomes	58	305
Monosomes	17	715

(for example, transferases and synthetases), tRNA and an energy source is critical for optimum peptide synthesis, and increased polysome purification often leads to decreased activity. Chain initiation and, probably, release, as opposed to chain extension, are usually severely reduced or absent in cell-free systems, but may be improved by supplementing the system with 'initiation' factors which are usually isolated from a salt wash of the ribonucleoprotein. In general, polyribosomes are more active than monoribosomes in endogenous protein synthesis but the converse is true when synthetic polynucleotides, for example, polyU, are added as exogenous messengers in the system (*Table 2*).

ISOLATION OF POLYSOMES FROM SPECIFIC CELLS AND TISSUES

The preparation of polysomes from two representative tissues is outlined below.

RABBIT RETICULOCYTES *(essentially according to Arnstein* et al., *1964)*

The rabbits are made anaemic by subcutaneous injection of neutralised 2·5% (w/v) phenylhydrazine (0·3 ml/kg body weight) once a day for four to six consecutive days. After one or two days' rest the animals are anaesthetised by intravenous injection of 1·0 ml of nembutal mixed with 1·0 ml of heparin, 5000 I.U./ml, and the blood is collected by cardiac puncture. The cells (approximately 20 ml for 100 ml of blood collected from a 2·5 kg rabbit), which have a reticulocytosis of about 70 to 90%, are sedimented by a low-speed centrifugation (10 min at 1200*g* or 2500 rev/min in an MSE Magnum centrifuge) and washed three times with cold buffered saline (0·154 M NaCl, 0·01 M phosphate buffer, pH 7·0) or TKM buffer (0·05 M tris-HCl buffer, pH 7·6, 0·025 M KCl, 0·005 M $MgCl_2$). The white cells are found in a layer on top of the reticulocytes and can be removed by aspiration. All subsequent operations are carried out at, or near, 0°C. The cells are lysed by osmotic shock, either with 1 vol of water when a concentrated lysate is required for incorporation of amino acids (Lamfrom and Knopf, 1964) or, when subcellular fractionation is required, with 4 vol of 5 mM $MgCl_2$ and, after gentle stirring in ice for 1 min, 1 vol of 1·5 M sucrose, 0·15 M KCl is added to the suspension to restore isotonicity. The lysate prepared by either method is centrifuged at 10 000*g* (12 000 rev/min in an MSE High-speed 18 centrifuge) for 10 min and the supernatant fluid is carefully decanted. Polysomes and ribosomes can now be isolated by differential centrifugation, acid precipitation or sucrose gradient fractionation. The yield is 2 to 3 mg of ribosomes per ml of packed cells.

Eighty per cent of the total ribonucleoprotein can be obtained by centrifuging at 100 000*g* (40 000 rev/min in an MSE 8 × 25 fixed-angle rotor) for 1 h. This first 'IX' pellet consists of over 60% polysomes, but is heavily contaminated with cytoplasmic proteins, notably haemoglobin. For complete sedimentation of all ribosomes and subunits centrifugation of the post-ribosomal supernatant fraction at 100 000*g* for a further 3 to 4 h is required (Bishop, 1966). Precipitation of a concentrated reticulocyte lysate by adjustment of the pH to 5·4 results in a pellet which can initiate peptide synthesis in a cell-free system for up to 10 min at 37°C (Fuhr *et al*., 1969), that is, almost to the same extent as the original lysate (for comparison of the activity of intact reticulocytes with lysates of various dilutions and fractionated polysomes see Lamfrom and Knopf, 1964; also Hunt *et al*., 1969). For sucrose-gradient fractionation, the sample (1 to 2 ml) is carefully layered onto a 20 ml 15 to 30% (w/v) linear sucrose gradient in RS buffer (0·01 M

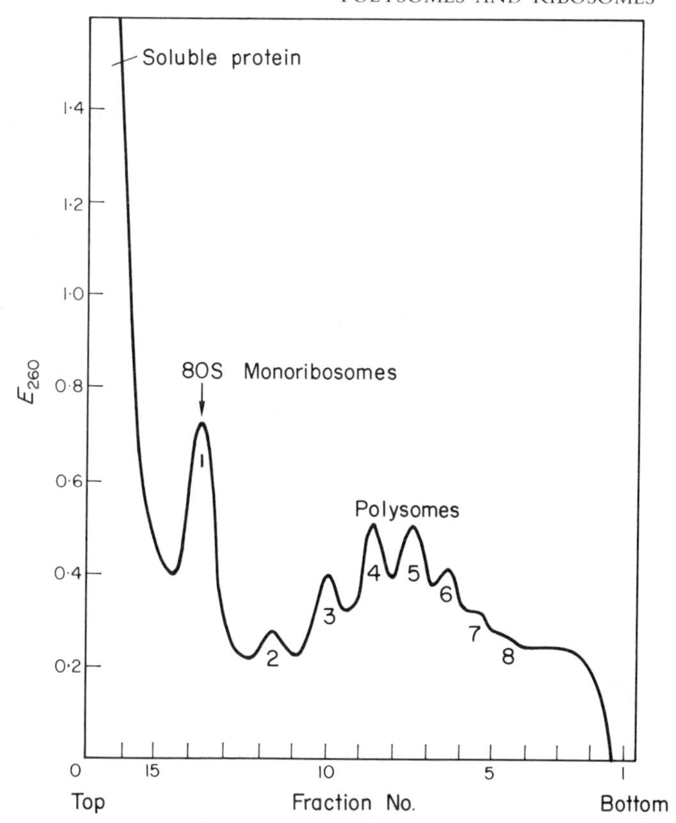

Figure 1. Sucrose-gradient sedimentation profile of rabbit-reticulocyte polysomes. Reticulocyte polysomes (0·3 mg) were suspended in 1 ml of RS buffer (10 mM KCl, 1·5 mM $MgCl_2$, 10 mM tris-HCl, pH 7·6) and layered onto an 18 ml 15 to 30% (w/v) linear sucrose gradient in RS buffer. The gradient was centrifuged for $2\frac{1}{2}$ h at 30 000 rev/min in an MSE 3 × 23 swing-out rotor at 4°C. The gradient was pumped out from the bottom, the extinction at 260 nm was monitored with a Uvicord absorptiometer and 1 ml fractions were collected. Resolution of monoribosomes and polysomes ranging from dimers to octamers was obtained

tris-HCl, pH 7·6, 0·01 M KCl, 0·0015 M $MgCl_2$) and centrifuged at 4°C for $2\frac{1}{2}$ h at 30 000 rev/min (100 000*g*) in an MSE 3 × 23 swing-out rotor (*Figure 1*). The concentration of sucrose, the buffer and the type of rotor employed can be varied, however, to suit different experiments.

RAT LIVER *(essentially according to Wettstein* et al., *1963)*

Rats weighing 200 to 250 g, usually starved to deplete glycogen stores, are killed by decapitation or ether anaesthesia, the livers are excised, blotted and quickly placed in ice-cold medium (0·25 M sucrose, 0·025 M KCl, 0·001 M $MgCl_2$, 0·05 M tris-HCl, pH 7·6). After weighing, more medium is added to give 2·5 ml per g of liver. The tissue is rapidly minced with scissors and homogenised in the cold in a Potter–Elvehjem glass vessel, with a Teflon pestle of about 0·01 in clearance, with 6 to 8 up-and-down strokes. The speed of the motor-driven pestle is regulated to avoid heating. The homogenate is centrifuged at 12 000*g* for 15 min (15 000 rev/min in an MSE High-speed 18 centrifuge). The post-mitochondrial supernatant fluid is filtered through gauze to remove excess fat, and sedimented at 100 000*g* for 1 h to obtain microsomes. The microsomes may be treated subsequently with detergent, in the presence of cell sap, to release polysomes and ribosomes which are pelleted by a further centrifugation. Alternatively, detergent is added directly to the post-mitochondrial supernatant fluid, which is centrifuged at 100 000*g* for 1 h or more, but the pellets obtained this way are heavily contaminated, especially with ferritin, and are normally washed before further use. In order to obtain a clean pellet which is relatively rich in polysomes it is preferable to layer 4 to 5 ml samples of post-mitochondrial supernatant fluid, made 1·3% (w/v) in DOC, directly over two layers of buffered sucrose, for example, 4 ml of 0·5 M sucrose layered over 3 ml of 2·0 M sucrose in an MSE 8 × 25 tube which is centrifuged at 100 000*g* for $2\frac{1}{2}$ h at 4°C. Under these conditions the pellet, which is colourless and transparent, is selectively enriched in polysomes since complete recovery of the ribosomes would require much longer centrifugation. The ferritin forms an intense yellow band below the interface of the two sucrose layers. The yield is about 0·9 mg of ribosomes per g of liver.

If separation of free and membrane-bound polysomes is required, the post-mitochondrial supernatant fluid, in the absence of detergent, is layered onto the 0·5 M/2·0 M sucrose solution and subjected to prolonged centrifugation at 100 000*g* in a fixed-angle rotor (Blobel and Potter, 1967; Loeb *et al.*, 1967). Free particles form a translucent pellet at the bottom of the centrifuge tube while the membrane-bound polysomes and ribosomes, which have a much lower density, accumulate at the interface and can be removed by aspiration. Starvation may affect the ratio of free to membrane-bound polysomes (Henshaw *et al.*, 1963), but it is usually accepted that in liver about 75% of the total RNA is found associated with membranes. It is essential to continue centrifugation long enough to allow complete sedimentation of free polysomes (for example, 20 h at 100 000*g*) or to employ a dilute post-mitochondrial supernatant fraction in order to minimise a concentration-dependent interaction of polysomes with non-sedimenting components (Lowe and Hallinan, 1970). The membrane-bound polysomes can be released from the membranes with detergent, in the presence of cell sap, and resedimented. Purified pellets or post-mitochondrial supernatant fluid (with or without detergent) can be further fractionated into ribosomes and polysomes of various sizes on sucrose gradients.

REFERENCES

ALLFREY, V. G. (1970). In *Aspects of Protein Biosynthesis*, (ed. Anfinsen, C. B.), vol. 1, p. 247, New York and London (Academic Press Inc.)

ARNSTEIN, H. R. V. (1961). *Biochem. J.*, **81**, 24P

ARNSTEIN, H. R. V. (1963). *Rep. Prog. Chem.*, **60**, 512

ARNSTEIN, H. R. V., COX, R. A., GOULD, H. and POTTER, H. (1965). *Biochem. J.*, **96**, 500

ARNSTEIN, H. R. V., COX, R. A. and HUNT, J. A. (1964). *Biochem. J.*, **92**, 648

ARORA, D. J. S. and DE LAMIRANDE, G. (1968). *Arch. Biochem. Biophys.*, **123**, 416

ASCIONE, R. and ARLINGHAUS, R. P. (1970). *Biochim. biophys. Acta*, **204**, 478

ATTARDI, G. and OJALA, D. (1971). *Nature New Biology, Lond.*, **229**, 133

BALIGA, B. S., PRONCZUK, A. W. and MUNRO, H. N. (1968). *J. molec. Biol.*, **34**, 199

BECKER, M. J. and RICH, A. (1966). *Nature, Lond.*, **212**, 142

BENEDETTI, E. L., BONT, W. S. and BLOEMENDAL, H. (1966). *Lab. Invest.*, **15**, 196

BENEDETTI, E. L., ZWEERS, A. and BLOEMENDAL, H. (1968). *Biochem. J.*, **108**, 765

BISHOP, J. O. (1966). *Biochim. biophys. Acta*, **119**, 130

BLOBEL, G. and POTTER, V. R. (1966). *Proc. natn. Acad. Sci. U.S.A.*, **55**, 1283

BLOBEL, G. and POTTER, V. R. (1967). *J. molec. Biol.*, **26**, 279

BLOEMENDAL, H., BONT, W. S. and BENEDETTI, E. L. (1964). *Biochim. biophys. Acta*, **87**, 177

BONANOU, S., COX, R. A., HIGGINSON, B. and KANAGALINGAM, K. (1968). *Biochem. J.*, **110**, 87

BREILLATT, J. and DICKMAN, S. R. (1969). *Biochim. biophys. Acta*, **195**, 531

BREGA, A. and VESCO, C. (1971). *Nature New Biology, Lond.*, **229**, 136

BREUER, E. N., FOSTER, L. B. and SELLS, B. H. (1969). *J. biol. Chem.*, **244**, 1389

BRITTEN, R. J. and ROBERTS, R. B. (1960). *Science, N.Y.*, **131**, 32

CAMPBELL, P. N. (1970). *FEBS Letters*, **7**, 1

CAMPBELL, P. N. and SARGENT, J. R. (1967). In *Techniques in Protein Synthesis* (ed. Campbell, P. N. and Sargent, J. R.), vol. 1, p. 299, New York and London (Academic Press Inc.)

CARTOUZOU, G., ATTALI, J. C. and LISSITZKY, S. (1968). *Eur. J. Biochem.*, **4**, 41

CHRISTMAN, J. K. and GOLDSTEIN, J. (1971). *Nature New Biology, Lond.*, **230**, 272

COHEN, B. B. (1968). *Biochem. J.*, **110**, 231

COLOMBO, B., VESCO, C. and BAGLIONI, C. (1968). *Proc. natn. Acad. Sci. U.S.A.*, **61**, 651

CRICK, F. H. C., BARNETT, L., BRENNER, S. and WATTS-TOBIN, R. J. (1961). *Nature, Lond.*, **192**, 1227

DAHLBERG, A. E., DINGMAN, C. W. and PEACOCK, A. C. (1969). *J. molec. Biol.*, **41**, 139

DAS, H. K. and GOLDSTEIN, A. (1968). *J. molec. Biol.*, **31**, 209

VON DER DECKEN, A. (1967). In *Techniques in Protein Synthesis* (ed. Campbell, P. N. and Sargent, J. R.), vol. 1, p. 65, New York and London (Academic Press Inc.)

DICKMAN, S. R. and BRUENGER, E. (1969). *Biochemistry*, **8**, 3295

DINTZIS, H. M. (1961). *Proc. natn. Acad. Sci. U.S.A.*, **47**, 247

EARL, D. C. N. and KORNER, A. (1965). *Biochem. J.*, **94**, 721

EARL, D. C. N. and MORGAN, H. E. (1968). *Arch. Biochem. Biophys.*, **128**, 460

EISENBERG, H. and FELSENFELD, G. (1967). *J. molec. Biol.*, **30**, 17

ELIASSON, E., BAUER, G. E. and HULTIN, T. (1967). *J. Cell Biol.*, **33**, 287

FALVEY, A. K. and STAEHELIN, T. (1970). *J. molec. Biol.*, **53**, 1
FIRTEL, R. A. and MONROY, A. (1970). *Devl. Biol.*, **21**, 87
FLORINI, J. R. and BREUER, C. B. (1966). *Biochemistry*, **5**, 1870
FORD, P. J. (1966). *Biochem. J.*, **101**, 369
FRIEDMAN, H., LU, P. and RICH, A. (1969). *Nature, Lond.*, **223**, 909
FUHR, J. E., LONDON, I. M. and GRAYZEL, A. I. (1969). *Proc. natn. Acad. Sci. U.S.A.*, **63**, 129
FULLER, W. and HODGSON, A. (1967). *Nature, Lond.*, **215**, 817
GASKILL, P. and KABAT, D. (1971). *Proc. natn. Acad. Sci. U.S.A.*, **68**, 72
GAYE, P. and DENAMUR, R. (1970). *Biochem. biophys. Res. Commun.*, **41**, 266
GERLINGER, P., LE MEUR, M. A., BECK, G. and EBEL, J. P. (1969). *Bull. Soc. Chim. Biol.*, **51**, 1157
GIERER, A. (1963). *J. molec. Biol.*, **6**, 148
GOLDBERG, B. and GREEN, H. (1967). *J. molec. Biol.*, **26**, 1
GOOD, N. E., WINGET, G. D., WINTER, W., CONNOLLY, T. N., ISAWA, S. and SINGH, R. M. M. (1966). *Biochemistry*, **5**, 467
GRAU, O. and FAVELUKES, G. (1968). *Arch. biochem. Biophys.*, **125**, 647
GROLLMAN, A. P. and HUANG, M. T. (1970). *Fed. Proc.*, **29**, 1624
HALLINAN, T., MURTY, C. N. and GRANT, J. H. (1968). *Arch. biochem. Biophys.*, **125**, 715
HAMADA, K., YANG, P., HEINTZ, R. and SCHWEET, R. (1968). *Arch. biochem. Biophys.*, **125**, 598
HARDESTY, B., HUTTON, J. J., ARLINGHAUS, R. and SCHWEET, R. (1963). *Proc. natn. Acad. Sci. U.S.A.*, **50**, 1078
HARRIS, H. (1970). In *Nucleus and Cytoplasm*, p. 47, Oxford (Clarendon Press)
HENSHAW, E. C., BOJARSKI, T. B. and HIATT, H. H. (1963). *J. molec. Biol.*, **7**, 122
HERRINGTON, M. D. and HAWTREY, A. O. (1971). *Biochem. J.*, **121**, 279
HEYWOOD, S. M., DOWBEN, R. and RICH, A. (1967). *Proc. natn. Acad. Sci. U.S.A.*, **57**, 1002
HEYWOOD, S. M. and NWAGWU, M. (1968). *Proc. natn. Acad. Sci. U.S.A.*, **60**, 229
HICKS, S. J., DRYSDALE, J. W. and MUNRO, H. N. (1969). *Science, N.Y.*, **184**, 584
HOLDER, J. W. and LINGREL, J. B. (1970). *Biochim. biophys. Acta*, **204**, 210
HORI, M. and RABINOWITZ, M. (1968). *Proc. natn. Acad. Sci. U.S.A.*, **59**, 1349
HUANG, A. S. and BALTIMORE, D. (1969). *J. molec. Biol.*, **47**, 275
HUEZ, G., BURNY, A., MARBAIX, G. and LEBLEU, B. (1967). *Biochim. biophys. Acta*, **145**, 629
HUNT, T., HUNTER, T. and MUNRO, A. (1969). *J. molec. Biol.*, **43**, 123
HUNTER, A. R. and KORNER, A. (1966). *Biochem. J.*, **100**, 73P
HUSTON, R. L., SCHRADER, L. E., HONOLD, G. R., BEECHER, G. R., COOPER, W. K. and SAUBERLICH, H. E. (1970). *Biochim. biophys. Acta*, **209**, 220
JACOB, F. and MONOD, J. (1961). *J. molec. Biol.*, **3**, 318
KAGEN, L. J. and LINDER, S. (1969). *Biochim. biophys. Acta*, **195**, 523
KAZAZIAN, H. H. JR. and FREEDMAN, M. L. (1968). *J. biol. Chem.*, **243**, 6446
KEDES, L. H., KUFF, E. L. and PETERSON, E. A. (1969). *Biochemistry*, **8**, 2923
KORNER, A. (1961). *Biochem. J.*, **81**, 168
KRAFT, N. and SHORTMAN, K. (1970). *Biochim. biophys. Acta*, **217**, 164
KUFF, E. L. and ROBERTS, N. E. (1967). *J. molec. Biol.*, **26**, 211
KÜNTZEL, H. and NOLL, H. (1967). *Nature, Lond.*, **215**, 1340
KUWANO, M., KWAN, C. N., APIRION, D. and SCHLESSINGER, D. (1969). *Proc. natn. Acad. Sci. U.S.A.*, **64**, 693
LABRIE, F. (1969). *Nature, Lond.*, **221**, 1217
LAGA, E. M., BALIGA, B. S. and MUNRO, H. N. (1970). *Biochim. biophys. Acta*, **213**, 391
LAMFROM, H. and KNOPF, P. M. (1964). *J. molec. Biol.*, **9**, 558
LAMFROM, H. and KNOPT, P. M. (1965). *J. molec. Biol.*, **11**, 589
LAWFORD, G. R. (1969). *Biochem. biophys. Res. Commun.*, **37**, 143
LAWFORD, G. R., LANGFORD, P. and SCHACHTER, H. (1966). *J. biol. Chem.*, **241**, 1835
LAWFORD, G. R., SADOWSKI, P. and SCHACHTER, H. (1967). *J. molec. Biol.*, **23**, 81
LA VIA, M. R., VATTER, A. E., HAMMOND, W. S. and NORTHRUP, P. V. (1967). *Proc. natn. Acad. Sci. U.S.A.*, **57**, 79

LEBLEU, B., MARBAIX, G., WERENNE, J., BURNY, A. and HUEZ, G. (1970). *Biochem. biophys. Res. Commun.*, **40**, 731

LING, V. and DIXON, G. H. (1970). *J. biol. Chem.*, **245**, 3035

LITTLEFIELD, J. W., KELLER, E. B., GROSS, J. and ZAMECNIK, P. C. (1955). *J. biol. Chem.*, **217**, 111

LOEB, J. N., HOWELL, R. R. and TOMKINS, G. M. (1967). *J. biol. Chem.*, **242**, 2069

LOWE, D. and HALLINAN, T. (1970). *Biochem. J.*, **117**, 64P

MANCHESTER, K. L. (1970). *Biochim. biophys. Acta*, **213**, 532

MANNER, G. and GOULD, B. S. (1965). *Nature, Lond.*, **205**, 670

MARKS, P. A., BURKA, E. R., CONCONI, F. M., PERL, W. and RIFKIND, R. A. (1965). *Proc. natn. Acad. Sci. U.S.A.*, **53**, 1437

MARTIN, R. G. and AMES, B. N. (1962). *Proc. natn. Acad. Sci. U.S.A.*, **48**, 2171

MARTIN, T. E. and WOOL, I. G. (1969). *J. molec. Biol.*, **43**, 151

MCCARTY, K. S., PARSONS, J. T., CARTER, W. A. and LASZLO, J. (1966). *J. biol. Chem.*, **241**, 5489

MEANS, R., HALL, P. F., NICOL, L. W., SAWYER, W. H. and BAKER, C. A. (1969). *Biochemistry*, **8**, 1488

MILLER, R. L. and SCHWEET, R. (1968). *Arch. Biochem. Biophys.*, **125**, 632

MOLDAVE, K. and SKOGERSON, L. (1967). In *Methods in Enzymology* (Ed. by Grossman, L. and Moldave, K.), vol. XII, p. 478, New York and London (Academic Press Inc.)

MUNRO, A. J., JACKSON, R. J. and KORNER, A. (1964). *Biochem. J.*, **92**, 289

MURTHY, M. R. V. and RAPPOPORT, D. A. (1965). *Biochim. biophys. Acta*, **95**, 132

NAIR, K. G., ZAK, R. and RABINOWITZ, M. (1966). *Biochemistry*, **5**, 2674

NEMER, M. and INFANTE, A. A. (1965). *Science, N.Y.*, **150**, 217

NOLL, H. (1969). In *Techniques in Protein Biosynthesis* (ed. by Campbell, P. N. and Sargent, J. R.), vol. 2, p. 101, New York and London (Academic Press Inc.)

NOLL, H., STAEHELIN, T. and WETTSTEIN, F. (1963). *Nature, Lond.*, **198**, 632

NORTHRUP, R. V., HAMMOND, W. S. and LA VIA, M. R. (1967). *Proc. natn. Acad. Sci. U.S.A.*, **57**, 273

O'BRIEN, T. W. and KALF, G. F. (1967). *J. biol. Chem.*, **242**, 2172

OLSNES, S. (1970*a*). *Biochim. biophys. Acta*, **213**, 149

OLSNES, S. (1970*b*). *Eur. J. Biochem.*, **15**, 464

OPPENHEIM, J., SCHEINBUCKS, J., BIAVA, C. and MARCUS, L. (1968). *Biochim. biophys. Acta*, **161**, 386

PALADE, G. E. and SIEKEVITZ, P. (1956). *J. biophys. biochem. Cytol.*, **2**, 171

PENMAN, S., SCHERRER, K., BECKER, Y. and DARNELL, J. E. (1963). *Proc. natn. Acad. Sci. U.S.A.*, **49**, 654

PERRY, R. P. (1967). *Progr. Nucleic Acid Res. molec. Biol.*, **6**, 219

PERRY, R. P. and D. E. KELLEY (1968). *J. molec. Biol.*, **35**, 37

PETERMANN, M. L. (1964). *The Physical and Chemical Properties of Ribosomes*, Amsterdam (Elsevier Publ. Corp.)

PETERSON, E. A. and KUFF, E. L. (1969). *Biochemistry*, **8**, 2916

PFUDERER, P., CAMMARANO, P., HOLLADAY, D. R. and NOVELLI, G. D. (1965). *Biochim. biophys. Acta*, **109**, 595

PIATIGORSKY, J. (1968). *Biochim. biophys. Acta*, **166**, 142

POGO, A. O., POGO, B. G. T., LITTAU, V. C., ALLFREY, V. G., MIRSKY, A. E. and HAMILTON, M. G. (1962). *Biochim. biophys. Acta*, **55**, 849

PRAGNELL, I. B. and ARNSTEIN, H. R. V. (1970). *FEBS Letters*, **9**, 331

PRICHARD, P. M., GILBERT, J. M., SHAFRITZ, D. A. and ANDERSON, W. F. (1970). *Nature, Lond.*, **226**, 511

PRIESTLEY, G. C., PRUYN, M. R. and MALT, R. A. (1969). *Biochim. biophys. Acta*, **190**, 154

REDMAN, C. M. (1969). *J. biol. Chem.*, **244**, 4308

RENDI, R. and HULTIN, T. (1960). *Expl Cell Res.*, **19**, 253

RICH, A., WARNER, J. and GOODMAN, H. M. (1963). *Cold Spring Harb. Symposia Quant. Biol.*, **28**, 269

RIFKIN, M. R., WOOD, D. D. and LUCK, D. J. L. (1967). *Proc. natn. Acad. Sci. U.S.A.*, **58**, 1025
RIFKIND, R. A., LUZZATTO, L. and MARKS, P. A. (1964). *Proc. natn. Acad. Sci. U.S.A.*, **52**, 1227
ROBERTSON, H. D., WEBSTER, R. E. and ZINDER, N. D. (1968). *J. biol. Chem.*, **243**, 82
ROTH, J. S. (1958). *J. biol. Chem.*, **23**, 1085
ROWLEY, P. T. and MORRIS, J. (1966). *Expl Cell Res.*, **45**, 494
SADOWSKI, P. D. and ALCOCK-HOWDEN, J. (1967). *J. Cell Biol.*, **37**, 163
SCHARFF, M. D. and UHR, J. W. (1965). *Science, N.Y.*, **148**, 646
SHELTON, E. and KUFF, E. L. (1966). *J. molec. Biol.*, **22**, 23
SLAYTER, H. S., WARNER, J. R., RICH, A. and HALL, C. E. (1963). *J. molec. Biol.*, **7**, 652
SPOHR, G., GRANBOULAN, N., MOREL, C. and SCHERRER, K. (1970). *Eur. J. Biochem.*, **17**, 296
STAEHELIN, T., VERNEY, E. and SIDRANSKY, H. (1967). *Biochim. biophys. Acta*, **145**, 105
STAEHELIN, T., WETTSTEIN, F. O., OURA, H. and NOLL, H. (1964). *Nature, Lond.*, **201**, 264
STAVY, L., FELDMAN, M. and ELSON, D. (1964). *Biochim. biophys. Acta*, **91**, 606
SUGANO, H., WATANABE, I. and OGATA, K. (1967). *J. Biochem., Tokyo*, **61**, 778
SWANSON, R. F. and DAWID, I. B. (1970). *Proc. natn. Acad. Sci. U.S.A.*, **66**, 117
TAKAGI, M. and OGATA, K. (1968). *Biochem. biophys. Res. Commun.*, **33**, 55
TAKAHASHI, Y., MASE, K. and SUGANO, H. (1966). *Biochim. biophys. Acta*, **119**, 627
TAKANAMI, M. (1960). *Biochim. biophys. Acta*, **39**, 318
TALAL, N. and KALTREIDER, H. B. (1968). *J. biol. Chem.*, **243**, 6504
TENG, C. S. and HAMILTON, T. H. (1967). *Biochem. J.*, **105**, 1091
VESCO, C. and COLOMBO, B. (1970). *J. molec. Biol.*, **47**, 335
VILLA-TREVINO, S., FARHER, E., STAEHELIN, T., WETTSTEIN, F. O. and NOLL, H. (1964). *J. biol. Chem.*, **239**, 3826
WARNER, J., RICH, A. and HALL, C. E. (1962). *Science, N.Y.*, **138**, 1399
WATTS, R. L. and MATHIAS, A. P. (1967). *Biochim. biophys. Acta*, **145**, 828
WETTSTEIN, F. O., STAEHELIN, T. and NOLL, H. (1963). *Nature, Lond.*, **197**, 430
WILLIAMSON, A. R. and ASKONAS, B. A. (1967). *J. molec. Biol.*, **23**, 201
WILLIAMSON, R., LANYON, G. and PAUL, J. (1969). *Nature, London*, **223**, 628
WILSON, S. H. and HOAGLAND, M. B. (1965). *Proc. natn. Acad. Sci. U.S.A.*, **54**, 600
ZAK, R., NAIR, K. G. and RABINOWITZ, M. (1966). *Nature, Lond.*, **210**, 169
ZIMMERMAN, E. F., HACKNEY, J., NELSON, P. and ARIAS, I. M. (1969). *Biochemistry*, **8**, 2636
ZOMZELY, C., ROBERTS, S., BROWN, D. M. and PROVOST, C. (1966). *J. molec. Biol.*, **19**, 455
ZYLBER, E. A. and PENMAN, S. (1970). *Biochim. biophys. Acta*, **204**, 230

ADDENDUM

Among the several papers which appeared while this article was in the press we would like to mention the following: on the basis of experiments with EDTA, puromycin and ribonuclease two classes of membrane-bound polysomes could be distinguished in HeLa cells (Rosbach and Penman, 1971); the optimal conditions for the isolation of polysomes from tissue culture cells were investigated (Gielkens *et al.*, 1971); free globin mRNA was released from active reticulocyte polysomes which were dissociated by puromycin at high ionic strength and fractionated on sucrose gradients (Blobel, 1971).

References

BLOBEL, G. (1971). *Proc. natn. Acad. Sci. U.S.A.*, **68**, 832
GIELKENS, A. L. J., BERNS, T. J. M. and BLOEMENDAL, H. (1971). *Eur. J. Biochem.*, **22**, 478
ROSBACH, M. and PENMAN, S. (1971). *J. molec. Biol.*, **59**, 227

10

SEPARATION OF POLYSOMES, RIBOSOMES AND RIBOSOMAL SUBUNITS IN ZONAL ROTORS

G. D. Birnie, Sylvia M. Fox and D. R. Harvey

The sedimentation coefficients of the major constituents of mammalian cells range from approximately 2*S* (proteins) to about 10^7S (nuclei) with considerable overlap between many of the different species of particles (*Table 1*). The isolation of a purified preparation of any one cell component by differential centrifugation is, if possible at all, both difficult and time-consuming. This method depends entirely on differences in the sedimentation rates of the particles and it is only successful when these differences are large. Even when this is so, preparation of relatively pure samples of most components of the mammalian cell requires repeated resuspension and resedimentation, a procedure inevitably accompanied by considerable loss of material and, in some cases, of structure and biological activity. Usually, to obtain preparations of reasonable purity, more sophisticated means of separating the desired particles from contaminants is required. Sedimentation of crude preparations through sucrose gradients (continuous and discontinuous) has often been used for this purpose since this technique enables small differences in sedimentation rate, or density, or both, to be exploited.

The isolation of a mixture of ribonucleoprotein particles and its fractionation is one case in point. The sedimentation coefficients of these particles form an almost continuous series from approximately 30*S* to 400*S* (*Table 1*) and they overlap to a significant extent at the upper end of the scale with the smaller fragments of microsomes and plasma membranes. A crude mixture of polysomes and ribosomes can be isolated by differential centrifugation and if this is followed by sedimentation through dense sucrose to remove contaminating membranous material a relatively pure preparation of the ribonucleoprotein particles can be obtained. The mixture of particles can then be resolved into its component polysomes and ribosomes by sedimentation through a continuous sucrose density gradient (see Bonanou-Tzedaki and Arnstein, this volume, *Chapter 9*). However, until quite recently this method

Table 1. APPROXIMATE SEDIMENTATION COEFFICIENTS OF MAJOR COMPONENTS OF MAMMALIAN CELLS

Cell component	*Sedimentation coefficient (S)*
Soluble proteins	2–25
RNA	4–50
DNA	up to 100
Ribosomal subunits	30–60
Ribosomes	70–80
Polysomes	100–400
Microsomes	$100–1{\cdot}5\times10^4$
Plasma membrane	$100–10^5$
Lysosomes	$1\times10^4–2\times10^4$
Mitochondria	$2\times10^4–7\times10^4$
Nuclei	$4\times10^6–10^7$

has been severely limited, mainly by the low capacity of the available high-speed swing-out rotors which were necessary for high-resolution density-gradient separations. The advent of the zonal ultracentrifuge (Anderson, 1966*a*) has greatly facilitated separations such as these. Not only do zonal rotors have capacities at least 100 times that of high-speed swing-out rotors of comparable separating capabilities but also their design is such that, when they are used under optimum conditions, their resolving powers are significantly greater than those of comparable swing-out rotors. Thus it is possible to separate with good resolution a variety of small subcellular particles from 5 to 10 g of tissue by rate-zonal centrifugation of tissue extracts in zonal centrifuge rotors.

This article describes conditions under which polysomes, ribosomes and ribosomal subunits can be isolated from extracts of mouse embryos and rabbit reticulocytes by rate-zonal centrifugation through sucrose density gradients in BIV, BXIV, BXV and BXXIX zonal rotors. Mouse embryos and rabbit reticulocytes were chosen as a source of material for a variety of reasons, not least because mouse-embryo and rabbit-reticulocyte cells are very readily lysed, so that extracts can be prepared with the minimum amount of damage to the subcellular structures. Another important reason was that the vast proportion of polysomes and ribosomes are not bound to membranes in these cells so that these particles can be isolated directly from homogenates without prior treatment with detergent.

EXPERIMENTAL PROCEDURES

PREPARATION OF MOUSE-EMBRYO EXTRACTS

The method of preparing mouse-embryo extracts was based on that described for rabbit-reticulocyte extracts by Arnstein *et al.* (1964). The embryos were removed from 13- or 14-days-pregnant mice (strain TO) and immediately chilled in ice-cold phosphate-buffered saline (PBS) (0·145 M NaCl, 10 mM sodium phosphate, pH 7·2). All further operations were carried out at 0 to 4°C. The embryos were minced by squeezing them through a fine stainless-steel gauze (22 × 22 mesh) in a glass syringe. To remove red blood cells, the minced material was washed three times by suspending it in 10 vol of PBS, allowing it to settle for 5 min and then removing the supernatant fluid. The minced tissue was finally collected by centrifugation at 1000*g* for 5 min. The pelleted material was suspended in 3 vol of 5 mM $MgCl_2$ and homogenised in an electrically-driven all-glass Potter homogeniser (20 to 25 strokes at approx. 1000 rev/min). The homogenate was mixed with one-fifth of its volume of 1·5 M sucrose, 0·15 M KCl, to give final concentrations of 0·25 M sucrose and 25 mM KCl. The post-nuclear extract was prepared by centrifuging the mixture at 800*g* for 15 min, and the post-mitochondrial extract by centrifuging the mixture at 10 000*g* for 10 min. The extracts thus contained the equivalent of approx. 0·2 g of tissue per ml.

PREPARATION OF RABBIT-RETICULOCYTE EXTRACTS

Extracts of rabbit reticulocytes were prepared as described by Arnstein *et al.* (1964). Blood from a rabbit which had been made anaemic by repeated injections of phenylhydrazine (Arnstein *et al.*, 1964) was chilled in an ice-bath and centrifuged at 1000*g* for 10 min. The pellet was washed three times by suspension in ice-cold PBS and centrifugation at 1000*g*. The cells were lysed by suspending the pellet in 4 vol of 5 mM $MgCl_2$ and, after 5 min, the lysate was mixed with one-fifth of its volume of 1·5 M sucrose, 0·15 M KCl. The post-mitochondrial extract was prepared by centrifuging the mixture at 10 000*g* for 10 min.

PREPARATION OF RABBIT-RETICULOCYTE POLYSOMES

A post-mitochondrial extract of rabbit reticulocytes was centrifuged at 105 000*g* for 1 h. The pellet was resuspended in 0.25 M sucrose buffered with 0·05 M tris-HCl, 25 mM KCl, 3 mM $MgCl_2$, pH 7·6. The suspension was layered on top of 0·5 M sucrose and 2·0 M sucrose (both containing the tris–KCl–$MgCl_2$ buffer) and centrifuged at 105 000*g* for 4 h. The transparent pellets were resuspended in sufficient 0·25 M sucrose containing the tris–KCl–$MgCl_2$ buffer to give a concentration of 6 to 7 mg/ml (Bonanou *et al.*, 1968).

CENTRIFUGATION IN ZONAL ROTORS

The separations were made in aluminium BXIV, BXV and BXXIX zonal rotors, using either an MSE Super-speed 40 ultracentrifuge, which had been adapted for zonal centrifugation, or an MSE Super-speed 50 ultracentrifuge. An aluminium BIV rotor in a Beckman model ZU ultracentrifuge was also used on some experiments. Before an experiment the rotor and all solutions were cooled to 4°C. The rotors were loaded with gradient, sample and overlay solutions, and unloaded at the end of the experiments, as described previously (Anderson, 1966*a,b*; Birnie, 1967, 1969). As the rotor contents were displaced during the unloading the optical density of the gradient solution was monitored continuously with a Perkin–Elmer model 137UV spectrophotometer fitted with a variable light-path flow-cell (Beckman-RIC, type BTF5) and a time-base chart recorder. Fractions of 20 or 40 ml were collected by hand and the concentration of sucrose in each fraction was measured with an Abbe refractometer.

INTERPRETATION OF DATA

Initial identification of particle zones was by calculation of sedimentation coefficients (kindly done by Mr J. M. Leach) using the Fortran IV programme developed at Oak Ridge National Laboratory and adapted to an 8K IBM 1130 computer (Leach,

1971). Sedimentation coefficients of zones in gradients linear with radius were also calculated manually, using the tables of McEwen (1967). For a variety of reasons discussed elsewhere (Birnie, 1971) significant errors can occur in sedimentation coefficients computed from separations in zonal rotors so that other criteria should be used to confirm the identity of zones of particles. In these experiments, electron microscopy and, for the ribosomal subunits, analytical centrifugation and measurement of the ratio of protein to RNA, were used.

FRACTIONATION OF TISSUE EXTRACTS

RESOLUTION IN ZONAL ROTORS

The significant increase in resolution which can be obtained in zonal rotors, as compared to that in swing-out rotors, is due to two features of these rotors. First, zonal rotors are loaded and unloaded while spinning, with the result that the initial sample zone and, later, the zones of separated particles are stabilised by a high gravitational field during the times when the likelihood of disturbance is greatest. Second, in zonal rotors the particles sediment in sector-shaped compartments so that the wall effects encountered in tubes are minimised. However, to obtain high-resolution rate-zonal separations in zonal rotors a number of factors require careful consideration. Probably the most important of these are the design of the gradient and the volume, density and concentration of the sample applied to it.

Density gradients which are linear with volume are popular since they are the simplest to prepare. In tubes, gradients linear with volume are also linear with radius over at least 80% of the tube. However, in zonal rotors, precisely that feature which is designed to eliminate the wall effects encountered in swing-out rotors introduces a complication in that gradients linear with volume are not linear with the radius of the rotor, but deviate significantly from it. Fortunately, computation of gradients linear with radius in a zonal rotor is simple. Their preparation is also simple, particularly with gradient engines which can be programmed to give complex gradients. Indeed, very good approximations can often be obtained using simple, home-made gradient engines of the types described by Anderson and Ruttenberg (1967), Birnie and Harvey (1968) and Hinton and Dobrota (1969).

One major attraction of using gradients linear with radius is the ease with which sedimentation coefficients can be calculated (McEwen, 1967). However, so far as resolution is concerned, these gradients are not always the most suitable. Except for very light sucrose gradients [for example, 5 to 10% (w/w)], resolution in linear sucrose gradients is inversely proportional to the radius. This is particularly unfortunate in the case of polysome fractionations, which ideally require the reverse. Better resolution can be obtained with very light sucrose gradients but only by sacrificing capacity. The problem can be overcome by using isokinetic (Steensgaard, 1970) or equivolumetric (Pollack and Price, 1971) gradients, but these are complex gradients which are difficult to compute and which require a sophisticated gradient engine to produce.

Since separations of polysomes, ribosomes and ribosomal subunits are made on a rate-zonal basis, the width of the initial sample zone has a critical effect on resolution. This width depends on the volume of the sample introduced into the rotor, its density and its concentration. The effect of sample volume on resolution is illustrated by *Figure 1*, in which the fractionation of a 24 ml sample in a BXIV rotor (total capacity 670 ml) is compared with that of the same volume of sample in a BXV rotor (total capacity 1670 ml). It is clear that the resolution obtained in the smaller rotor (*Figure 1a*), in which the volume of the sample is 6% of that of the gradient, is significantly poorer than that in the larger rotor (*Figure 1b*), in which the sample volume is only 2% of the gradient volume. The same improvement in resolution can be achieved with the BXIV rotor if the sample volume is reduced to 8 ml.

The density of the sample zone must not exceed that of the light end of the gradient since, if it does, the local instability so introduced is spontaneously eliminated by diffusion, which causes broadening of the sample zone to an extent dependent on the difference between the density of the sample and that of the top of the gradient. In addition, the density of the sample zone must not be much less than that of the top of the gradient nor much greater than that of the overlay solution. Density discontinuities are rapidly smoothed out in a zonal rotor by diffusion of sucrose from the gradient and this diffusion causes back-diffusion of the sample material, which again results in broadening of the sample zone. The occurrence of such sample-band diffusion is clearly seen in *Figure 1*, in which there are relatively large differences in density between the top of the gradient and the sample and between the sample and the overlay. Ideally, there should be no discontinuities in the density gradient in these regions.

Broadening of the sample zone also occurs as a result of con-

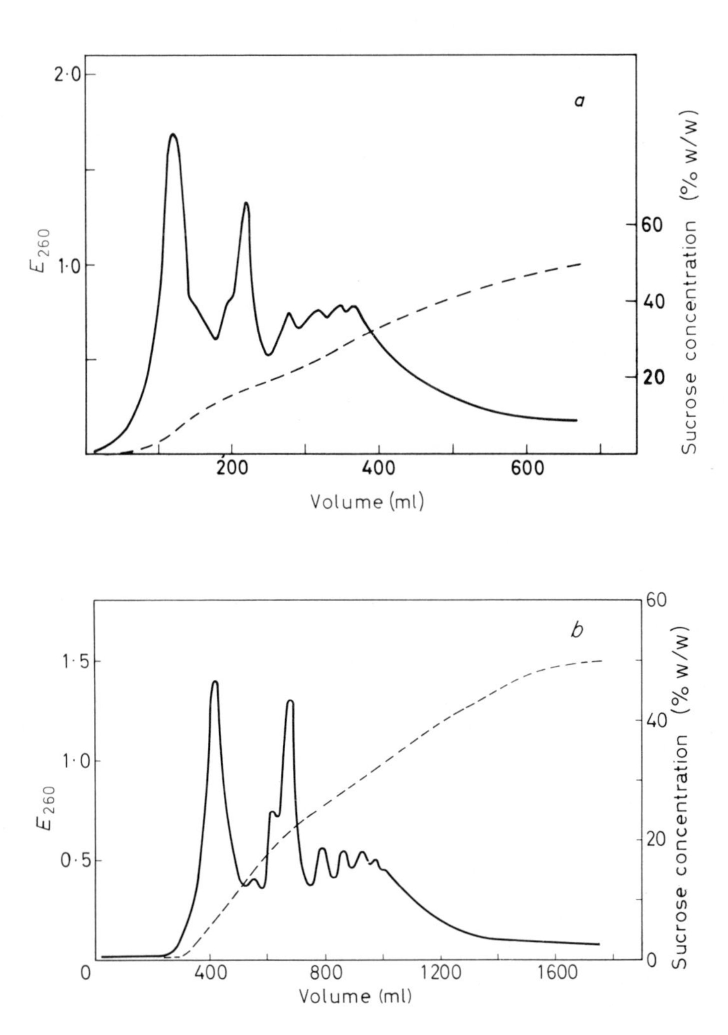

The BXIV rotor was loaded with 120 ml of overlay solution (10 mM tris-HCl, 5 mM $MgCl_2$, 10 mM KCl, pH 7·5), 24 ml of a post-mitochondrial extract of mouse embryos, a 400 ml 10–50% (w/w) sucrose gradient and a 115 ml cushion of 50% (w/w) sucrose. The BXV rotor was loaded with the same solutions as follows: overlay, 400 ml; embryo extract, 24 ml; gradient, 1200 ml; cushion, 40 ml. All sucrose solutions contained the tris–$MgCl_2$–KCl buffer.

Centrifugation conditions: (a) *30 000 rev/min for 140 min at 4°C (ω^2t = 8·3 × 10^{10} rad²/sec), RCF at initial sample position, g_{sample} = 34 500g, RCF at periphery of rotor, g_{max} = 67 100g;* (b) *21 000 rev/min for 4·5 h at 4°C (ω^2t = 7·8 × 10^{10} rad²/sec), g_{sample} = 24 000g, g_{max} = 43 800g. ———, E_{260} at 1 cm; — — —, concentration of sucrose in gradient*

Figure 1. Effect of sample volume on resolution in zonal rotors: separation of ribosomes and polysomes in (a) *a BXIV rotor and* (b) *a BXV rotor (from Birnie, 1971)*

vective diffusion if the concentration of particles in it exceeds a critical level. Hitherto it has proved impossible to calculate critical loads accurately since these depend on the difference between the molecular weight of the gradient solute and that of the particles (Meuwissen, 1971). For complex mixtures of particles the critical load must be determined empirically; in the case of mixtures of polysomes and ribosomes loadings of up to 10 mg per ml have been found to be subcritical.

BIV, BXIV, BXV and BXXIX rotors have been used interchangeably to fractionate extracts of mouse embryos and rabbit reticulocytes into their component subcellular particles. Although these rotors differ in capacity and there are large differences in shape between the BIV rotor on the one hand and the BXIV, BXV and BXXIX rotors on the other (Birnie, 1969), direct comparison has shown that there is no significant difference in the degree of resolution obtainable with each of them under comparable conditions. However, it is essential that these differences are borne in mind when attempts are made to duplicate the separation achieved in, say, a BIV rotor with, for example, a BXIV rotor, since the shape of the rotor influences the choice of both the gradient and the speed and time of centrifugation. For this reason it has been found useful to record the centrifugal force at the initial position of the sample zone in the rotor as well as that at the periphery of the rotor.

SEPARATION OF POLYSOMES AND RIBOSOMES

Figure 2 shows the distribution of u.v.-absorbing material in the gradient after centrifugation of a post-nuclear extract of mouse embryos in a BIV rotor at 40 000 rev/min for 1 h. The rotor contained a 1200 ml sucrose density gradient varying linearly with volume from 20 to 50% (w/w). The first peak (A), the 'soluble' fraction, contains soluble proteins and other subcellular components and particles of sedimentation coefficients up to about 20*S*, while the second peak (B) consists of the free 70*S* ribosomes. In the remainder of the gradient there has been partial separation of the other subcellular constituents of the post-nuclear fraction. In region C, 5 peaks can be clearly distinguished, corresponding to polysomes containing from 2 to 6 ribosomes. The polysome peaks merge with the regions containing membraneous material (D and E), while mitochondria are concentrated at the bottom of the gradient, in the fastest-sedimenting zone (E) which mainly consists of large pieces of cell debris. In fact, the major part of this

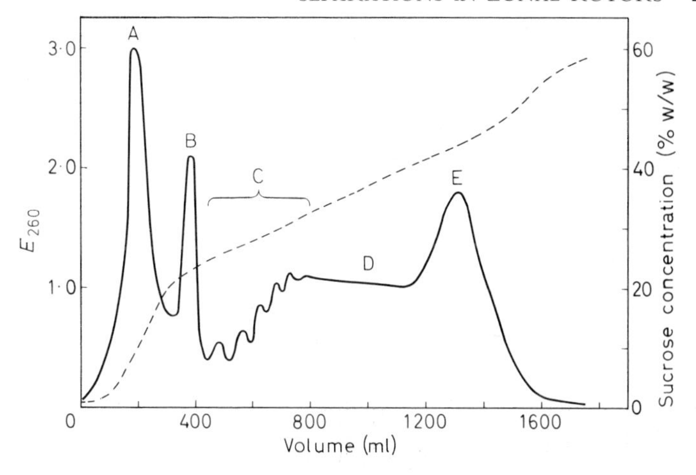

The rotor was loaded with 200 ml of overlay solution (tris–$MgCl_2$–KCl buffer as Figure 1*), 35 ml of embryo extract, a 1200 ml 20–50% (w/w) sucrose gradient and a 285 ml cushion of 60% (w/w) sucrose; the sucrose solutions contained tris–$MgCl_2$–KCl buffer. Centrifugation conditions: 40 000 rev/min for 60 min at 4°C ($\omega^2 t$ = 6·3 × 10^{10} rad²/sec), RCF at initial sample position, g_{sample} = 46 000g, RCF at periphery of rotor, g_{max} = 90 900g. ——— and — — — as* Figure 1

Figure 2. Fractionation of a post-nuclear extract of mouse embryos in a BIV zonal rotor.

debris, the mitochondria and the membranes have already reached their isopycnic points and these contaminants are in the process of being overtaken by the still-sedimenting zones of polysomes.

A much clearer picture emerges if the fast-sedimenting material is first removed from the mouse-embryo extracts by a brief preliminary centrifugation (10 000*g* for 10 min). This is shown by *Figure 3*, which depicts the distribution of material after centrifugation of a post-mitochondrial extract of mouse embryos in a BXXIX rotor at 25 000 rev/min for 3·3 h. In this experiment a 1200 ml sucrose gradient, which varied linearly with radius from 10 to 50% (w/w), was used. Debris, mitochondria and much of the membraneous material (*Figure 2*, D and E) is absent in *Figure 3*, while the soluble fraction (A), ribosomes (B) and polysomes (C) are well separated from each other. Five polysome peaks are clearly distinguishable and a sixth peak can just be discerned sedimenting just ahead of the fastest of these. The most rapidly-sedimenting material in this gradient consists of a mixture of larger polysomes and residual small fragments of membrane. In addition, two shoulders are clearly seen on the ribosome peak (B); that on the trailing edge is due to the 60*S* native ribosomal subunit while the one on the leading edge probably represents 80*S* monosomes.

Post-mitochondrial extracts of rabbit reticulocytes have also

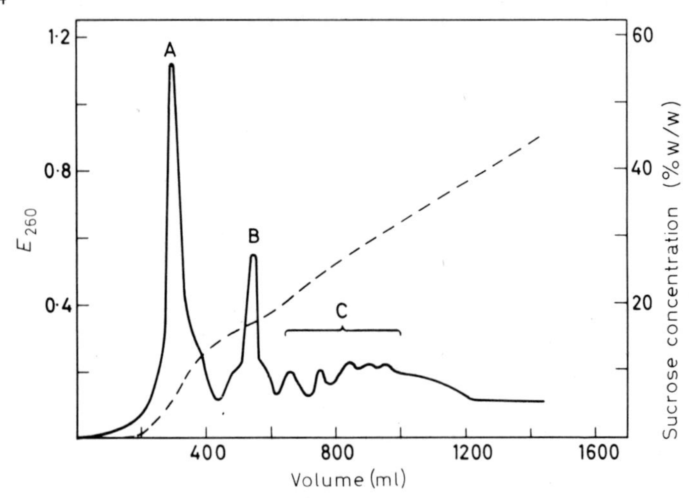

The rotor was loaded with 300 ml of overlay solution (tris–$MgCl_2$–KCl as Figure 1*), 30 ml of embryo extract and the first 1100 ml of a 1200 ml 20–50% (w/w) sucrose gradient containing tris–$MgCl_2$–KCl buffer. Centrifugation conditions: 25 000 rev/min for 200 min at 4°C (ω^2t = 8·2 × 10^{10} rad²/sec), g_{sample} = 30 300g, g_{max} = 58 000g. ——— and — — — as* Figure 1

Figure 3. Fractionation of a post-mitochondrial extract of mouse embryos in a BXXIX zonal rotor.

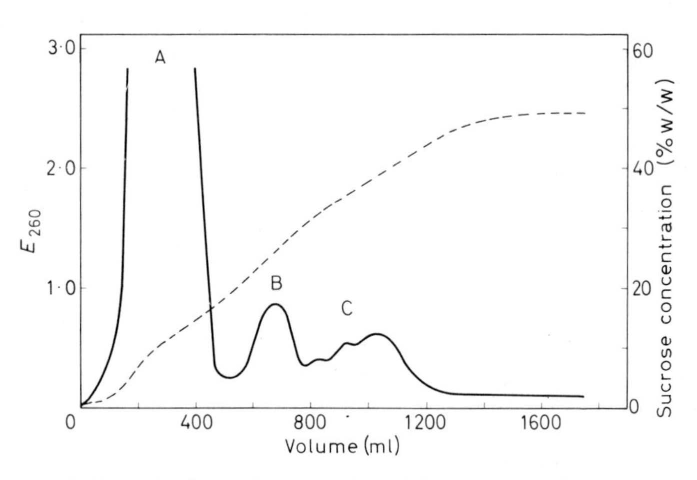

The rotor was loaded with 200 ml of overlay solution (tris–$MgCl_2$–KCl buffer as Figure 1*), 40 ml of reticulocyte extract, a 1200 ml 10–50% (w/w) sucrose gradient and a 280 ml cushion of 50% (w/w) sucrose; the sucrose solutions contained tris–$MgCl_2$–KCl buffer. Centrifugation conditions: 40 000 rev/min for 90 min at 4°C (ω^2t = 9·5 × 10^{10} rad²/sec), g_{sample} = 46 000g, g_{max} = 90 900g. ——— and — — — as* Figure 1

Figure 4. Fraction of a post-mitochondrial extract of rabbit reticulocytes in a BIV zonal rotor

been fractionated on sucrose density gradients under similar conditions in the BIV rotor. As shown in *Figure 4*, these extracts are separated into a broad 'soluble' peak (A), ribosomes (B) and at least three peaks of polysomes (C). It is also clear from *Figure 4* that the rabbit-reticulocyte extracts contain a much higher proportion of soluble and slowly-sedimenting material than do post-mitochondrial extracts of mouse embryos prepared by a similar method.

Mouse-embryo post-mitochondrial extracts are quite stable to storage at −20°C for considerable periods of time. As judged by the relative amounts of material in the polysome and ribosome regions, there is no significant breakdown of the polysomes during storage at −20°C for 30 days (*Figure 5*). However, *Figure 5* does show that, after they have been kept at −20°C for a considerable period, post-mitochondrial extracts contain a small amount of

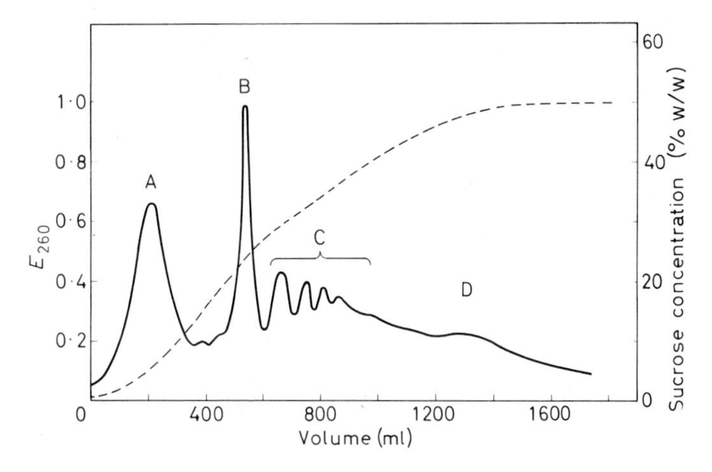

The rotor was loaded with 200 ml of overlay solution (tris–$MgCl_2$–KCl buffer as Figure 1*), 25 ml of embryo extract, a 1200 ml 10–50% (w/w) sucrose gradient and a 295 ml cushion of 50% (w/w) sucrose; the sucrose solutions contained tris–$MgCl_2$–KCl buffer. Centrifugation conditions: as* Figure 4 ——— *and* — — — *as* Figure 1

Figure 5. Fractionation of a post-mitochondrial extract of mouse embryos in a BIV zonal rotor after storage at −20°C for 30 days

fast-sedimenting material which corresponds to the membraneous material in peaks D and E shown in *Figure 2*. This material is probably small fragments of cell membrane which have aggregated during storage.

In most mammalian cells the majority of polysomes are attached to membranes which must be solubilised by treatment with a suitable detergent before free polysomes can be isolated (see

Bonanou-Tzedaki and Arnstein, this volume, *Chapter 9*). Treatment of microsomal pellets with detergents such as sodium deoxycholate (DOC), a method frequently used in the isolation of ribosomes from tissues such as rat liver (Littlefield *et al.*, 1955) often yields only monomers and dimers as a result of the release of ribonuclease from the membranes (Blobel and Potter, 1966). However, liver-cell sap contains a potent inhibitor of ribonuclease so that the breakdown of polysomes can be prevented by treating liver post-mitochondrial extracts directly with DOC (Blobel and Potter, 1966), or other detergents. Norman (1971) has resolved ribonucleoprotein particles from a rat-liver homogenate, which has been treated with Triton X-100, into two zones of ribosomal subunits and no fewer than 11 zones of polysomes.

Since most of the polysomes in embryonic cells are not membrane-associated, it was of interest to determine the effect of detergent on such polysomes. As shown by *Figure 6*, addition of

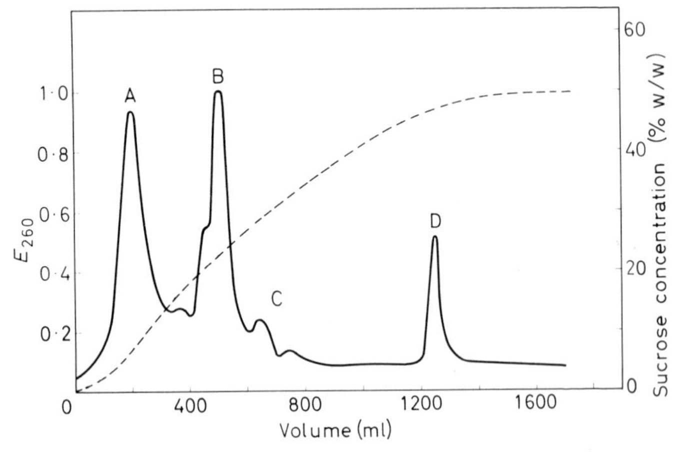

The rotor was loaded with 200 ml of overlay solution (tris–$MgCl_2$–KCl buffer as Figure 1*), 20 ml of embryo extract containing 1% (w/v) sodium deoxycholate, a 1200 ml 10–50% (w/w) sucrose gradient and a 300 ml cushion of 50% (w/w) sucrose; the sucrose solutions contained tris–$MgCl_2$–KCl buffer. Centrifugation conditions: as* Figure 4. ——— *and* — — — *as* Figure 1

Figure 6. Fractionation of a deoxycholate-treated post-mitochondrial extract of mouse embryos in a BIV zonal rotor

1% (w/v) DOC to a post-mitochondrial extract of mouse embryos causes extensive degradation of the polysomes since only monomers (B), together with small amounts of dimers and trimers (C) can be isolated after this treatment. Peak D in the gradient is magnesium deoxycholate, which is sparingly soluble in the cold. This result suggests that either mouse-embryo extracts are deficient in the inhibitor of ribonuclease or DOC inactivates it.

ISOLATION OF NATIVE AND DERIVED RIBOSOMAL SUBUNITS

Mouse-embryo cells contain only small quantities of 45*S* and 60*S* native ribosomal subunits. On occasion, fractionation of post-mitochondrial extracts reveals zones of these subunits as small peaks appearing between the soluble and ribosomal peaks (see *Figure 1b*). They can readily be separated from both the soluble material and the ribosomes by more prolonged centrifugation through a shallow sucrose density gradient (Kedes *et al.*, 1966). *Figure* 7 shows the distribution of material from a mouse-embryo

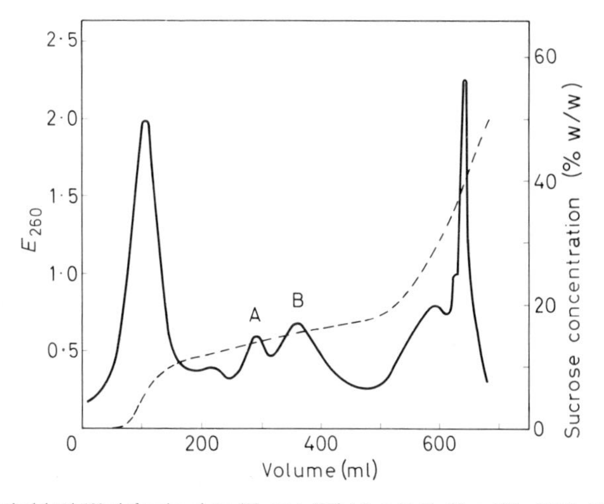

The rotor was loaded with 100 ml of overlay solution (10 mM tris-HCl, 1·5 mM $MgCl_2$, 10 mM KCl, pH 7·5), 20 ml of embryo extract, a 430 ml 10–20% (w/w) sucrose gradient and a 110 ml cushion of 50% (w/w) sucrose; the sucrose solutions contained tris–$MgCl_2$–KCl buffer. Centrifugation conditions: 30 000 rev/min for 4 h at 4°C ($\omega^2 t$ = 14·2 × 10^{10} rad²/sec), g_{sample} = 34 500g, g_{max} = 67 100g. ——— and — — — as Figure 1

Figure 7. Separation of native ribosomal subunits from a post-mitochondrial extract of mouse embryos in a BXIV zonal rotor

extract after centrifugation in a BXIV rotor at 30 000 rev/min for 4 h. The rotor contained a 430 ml sucrose density gradient which varied linearly with volume from 10 to 20% (w/w) and a 110 ml cushion of 50% (w/w) sucrose. Under these conditions, ribosomes and polysomes sediment to the bottom of the gradient while the two zones of ribosomal subunits (A, 45*S*; B, 60*S*) are separated in the shallow part of the gradient.

Rate-zonal centrifugation has also been found to be a rapid and simple method of preparing large amounts of 30*S* and 50*S* subunits from disrupted ribosomes (Kedes *et al.*, 1966; Klucis and Gould, 1966; Eikenberry *et al.*, 1970). The separation of the 30*S* (A) and

50*S* (B) subunits from 125 mg of reticulocyte ribosomes by centrifugation through a 15 to 30% (w/w) sucrose gradient at 40 000 rev/min for 4 h in a BIV rotor is shown in *Figure 8*. The gradient solutions contained 0·5 M KCl so that, when the suspension of reticulocyte polysomes was loaded on top of the gradient, the ribosomes were disrupted *in situ* into their constituent subunits by the high salt concentration. It is clear from *Figure 8* that, compared

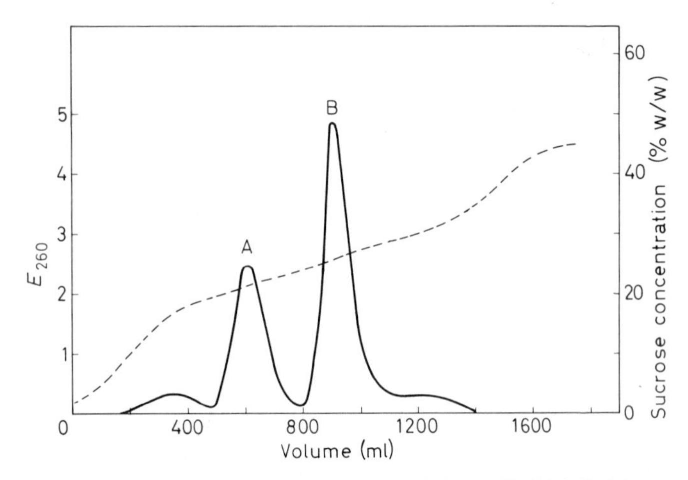

The rotor contained 200 ml of overlay solution (50 mM tris-HCl, 3 mM $MgCl_2$, 25 mM KCl, pH 7·6), 20 ml of a suspension of polysomes (125 mg), a 1200 ml 15–30% (w/w) sucrose gradient and a 300 ml cushion of 45% (w/w) sucrose; the sucrose solutions contained 0·5 M KCl and 10 mM tris-HCl, pH 7·6. Centrifugation conditions: 40 000 rev/min for 4 h at 4°C ($\omega^2 t$ = 25·2 × 10^{10} rad²/sec), g_{sample} = 46 000g, g_{max} = 90 900g. ——— *and* — — — *as* Figure 1

The polysomes preparation was kindly supplied by Miss Sophia Bonanou and Dr R. A. Cox, National Institute for Medical Research, London, NW7

Figure 8. Separation of derived ribosomal subunits from rabbit-reticulocyte polysomes in a BIV zonal rotor

to high-speed swing-out rotors, large-capacity zonal rotors not only facilitate the isolation of much larger quantities of derived subunits but also the resolution achieved is considerably better. In addition, this superior resolution should allow significant amounts of important minor constituents of ribosomes to be isolated (see, for example, Huez *et al.*, 1967).

CONCLUSIONS

The conditions under which these experiments were done were not ideal. The density gradients used throughout were simple ones linear with volume or with radius. It is now clear from some recent work that significantly better resolution can be achieved by the use of isokinetic (Steensgaard, 1970) or equivolumetric (Pollack and Price, 1971) gradients. In addition, the initial width of the sample zone was not as closely controlled as it might have been. In most experiments the difference in density between the top of the gradient and the sample, or between the sample and the overlay, was too great so that significant broadening of the sample zone took place. Nevertheless, the separations of polysomes, ribosomes and ribosomal subunits described here demonstrate the usefulness of the B-series of zonal rotors for large-scale rate-zonal fractionation of tissue extracts and the isolation of subcellular components ranging in sedimentation coefficient from 30*S* to approximately 400*S*. The resolution of the zones of particles compares very favourably with that achieved using preparative swing-out rotors (see, for example, Wettstein *et al.*, 1963; Arnstein *et al.*, 1965; Blobel and Potter, 1966; Drysdale and Munro, 1967).

The success of these fractionations of reticulocyte and mouse-embryo extracts is due in no small measure to the ease with which both the reticulocytes and the embryo cells can be lysed. Gentle homogenisation of mouse-embryo tissue breaks more than 95% of the cells so that mouse-embryo homogenates containing relatively undamaged subcellular components can be obtained readily and reproducibly. Microscopic examination of the stained homogenates shows that the nuclei are morphologically intact and free of cytoplasmic tags. The situation is very different for many other tissues, particularly tumour tissue, the cells of which are more refractory to the gentler methods of homogenisation. In these cases, reproducible fractionation patterns are difficult to obtain since the harsh methods required to lyse the cells damage subcellular components to a considerable and variable extent (Negroni, Harvey and Birnie, unpublished). Indeed, work with such tissues provides compelling evidence for the need to develop a reproducible method for breaking cells which leaves the cell organelles intact.

REFERENCES

ANDERSON, N. G. (1966*a*). Ed., *Natn. Cancer Inst. Monog.*, **21**

ANDERSON, N. G. (1966*b*). *Science, N.Y.*, **154**, 103

ANDERSON, N. G. and RUTTENBERG, E. (1967). *Analyt. Biochem.*, **21**, 259

ARNSTEIN, H. R. V., COX, R. A., GOULD, H. and POTTER, H. (1965). *Biochem. J.*, **96**, 500

ARNSTEIN, H. R. V., COX, R. A. and HUNT, J. A. (1964). *Biochem. J.*, **92**, 648

BIRNIE, G. D. (1967). *Lab. Practice*, **16**, 705

BIRNIE, G. D. (1969). *Lab. Equip. Digest*, **7**, 59

BIRNIE, G. D. (1971). *Vierteljahrssch. Naturforsch. Geselt. Zürich*, **116**, 355

BIRNIE, G. D. and HARVEY, D. R. (1968). *Analyt. Biochem.*, **22**, 171

BLOBEL, G. and POTTER, V. R. (1966). *Proc. natn. Acad. Sci., U.S.A.*, **55**, 1283

BONANOU, S., COX, R. A., HIGGINSON, B. and KANAGALINGAM, K. (1968). *Biochem. J.*, **110**, 87

DRYSDALE, J. W. and MUNRO, H. N. (1967). *Biochim. Biophys. Acta*, **138**, 616

EIKENBERRY, E. F., BICKLE, T. A., TRAUT, R. R. and PRICE, C. A. (1970). *Eur. J. Biochem.*, **12**, 113

HINTON, R. H. and DOBROTA, M. (1969). *Analyt. Biochem.*, **30**, 99

HUEZ, G., BURNY, A., MARBAIX, G. and SCHRAM, E. (1967). *Eur. J. Biochem.*, **1**, 179

KEDES, L. H., KOEGEL, R. J. and KUFF, E. L. (1966). *J. molec. Biol.*, **22**, 359

KLUCIS, E. S. and GOULD, H. J. (1966). *Science, N.Y.*, **152**, 378

LEACH, J. M. (1971). In *Separations with Zonal Rotors* (ed. Reid, E.), p. Z-4.1, Guildford (Wolfson Bioanalytical Centre, University of Surrey)

LITTLEFIELD, J. W., KELLER, E. B., GROSS, J. and ZAMECNIK, P. C. (1955). *J. biol. Chem.*, **217**, 111

MCEWEN, C. R. (1967). *Analyt. Biochem.*, **20**, 114

MEUWISSEN, J. A. T. P. (1971). In *Separations with Zonal Rotors* (ed. Reid, E.), p. B-2.1, Guildford (Wolfson Bioanalytical Centre, University of Surrey)

NORMAN, M. (1971). In *Separations with Zonal Rotors* (ed. Reid, E.), p. P-6.1, Guildford (Wolfson Bioanalytical Centre, University of Surrey)

POLLACK, M. S. and PRICE, C. A. (1971). *Analyt. Biochem.*, **42**, 38

STEENSGAARD, J. (1970). *Eur. J. Biochem.*, **16**, 66

WETTSTEIN, F. O., STAEHELIN, T. and NOLL, H. (1963). *Nature, Lond.*, **197**, 430

11

ISOLATION AND FRACTIONATION OF RNA

J. H. Parish

Most RNA occurs as a component of ribonucleoprotein (in, for example, ribosomes, polysomes and RNA viruses); in eukaryotic cells, cellular ribonucleoprotein is localised in various subcellular fractions and organelles. In general, RNA of functionally-distinct classes occurs in these fractions and consequently many studies on eukaryotic RNA require a preliminary subcellular fractionation to yield, for example, nuclei, nucleoli, microsomes (vesicles derived from endoplasmic reticulum), mitochondria and chloroplasts. There is an important exception to these generalisations; transfer RNA (tRNA) exists partly as a component of polysomes (as polypeptidyl-tRNA and aminoacyl-tRNA) and partly as free cytoplasmic tRNA.

There are two general methods available for the isolation and fractionation of RNA. Either the whole tissue can be broken up in a medium in which all ribonucleoprotein is disaggregated and total cell RNA can be subsequently fractionated, or the subcellular ribonucleoprotein or the organelles can be purified first and the RNA obtained by deproteinisation of these. The disadvantages of the latter approach are largely derived from the difficulties encountered in subcellular fractionation in general with the added problem of ensuring that the fractionation procedures used are ones in which ribonucleases (released during disruption of the cells) do not have a chance to degrade the RNA. The problems encountered in the former approach arise partly from the added difficulties in separating RNA from other macromolecules (such as DNA or polysaccharide) which are largely eliminated during subcellular fractionation, and partly from the increased complexity of the RNA mixture which has to be analysed. If RNA is to be prepared on a small scale, or if RNA from a particular organelle is required, it is usually preferable to start with some sort of subcellular fractionation. On the other hand, if large quantities of one of the major fractions of RNA (such as cytoplasmic rRNA or tRNA) are required, there is a lot to be said for deproteinising the whole tissue or cells. For example, it is within the capacity of the average laboratory to isolate rRNA from several hundred grammes of liver or *Esch. coli* paste while preparing ribosomes on this scale is a fairly formidable task.

Fractionation of RNA is based on differences either in the molecular weight or in the nucleotide composition of the components of the mixture. In practice it is impossible to dissociate these two properties satisfactorily. Thus, differences in nucleotide composition can be exploited in the ion-exchange chromatography of RNA (on the basis that the ionic properties of the nucleic acids contain contributions from the charge-distributions on the bases). However, the binding of RNA to the ion-exchange resin is due primarily to the phosphate groups. Consequently, the higher the molecular weight of the RNA, the more extensive will be the binding of its phosphate residues and the harder it will be to elute it. Similarly, separations on the basis of the sedimentation (or other hydrodynamic) properties of nucleic acids are partly dependent on the partial specific volume and conformational parameters of the molecules, the values of which depend to some extent on nucleotide composition.

In general, gel electrophoresis is the most versatile fractionation procedure, in that it will fractionate molecules varying in molecular weight from 1000 or so (small oligomers) to several million. The technique can be scaled up only with considerable difficulty. For the large-scale fractionation of large RNA molecules, zonal centrifugation and ion-exchange chromatography are used extensively. Both methods have severe practical limitations, as will be discussed. Low molecular-weight RNA can be fractionated on a preparative scale by ion-exchange or gel-exclusion chromatography.

ISOLATION OF RNA

Whether RNA is to be isolated from whole cells (or tissue) or from ribonucleoprotein particles or subcellular organelles, the essential requirements are the same. The procedure must be one in which endogenous ribonucleases are not released in an active form and in which protein, DNA and polysaccharide are completely removed. Other contaminants are not worth specific consideration since low molecular-weight tissue components, lipids and the like, have solubility properties so different from those of RNA that their removal is more or less automatic during any isolation procedure. Another point that must not be neglected is that there is no point taking great precautions to inactivate endogenous nucleases if exogenous nucleases are allowed to contaminate the preparation. In our experience, the alleged contamination of laboratory reagents with ribonuclease is much overrated; in any case, there is no need for idle (and gloomy) speculation about this since ribonuclease must be about the easiest enzyme to assay in small amounts (using as substrate highly-labelled RNA). The main danger comes from fingers which must be strictly kept out of all reagents that are to be used in the preparation of RNA.

There are numerous recipes for the isolation of RNA. Here I have concentrated on the phenol procedures as they exemplify the principles underlying the isolation of RNA and can be applied to a wide variety of different preparations. Equally, I have chosen to discuss the isolation of RNA from whole tissue (or cells) first, and in greatest detail, as the problems encountered in isolating nucleic acids are greatest in these cases.

ISOLATION OF RNA FROM WHOLE CELLS OR TISSUE

The procedures described here are based on the methods of the late K. S. Kirby (Parish and Kirby, 1966, 1967; Kirby *et al.*, 1967). They all rely on the fact that when the cells are disrupted in the presence of a two-phase mixture of aqueous salt solution and phenol, protein is denatured by phenol which has partitioned into the aqueous phase and the denatured protein is either insoluble or dissolves in the phenol layer. Nucleic acids are precipitated from the aqueous phase with an organic solvent such as ethanol. The following is a recipe for the isolation of total nucleic acid from *Esch. coli*, together with a discussion of the roles of the reagents that are employed.

Two solutions are prepared. One is a mixture of 1 kg of phenol (either 'detached crystal' grade or redistilled 'technical' grade) 140 ml of *m*-cresol, 110 ml of water and 0·2 g of 8-hydroxyquinoline. If the cresol looks pink or red it is worth distilling it before using it in this mixture. There is no point in doing this unless the distillation is done fairly slowly and the first tenth or so of the distillate is discarded, otherwise it will simply go red again (the red colour is due to auto-oxidation of *o*-cresol which has a lower boiling point than *m*-cresol). In the following, this mixture (which liquifies on standing overnight) is referred to as 'phenol mixture'. The cresol is present mainly as an anti-freeze but it also improves the deproteinisation; the 8-hydroxyquinoline is present partly to inhibit the oxidation of the phenol and cresol, and also to remove Mg^{2+} ions (which contribute to the stability of nucleoprotein complexes) as a phenol-soluble chelate. The second solution is prepared by dissolving 6 g of sodium 4-aminosalicylate (PAS), 1 g of NaCl and 1 g of tri-isopropylnaphthalene sulphonate (TIPNS) in 100 ml of water. TIPNS is salted out by the NaCl so the solution should also contain a solvent to solubilise the TIPNS (3 ml of phenol mixture is suitable). TIPNS is a powerful inhibitor of nucleases, the aminosalicylate is a chelating agent effective for deproteinisation and the NaCl is present to maintain the ionic strength and so prevent denaturation of the nucleic acids. This solution is referred to as 'PAS-TIPNS'. Unlike phenol mixture, which is stable indefinitely, PAS-TIPNS should be prepared freshly before use.

The *Esch. coli* cells are harvested and suspended in 15 vol of PAS-TIPNS. An equal amount of phenol mixture is added and the mixture is shaken or stirred at room temperature for about 30 min. The mixture is separated into two phases by centrifugation. It is advantageous to operate the centrifuge at a temperature somewhat lower than that at which the deproteinisation was performed as the emulsion breaks more easily if the mutual solubilities of phenol and water are reduced; 5°C is a suitable temperature. A speed of around 12 000 rev/min (20 000*g*) in a medium speed centrifuge (such as the MSE High-speed 18) for 20 min is sufficient. After centrifugation the upper (aqueous) phase is decanted. The DNA in the solution makes this phase very viscous and hence difficult to pipette. The deproteinisation is repeated by dissolving NaCl (2 g per 100 ml) in the aqueous phase and extracting it with half its volume of phenol mixture. (If an equal volume of phenol mixture is used, the upper phase contracts extensively due to re-partitioning of phenol and water.) The phases are separated by centrifugation as before. The upper phase is removed with great care; it is essential that none of the phenol layer or the insoluble 'interface' contaminates the preparation, since protein (including ribonuclease) remaining in the supernatant fluid at this stage will not be removed during subsequent processing. The preparation at this stage consists of a viscous, slightly cloudy, liquid containing DNA, RNA and polysaccharide. Methods of separating these components are discussed in the following Section.

This procedure can be adapted to the deproteinisation of most kinds of cells and tissues. With certain bacteria (notably many Gram-positive organisms such as *Bacilli*) it is necessary to make protoplasts by pre-treating the cells with lysozyme before deproteinisation. Most Gram-negative bacteria are lysed by the method as described. If difficulties are encountered they can usually be overcome by modifying the phenol treatment slightly. For example, *Serratia marcescens* is lysed by stirring the cells with a slurry consisting of PAS, TIPNS and NaCl (in the same proportions as in PAS-TIPNS) in a little water and phenol mixture. It is obvious when the cells have lysed since the mixture becomes sticky due to the release of DNA. Water and more phenol mixture are added to bring the final composition of the two-phase mixture to that described above, after which the procedure is continued as before (M. Brown, unpublished observation).

The procedure is equally suitable for the isolation of nucleic acids from tissue homogenates. In order to reduce the action of endogenous nucleases, it is wise to homogenise the tissue in the upper phase of the deproteinisation mixture. For example, liver is roughly chopped or minced, dropped into ice-cold PAS-TIPNS and homogenised in a Teflon-glass Potter–Elvehjem homogeniser, In this case only 10 vol of each phase are required. For tissues rich in DNA (such as lymphoid tissue or tumours) the proportions given above for *Esch. coli* should be used. After the initial homogenisation, the suspension is shaken with phenol mixture at room temperature as described above. Examples of the use of this procedure for the deproteinisation of a wide variety of cells and tissues (from animals, fungi and plants) are given by Loening (1968*a*).

Sometimes it is useful to isolate RNA preferentially and leave the bulk of the DNA behind. With mammalian tissues, it is easy to isolate cytoplasmic RNA by using less vigorous conditions of deproteinisation. The tissue is homogenised in a Waring Blendor with 12 vol of dilute salt (ice-cold 0·5% (w/v) disodium naphthalene-1,5-disulphonate (NDS) is an excellent choice, as ribonuclease is effectively inhibited in this solution) and 10 vol of phenol mixture. The mixture is then shaken or stirred at room temperature for 30 min and the phases are separated by centrifugation as before. There is a bulky insoluble 'interface' containing the nuclear material. The upper phase is removed and NaCl (3 g per 100 ml) and TIPNS (5 g per 100 ml) are dissolved in it. This mixture is extracted with half its volume of phenol mixture and the phases are separated as before. The final upper phase, a cloudy but limpid liquid, contains cytoplasmic RNA and polysaccharide (including a great deal of glycogen if liver is the tissue used). The interface from the first deproteinisation may be extracted with PAS–TIPNS and phenol mixture to yield DNA and nuclear RNA after cytoplasmic contaminants have been removed by extracting the interface with fresh NDS.

The NDS procedure does not selectively release RNA from bacterial preparations. However, it is possible to 'soak' RNA out of *Esch. coli*. The cells are suspended as gently as possible in PAS–TIPNS (about 200 ml per g wet weight) and one-third vol of phenol mixture is added. A roughly homogeneous emulsion is obtained by inverting the container once—shaking must be avoided. This mixture is left to stand at room temperature for 45 min; then the phases are separated by centrifugation as before. The upper phase is removed and used for the second deproteinisation exactly as described already. The result is a solution containing RNA and extremely little DNA.

SEPARATION OF HIGH AND LOW MOLECULAR-WEIGHT RNA, DNA AND POLYSACCHARIDE

High molecular-weight RNA is insoluble in strong salt; DNA and high molecular-weight RNA can both be selectively precipitated with organic solvents. Either strong solutions of salts or organic solvents may be used to separate polysaccharide and RNA. Combinations of these methods can be used to separate RNA from DNA and polysaccharide and to achieve a crude fractionation of RNA. Two methods for purifying RNA are summarised in *Chart 1*.

The first of these methods has been devised for the PAS–TIPNS supernatant fluid. The nucleic acids are precipitated with ethanol and collected by low-speed centrifugation. The precipitate is dissolved in the minimum volume of dilute sodium acetate. If a lot of DNA is present it is sometimes necessary to allow the precipitate to soak in the acetate solution overnight at 4°C and then shake it vigorously next morning. Solid NaCl is added to a final concentration of 4 M and the RNA is salted out at near (or preferably just below) 0°C. The precipitate is very finely divided and has to be obtained by centrifugation at 17 000 rev/min (40 000*g*) for 5 min. The excess NaCl, DNA and residual polysaccharide is removed

Chart 1. Use of organic solvents and solutions of salts for the purification of RNA

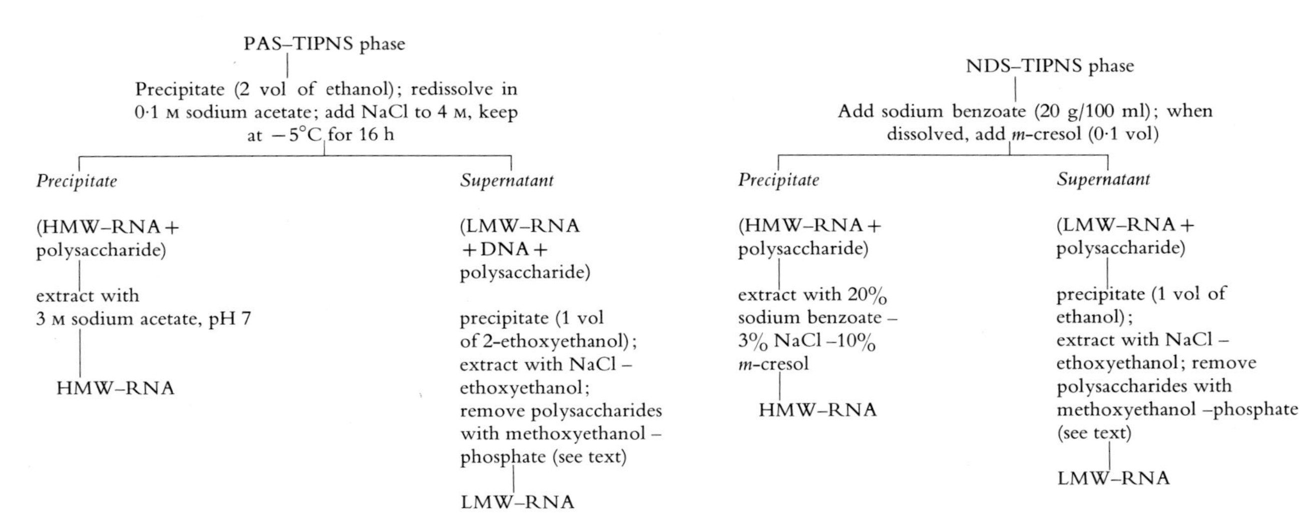

HMW, high molecular-weight; LMW, low molecular-weight.

from this pellet by repeated extraction with 3 M sodium acetate (pH adjusted to neutrality with glacial acetic acid). The number of extractions required depends on the amount of DNA and polysaccharide (glycogen, etc.) present in the preparation. If RNA has been extracted from bacteria by the 'soaking' procedure described above, one extraction is enough. If there is a lot of DNA or glycogen present (for example, in a preparation from the liver of an unfasted animal), three or four extractions are required before the supernatant fluid is limpid and clear. The precipitate is finally washed with 75% aqueous ethanol containing 2% (w/v) sodium acetate and can be stored as a slurry under this liquid at 4°C. The supernatant fluid from the treatment with 4 M NaCl contains DNA and tRNA (together with polysaccharides). The nucleic acids are precipitated with 2-ethoxyethanol (ethanol would precipitate the NaCl) and the precipitate is extracted with a mixture of M NaCl and 2-ethoxyethanol (2: 1 v/v). Only low molecular-weight RNA (and some polysaccharide) is extracted. The RNA is precipitated with 1 vol of ethanol and redissolved in a small volume of water. Polysaccharides are removed from this solution by adding to it 1 vol of 2·5 M potassium phosphate buffer, pH 7, and 1 vol of pre-cooled (−20°C) 2-methoxyethanol. The mixture is shaken for a few minutes and the phases are separated by centrifugation [say 15 000 rev/min (30 000*g*) at 5°C for 5 min]. The phases redistribute to produce a large upper phase, which contains the RNA, and a small lower phase (saturated potassium phosphate) with the polysaccharide floating at the interface. The upper phase is carefully removed and the RNA is precipitated (together with a lot of inorganic phosphate) from it by mixing it with an equal volume of ethanol. The precipitate is collected and redissolved in water, and the phosphate is removed by dialysis or by passing the solution through a column of Sephadex G-25. Low molecular-weight RNA is reprecipitated with ethanol. It is conveniently stored as a dried powder, obtained by washing the precipitate with ethanol and ether.

The method devised for the NDS-TIPNS supernatant fluid is explained in *Chart 1*. The high molecular-weight RNA is precipitated by adding *m*-cresol (sodium benzoate is present to solubilise the cresol) and polysaccharides are extracted with a solution of 20% (w/v) sodium benzoate, 3% (w/v) NaCl, 10% (v/v) *m*-cresol. The precipitated RNA forms a clear gel which changes to its familiar powdery form on being washed with 75% ethanol containing 2% (w/v) sodium acetate as above. The low molecular-weight RNA is obtained from the cresol supernatant fluid by the procedures described for the PAS-TIPNS phase (see *Chart 1*).

TRANSFER RNA

As tRNA is the one 'free' nucleic acid present in cells, it is possible, in favourable circumstances, to leach it out selectively. Moreover, it is frequently necessary to isolate large amounts of tRNA for use in experiments on protein biosynthesis, characterisation of minor nucleosides and sequence determination. Phenol and water may be used for the isolation of tRNA from yeast (Holley, 1963) and the NDS method described above is a good starting point for isolating large amounts of mammalian tRNA.

The following method is employed in our laboratory for the isolation of *Esch. coli* tRNA. About 500 g of *Esch. coli* paste is frozen and broken into pieces with a hammer. It is homogenised in portions in a Waring Blendor with solid phenol and 3% (w/v) NaCl (the amounts do not matter). When all the paste has been blended, more phenol and NaCl solution is added until there is a total of 3 l of 3% (w/v) NaCl solution and 2 kg of phenol. The mixture is stirred at room temperature for about 2 h and the phases are separated in a low-speed centrifuge [MSE Mistral 6L at 2400 rev/min (2500*g*) for 2 h at 5°C]. The upper phase is poured off and mixed with 5 l of industrial alcohol. The precipitate (which contains surprisingly little DNA) settles overnight at 4°C and is collected by centrifugation at low speed. The pellet is processed by the whole PAS-TIPNS procedure as though it was tissue, using 300 ml each of PAS-TIPNS and phenol mixture.

Transfer RNA isolated by these methods contains small amounts of both 5*S* and high molecular-weight RNA. Chromatographic methods of separating these materials are described in the next Section. However, if the tRNA is to be used for cell-free protein synthesis, it is not necessary to remove the contaminants. However, it is necessary to remove endogenous amino acids from the 3′ ends of the tRNA. This is achieved by incubating a solution of tRNA at pH 10 for 30 min at 37°C. The RNA can be recovered by precipitation from the neutralised solution with 2 vol of ethanol.

ISOLATION OF RNA FROM ORGANELLES

The PAS-TIPNS procedure has been applied successfully to the isolation of RNA from nuclei, nucleoli, chloroplasts and mitochondria (see Loening, 1969). It can be equally well applied to polysomes or to post-mitochondrial supernatant fractions (or their bacterial equivalents). In the last case, the PAS, TIPNS and

NaCl are simply added as solids to the supernatant fluid, which is then extracted with phenol mixture. Difficulties sometimes arise if the components of the buffer result in the phenol and aqueous phases becoming similar in density. The phenol can be made to sink by adding a few drops of chloroform to the emulsion. Alternatively, the aqueous phase may be made heavier with sucrose. In the latter case, the deproteinisation is performed in exactly the same way except, of course, the aqueous phase is always the *lower* of the two.

Less vigorous methods of deproteinising ribonucleoprotein are also effective. Thus, RNA is released from ribosomes or polysomes by treating them with a detergent such as sodium dodecyl sulphate (SDS). This is a particularly suitable method if the deproteinisation is to be effected directly prior to fractionation of the RNA by sucrose density-gradient centrifugation. This, and an alternative method of effecting deproteinisation on top of a gradient, are described later.

FRACTIONATION OF RNA

COLUMN CHROMATOGRAPHY

Transfer RNA is fractionated by column chromatography, the most familiar method being ion-exchange chromatography. Unlike high molecular-weight RNA, tRNA can be eluted from ion-exchange resins such as DEAE-cellulose with conventional salt solutions. A convenient method for removing high molecular-weight RNA from a tRNA preparation is to load the mixture on to a column of DEAE-cellulose in 0·1 M NaCl (in a convenient buffer of pH 7). All the RNA is bound on the column under these conditions and, if necessary, the column may be washed with this salt solution to remove contaminants such as radioactive nucleotides. The tRNA is eluted with 1 M NaCl. The high molecular-weight RNA cannot be recovered from such a column. The column may be re-used after it has been washed with 0·1 N NaOH whereupon RNA fragments are discharged. If the column is eluted with a salt gradient it is possible to achieve a fractionation of tRNA into various acceptor species. A more satisfactory resolution is obtained if the buffer contains urea. The various procedures are extremely versatile since the gradient can either be a salt gradient at constant urea concentration or a urea gradient at constant ionic strength; the conditions can also be modified by varying the pH of the eluting buffer. A valuable survey of these procedures has been made by Bock and Cherayil (1967). Other similar procedures for the fractionation of tRNA include the use of arylated DEAE-cellulose (Gillam *et al.*, 1967) and hydroxyapatite (Pearson and Kelmers, 1966).

Another technique applicable to tRNA is reverse-phase chromatography. This technique is based on the different partition coefficients of RNA species between two partly-miscible liquid phases. There are two versions of reverse-phase chromatography depending on which phase is chosen as stationary and which as moving. The more versatile system employs a column impregnated with the less aqueous (more hydrophobic) phase and the hydrophilic phase as moving phase. Using this system it is possible to introduce a new parameter into the fractionation by using a salt gradient for the elution. A number of materials have been used as supports for the stationary phase of reverse-phase chromatography, Sephadex or kieselguhr being the most common. However, for the fractionation of tRNA, the synthetic polymer Plaskon CTFE powder is most effective (Pearson *et al.*, 1971). An example of a fractionation employing this technique is shown in *Figure 1*.

The ion-exchange chromatography of high molecular-weight RNA requires an ion-exchanger in which the affinity between nucleotide residues and the exchanger is weak. The most familiar ion-exchanger of this type is methylated albumin, supported on kieselguhr (MAK). The main disadvantage of MAK column chromatography for the fractionation of high molecular-weight RNA is that the salt concentrations required for the elution are so high, and so close to the concentration at which precipitation occurs, that there is considerable danger of aggregation. On the whole, the method is more suitable for the fractionation of bacterial rather than eukaryotic RNA. Despite its limitations, MAK chromatography remains the one well-tried procedure which can be easily operated on a preparative scale and in which the major classes of cellular nucleic acids can be fractionated in one experiment. As an example of the techniques employed and the resolving powers of the columns, see *Figure 2*.

The applications of gel-exclusion chromatography to nucleic acids are rather limited. There are no gels whose exclusion limits conveniently bracket, for example, the two rRNA species (16*S* and 23*S* or 18*S* and 28*S*, etc.). Moreover, all tRNA species have approximately the same molecular weight and the procedure is clearly of no help in resolving them. However, there is one problem for which gel-exclusion chromatography is ideally suited, namely, the isolation on a preparative scale of 5*S* rRNA. The method of

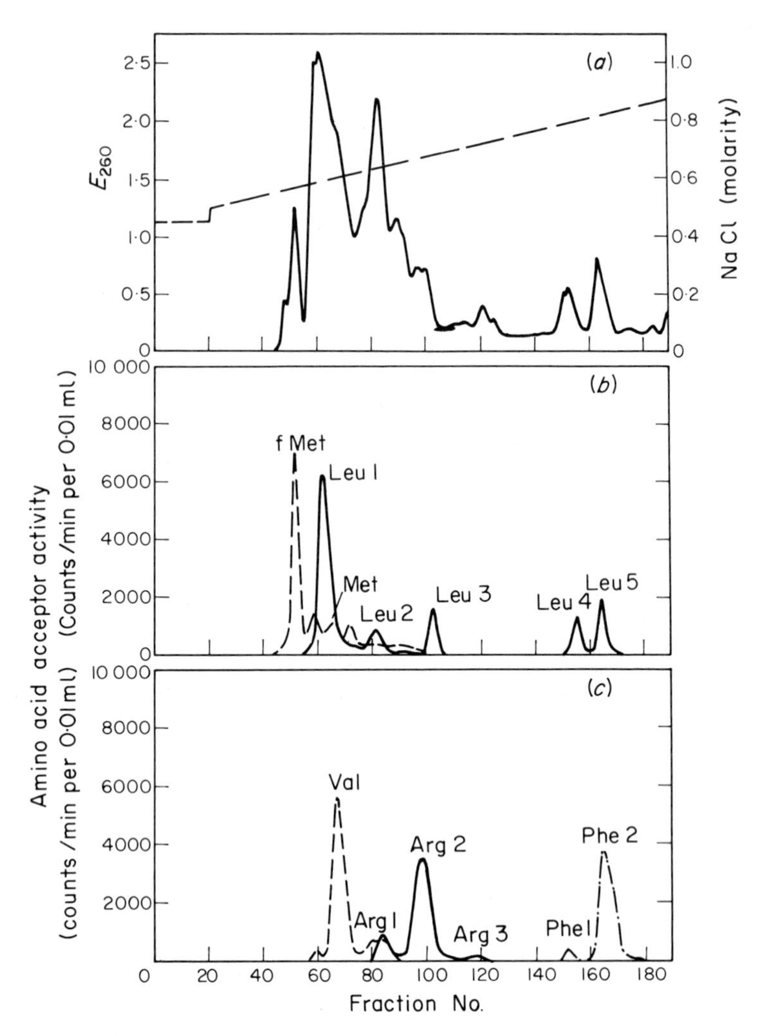

Plaskon CTFE 2300 powder (Allied Chemical Corp.; 300 g) was agitated for 2 h with a solution of tri-octylpropylammonium bromide (15 g) in chloroform (600 ml); the chloroform was removed by drying in air. A jacketed glass column (37°C ± 0·1%; 1 × 240 cm) was packed at 30–50 lb/in² with this material in 0·25 M NaCl, 10 mM magnesium acetate, 2 mM $Na_2S_2O_3$, 10 mM sodium acetate, pH 4·5; several column volumes of this buffer were used to equilibrate the column; 45 mg of RNA was loaded onto the column and eluted with 2 l of a linear gradient from 0·32 to 1·0 M NaCl (in buffer otherwise identical to the above)

Figure 1. Fractionation of tRNA by reverse-phase chromatography; (a), *absorbance profile of RNA;* (b) *and* (c), *the results of assays for various acceptor activities in the fractions (from Pearson* et al., *1971)*

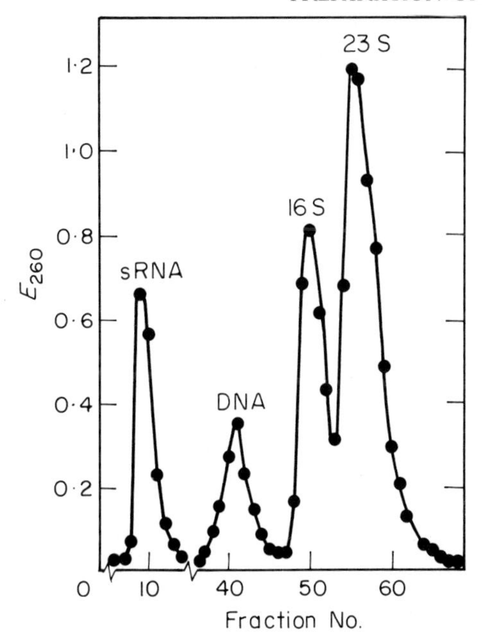

Methylated bovine albumin was prepared by the procedure of Mandell and Hershey (1960). Hyflo Supercel was washed in alkali and acid, and dried; 10 g was dispersed in 50 ml of 0·1 M NaCl buffered with 50 mM potassium phosphate, pH 6·7, boiled and cooled. Methylated albumin (2·4 ml of a 1% solution in water) was added and stirred; the mixture was poured into a column (1·8 × 15 cm) and washed with 300 ml of buffered NaCl (0·1 M); 1 g of Supercel, boiled and cooled in buffered NaCl (0·1 M) was added to the top. The column was loaded with 7 mg of total Esch. coli *nucleic acids and washed with 50 ml of buffered NaCl (0·2 M); the column was eluted with a 300 ml gradient of buffered NaCl (0·4 to 1·0 M)*

Figure 2. Fractionation of Esch. coli *nucleic acids by MAK column chromatography (from Osawa and Sibatani, 1967)*

Galibert *et al.* (1965) is universally suited to the isolation of this material. A column (1·8 × 190 cm) of Sephadex G-100 is equilibrated in 10 mM ammonium acetate, pH 5·1 and 2 mg of RNA is loaded on to it. High molecular-weight rRNA is eluted with the exclusion volume followed by 5*S* RNA and, finally, 4*S* RNA.

RATE-ZONAL CENTRIFUGATION

Centrifugation through sucrose gradients is a long-established procedure for the fractionation of high molecular-weight RNA. The advantages of the technique are that it can readily be scaled up, recovery of the sample is very simple, and the pattern of the fractionation may be used to estimate the sedimentation coefficients of the RNA species. The disadvantages are that it is time-consuming, it involves the use of expensive equipment, and it is of extremely limited resolving power. Thus for the routine analysis of mixtures of RNA it is inferior to gel electrophoresis (see later).

The basic techniques of sucrose density-gradient centrifugation are part of everyday laboratory lore. For fractionation in the tubes of a swing-out rotor of an ultracentrifuge, a continuous gradient of sucrose is estabished in the tube. The simplest method of preparing a linear sucrose gradient employs two test tubes with a little side-arm blown on to the bottom of one and two such arms on the other, as illustrated in *Figure 3*. A simple device (Noll, 1967)

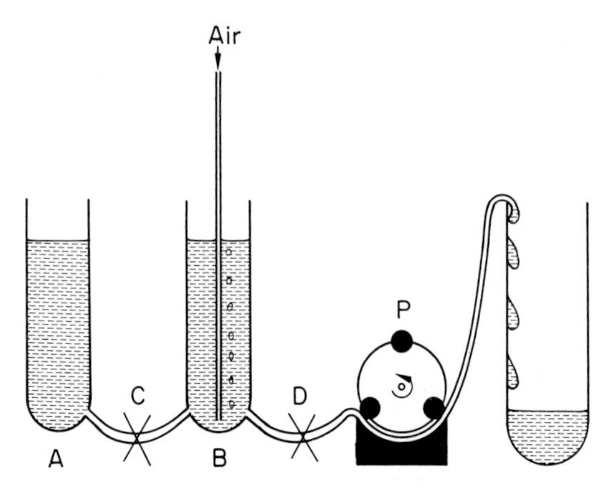

A and B are test tubes with small side-arms joined with flexible tubing, C and D are artery clips and P is a peristaltic pump. The heavier solution is placed in B and the lighter in A; air is bubbled slowly into B (as a stirrer). The pump is turned on and C and D are opened. As shown, the device is suitable for filling cellulose nitrate tubes; polypropylene tubes are poorly wetted by aqueous solutions and for them it is better to run the gradient down a spill of glass (or cellulose nitrate) clamped vertically so that it reaches the bottom of the tube; the spill is carefully removed when the tube is full

Figure 3. Device for creating a linear sucrose density gradient.

for creating an exponential gradient is illustrated in *Figure 4*. The advantages of these various arrangements will become clear shortly. Large-scale sucrose density-gradient centrifugation is performed in a zonal rotor, in which the gradient is established radially in a cylindrical rotor while it is spinning on its axis. Details of methods for loading and unloading zonal rotors appear in the technical literature published by ultracentrifuge manufacturers.

Some users of sucrose gradients are preoccupied with the possibility that the sucrose is contaminated with ribonuclease. It is certainly wise to keep a bottle of sucrose 'for preparing gradients only', to avoid contamination. If there is genuine reason to believe that the sucrose or buffer is contaminated, the simplest method is to sterilise the stock solutions. According to the procecure of Fedorcsák *et al.* (1969) the solutions are made 0·6% (v/v) with respect to diethyl pyrocarbonate ('Baycovin'). This reagent inactivates all proteins. Diethyl pyrocarbonate is destroyed to yield ethanol and CO_2 by heating at 100°C for 5 min.

Sucrose gradients can be monitored by pumping the contents of the tube, or zonal rotor, through a continuous-recording spectrophotometer. For a zonal rotor, this is achieved by pumping heavy sucrose solution to the edge of the rotor. The same technique can be employed with a centrifuge tube, in which case the heavy solution is pumped to the bottom of the tube. However, this is not necessary. A simpler method is to introduce to the bottom of the

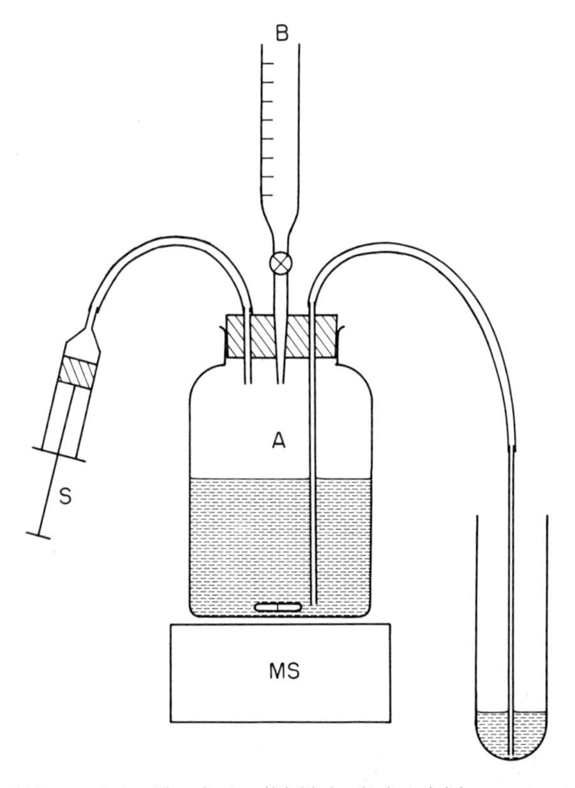

B is a burette and MS a magnetic stirrer. The gradient is established 'backwards', that is, the lighter sucrose is put into the bottle A and the heavier in the burette B; the increasingly heavy solution displaces the lighter sucrose as the centrifuge tube is filled. The syringe (S) is used to force the light solution into the filling tube initially; once the burette tap is opened, the tube is filled under gravity. The shape of the gradient is determined by the volume of solution in the mixing bottle and the volume of gradient; if V_m is the mixing volume, C_t the initial sucrose concentration in the mixing bottle and C_R the sucrose concentration in the reservoir (burette), then concentration (C_V) is a function of volume (V) as follows:

$$C_V = C_R - (C_R - C_t)e^{-(V/V_m)}$$

Figure 4. Device for creating an exponential sucrose density gradient (after Noll, 1967).

centrifuge tube a narrow tube (or long syringe needle) attached to a piece of narrow-bore flexible tubing which is passed through a peristaltic pump. The pump will pull the solution through a spectrophotometer cell quite readily. Fractions can be collected and assayed for radioactivity (see legend to *Figure 5*). Alternatively, it may be required to recover samples for further analysis, in which case the simplest procedure is to add two vol of ethanol. If the RNA solution is very dilute, the precipitation should be performed at −20°C or in the presence of a 'carrier'. For example, if fractions are to be used subsequently for studies of their capacity to stimulate the incorporation of labelled amino acids into polypeptides (in a cell-free system containing ribosomes and necessary co-factors),

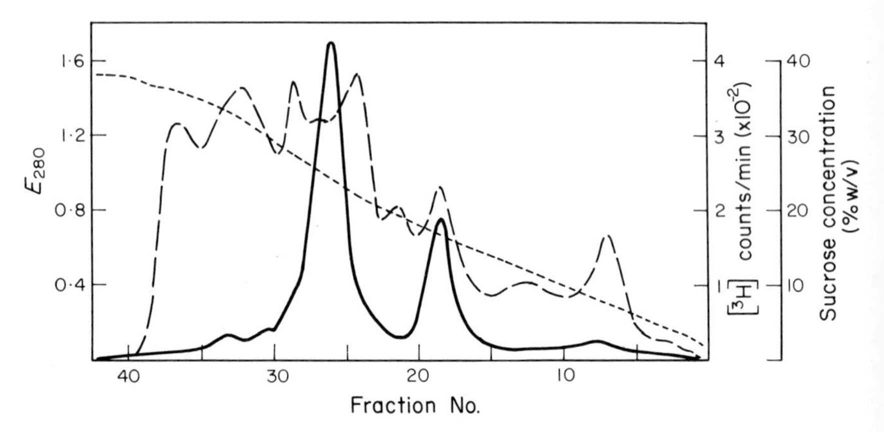

Rat-liver total RNA (40 mg), labelled in vivo *for 20 min with [³H] orotic acid, was dissolved in 40 ml of 0·1* M *sodium acetate, pH 6·5, containing 2·5% (w/v) sucrose and centrifuged at 40 000 rev/min at 5°C through a 5 to 30% (w/v) sucrose gradient (same buffer) in the BIV zonal rotor of the Beckman model L4 ultracentrifuge for 6 h. The contents of the rotor were pumped through a continuous-recording spectrophotometer and 40 ml fractions were collected. The concentration of sucrose in each fraction was determined by measuring the refractive index; the RNA in 2·0 ml samples was extracted into butanol containing a five-fold molar excess (over the acetate) of a long-chain aliphatic diammonium acetate and portions of the butanol layers were assayed by liquid-scintillation counting. Direction of sedimentation, right to left. ———, E_{280}; — — —, radioactivity; ...., sucrose concentration*

Figure 5. Example of sucrose density-gradient centrifugation of mammalian RNA in a zonal rotor (J. H. Parish and E. S. Klucis, unpublished data)

it is often convenient to add the tRNA at this stage where it acts as a co-precipitant for the mRNA. An alternative method of isolating small amounts of RNA from large volumes of sucrose solution (such as pooled fractions from a zonal rotor) consists of adding one-third vol of 3 M sodium acetate (pH adjusted to 7·0 with glacial acetic acid), then ice-cold acetone until the solution just becomes cloudy (approx. 1·5 to 2 vol is usually sufficient). When this mixture is kept at 4°C a small lower phase settles out and a precipitate of RNA collects at the interface. The upper phase is poured off and 1 drop of ethanol is added to the mixture of interface and lower phase. The RNA is collected by centrifugation (Hastings *et al.*, 1965).

An example of the fractionation of mammalian RNA by sucrose density-gradient centrifugation in a zonal rotor is shown in *Figure 5*. In addition to the 18*S* and 28*S* rRNA components, fast-sedimenting material, including the rRNA precursors, are present together with heterodisperse 'rapidly-labelled' RNA. The nature of this last material still remains obscure but it includes mRNA species.

For many purposes, it is useful to perform the deproteinisation step on top of the gradients. For the deproteinisation of ribosomes (or polysomes) the simplest procedure is to disperse the ribonucleoprotein in 10 mM NaCl, 5 mM tris-HCl, pH 7·5, and add SDS as a 10% (w/v) solution in water. For bacterial ribosomes, the final concentration of SDS is 0·5% (w/v) and the mixture is

0·8 ml	Sample
0·5 ml	5% PAS, 1% SDS, 3% PM
0·5 ml	6·5% PAS, 1% SDS, 10% PM
0·5 ml	8% PAS, 0·5% SDS, 15% PM
0·5 ml	12% PAS, 20% PM
0·5 ml	14% PAS, 15% PM
0·5 ml	16% PAS, 10% PM
1·0 ml	20% PAS
0·5 ml	10% PAS, 10% sucrose
14·0 ml	20–40% (w/v) linear sucrose gradient in 0·15 M sodium acetate

The gradient is established in a propropylene tube of the 3 × 23 swing-out rotor of the MSE Super-speed 50 ultracentrifuge by layering the solutions shown on the top of a pre-formed sucrose gradient; tubes made of materials such as cellulose nitrate cannot be used. The sample is prepared by mixing a suspension of the cells with 3 vol of 2% PAS, 1·3% SDS and 4% PM; it is kept at −4°C for 15 h prior to layering on top of the gradient. The gradient is centrifuged at 10°C, first for 1 h at 1000 rev/min and then for 15 h at 20 000 rev/min.

PAS, sodium 4-aminosalicylate; SDS, sodium dodecyl sulphate; PM, phenol-cresol-hydroxyquinoline mixture (p. 253)

Figure 6. Discontinuous gradient for the deproteinisation of cells and fractionation of nucleic acids (J. R. B. Hastings, unpublished work)

incubated at 37°C for 10 s. For eukaryotic ribosomes, the final concentration of SDS is 2% (w/v) and the incubation is for 2 min at 37°C. After the incubation, the solution is cooled and layered over the gradient. The SDS may precipitate (especially in the latter case) but it does not interfere with the fractionation in any way. These methods are due to Noll and are exemplified by the separation shown in *Figure 8*.

An alternative method of achieving deproteinisation *in situ* over a gradient is due to J. R. B. Hastings (unpublished). This method involves the use of phenol mixture and PAS in layers on top of a sucrose gradient and has the advantage that the deproteinisation conditions are strong enough to release RNA (and DNA) from cells, provided they can be lysed in the PAS-TIPNS-phenol procedure described earlier. The gradient employed (and the method of using it) is shown in *Figure 6*. An example of the use of this technique is illustrated in *Figure* 7, in which the fractionation

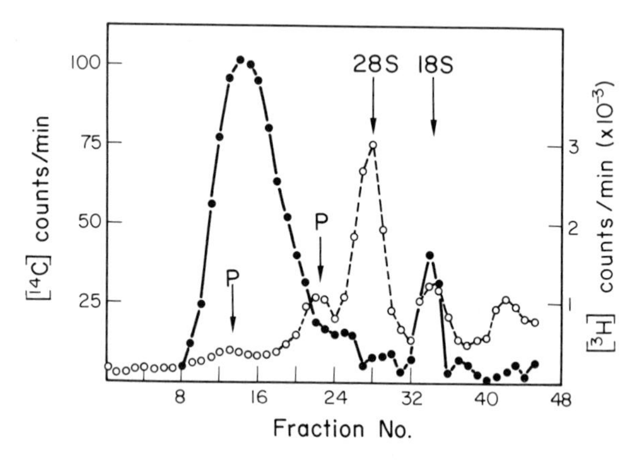

Polyoma virus-transformed BHK21 cells were labelled with [2–^{14}C]thymidine for 24 h and with [5–^{3}H]uridine for 2 h; the cells were washed, then lysed as described for Figure 6 *and the equivalent of 3 × 10^{4} cells was transferred (with a wide-bore pipette to minimise shear of DNA) on to the gradient. The conditions for centrifugation were as described in* Figure 6; *direction of sedimentation, right to left. In addition to 18S and 28S RNA (the positions of which were established from a similar gradient containing long-term labelled [^{14}C]RNA), two peaks (P) of ribosomal precursor RNA can be seen. Note the small peak of low molecular-weight DNA; the significance of this material is not yet clear.* ● —— ●, [^{14}C]; ○ — — ○, [^{3}H]

Figure 7. Example of the use of the gradient shown in Figure 6 *(I. H. Maxwell, unpublished data)*

of total nucleic acids from polyoma virus-transformed BHK21 cells is shown. Note that this method is only applicable to minute amounts of nucleic acid detectable only by radioactivity measurements. The RNA in this preparation was labelled for a short time so that rRNA and its precursors are labelled. If this technique is applied to the deproteinisation of bacteria, up to 10^7 cells per ml may be used in the initial suspension.

RATE-ZONAL CENTRIFUGATION AND MOLECULAR SIZE

The position to which a zone of RNA sediments in a sucrose gradient is determined by the ionic strength and temperature of the gradient, and the concentration and the value of the extrapolated sedimentation coefficient ($s^{\circ}_{20,w}$) of the RNA. This last value is the sedimentation coefficient at 20°C and infinite dilution, in water as solvent. It is possible to estimate the sedimentation coefficient for the components in a zone if the gradient is calibrated by including RNA standards in the sample. The estimation is most readily and reliably made if an isokinetic gradient is used. An isokinetic gradient is one in which the rate at which a molecule sediments is constant, that is, a gradient in which at all points there

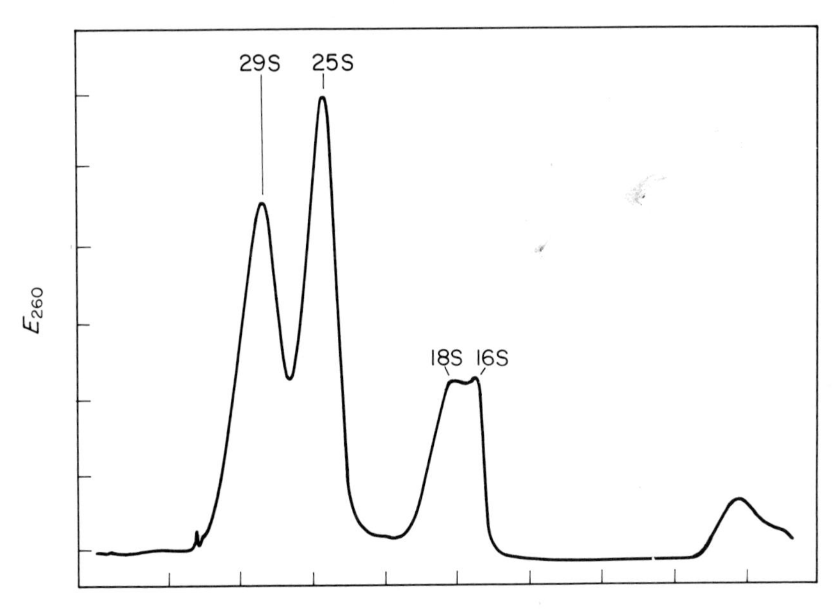

Ribosomes were deproteinised on the top of the gradient as described in the text. The gradient was monitored in an automatic device in which the bottom of the tube is pierced and the gradient pumped out by layering water over the top of it. Direction of sedimentation, right to left

Figure 8. *Separation of rRNAs from rat-liver (18S and 29S) and bean cytoplasm (16S and 25S) in an isokinetic gradient (from Noll and Stutz, 1968)*

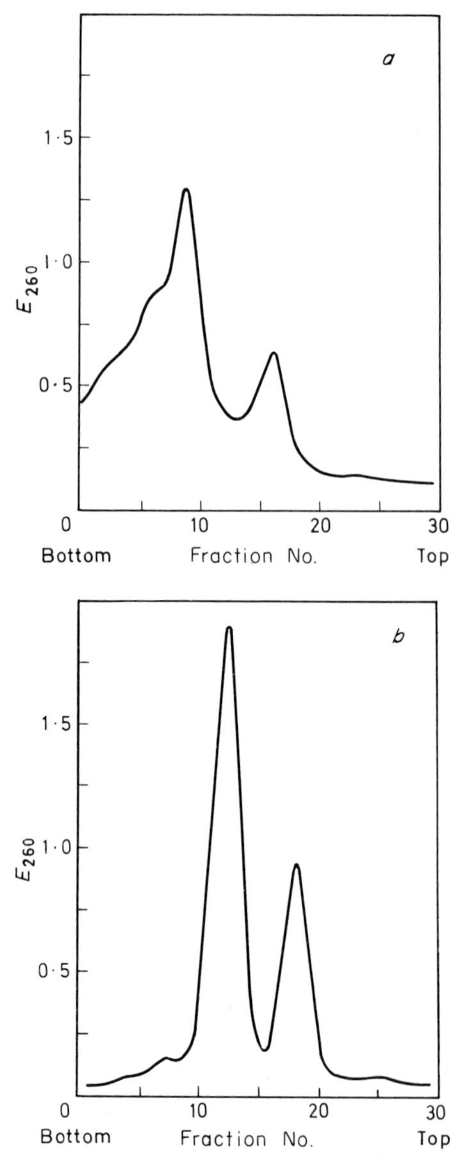

RNA was dissolved in 0·1 M sodium acetate, pH 6·5 and centrifuged through a 5–20% (w/v) sucrose gradient (containing 0·1 M sodium acetate, pH 6·5) in the SW25·1 rotor of a Beckman model L ultracentrifuge for 14 h (a) *at 22 000 rev/min and 4°C; and* (b) *at 19 000 rev/min and 18°C. Direction of sedimentation, right to left*

Figure 9. Sucrose density-gradient centrifugation of rat-kidney tumour RNA: (a) *at 4°C;* (b) *at 18°C*

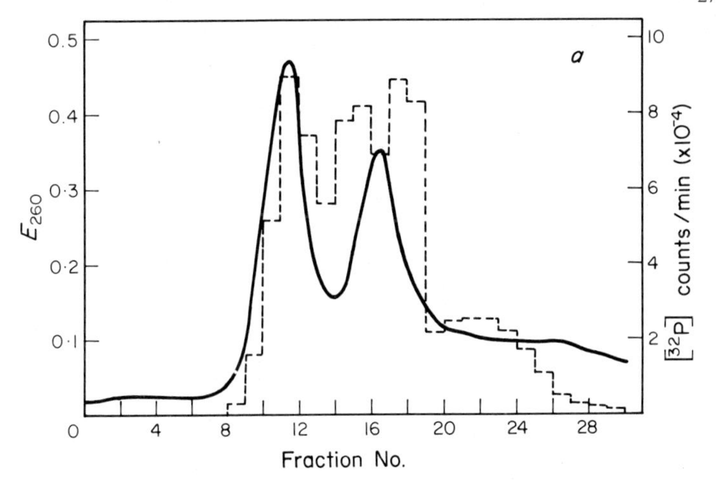

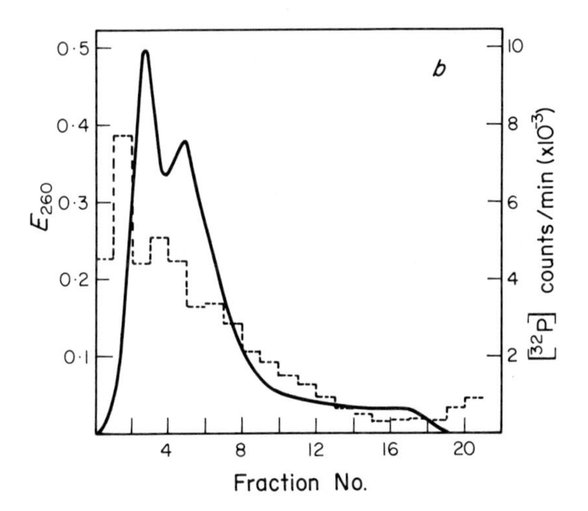

The sample was RNA isolated from Myxococcus xanthus *by the method described (p. 253) for* Esch. coli*; the RNA was salted out with 4* M *NaCl. The cells had been labelled* in vivo *with* [$^{32}PO_4^{3-}$] *for 5 min. In* (b), *the rRNA has been sedimented well down the gradient to demonstrate that none of the radioactivity associated with rRNA in the sucrose gradient* (a) *is in low molecular-weight RNA (associated with rRNA or aggregated); ———,* E_{260}*; — — —, radioactivity.*

(a) *Linear 7·5 to 30% (w/v) sucrose gradient in 0·15* M *sodium acetate (unbuffered) centrifuged at 4°C for 15 h at 22 000 rev/min in the SW25 rotor of the Beckman model L ultracentrifuge; direction of sedimentation, right to left.*

(b) *Linear gradient of 90% (w/v) methanol to 90% (w/v) 2-methoxyethanol in 1% glycerol, 50* mM *triethylammonium acetate. The gradient was prepared* (as in Figure 3) *in a tube of the 3 × 23 swing-out rotor of the MSE Super-speed 50 ultracentrifuge. RNA was dissolved in a small volume of water; 2* M *triethylammonium acetate and methanol were added to make the solution identical in composition to the top of the gradient except that the glycerol (included solely to create a 'density shelf' at the top of the gradient) was omitted. Centrifugation was at 15°C for 15·5 h at 30 000 rev/min; direction of sedimentation, right to left.*

Figure 10. Comparison of the sedimentation of RNA through a conventional linear sucrose gradient (a) *with that through a gradient in which RNA is denatured to a random coil conformation* (b) *(M. Brown, unpublished data)*

is a precise balance between the opposing forces of gravity (which increases linearly with distance down the centrifuge tube) and viscous drag and buoyancy (which increasingly retard the molecules as they penetrate into the gradient). The parameters for calculating the amounts required in the gradient-making device shown in *Figure 4* to establish gradients isokinetic for different classes of RNA are described by Noll (1967). Improved separation is obtained in these gradients so that standards can be used and resolved from the test material readily. An example of an isokinetic separation of four rRNA species is shown in *Figure 8*.

It is not possible reliably to convert sedimentation coefficients into molecular weights. There are two factors which make such conversions unreliable. The first is that the mean radius of gyration of a molecule (which contributes to the magnitude of $s^{\circ}_{20,w}$) is affected by the stability of its tertiary structure, which is dependent upon the nucleotide composition and sequence of the molecule. The second problem is that aggregates form in sucrose gradients. With rRNA components, aggregates are easy to identify but with minor 'rapidly-labelled' components these are difficult to identify as we do not have any real reference in many cases to establish whether aggregates are present or not. This is not to say that aggregates of rRNA never form. A striking example of aggregation was found by Hastings in the rRNA isolated from a spontaneous rat-kidney tumour (*Figure 9a*). The aggregate(s) are disrupted at higher temperatures (*Figure 9b*). It is to be hoped that further work on this very interesting RNA will continue in order to determine what features of its primary structure are responsible for specific association of 18*S* and 28*S* molecules.

Both problems can be overcome if totally-denatured RNA is centrifuged. In this case aggregates do not form and the mean radius of gyration is not dependant upon conformation differences between different RNA species as they all sediment as random coils. A technique has been developed by Sedat and Sinsheimer (1970) in which the RNA is sedimented through a gradient consisting of dimethyl sulphoxide and hexadeuterodimethyl sulphoxide. A simpler system has been proposed by Hastings (personal communication). In this case the gradient consists of methanol to methoxyethanol, in which RNA shows maximum hyperchromicity. Details of this system, and an example of its use, are given in *Figure 10*.

In these systems, the position of the RNA zone is proportional to the logarithm of the molecular weight of the RNA.

GEL ELECTROPHORESIS

The techniques involved in gel electrophoresis of RNA have been perfected by Loening (1968*b*). These proposals are all based on his methods.

Polyacrylamide gels are made by co-polymerising acrylamide ($CH_2=CHCONH_2$) and methylenebisacrylamide ($NH_2CO-CH=CH-CH_2-CH=CHCONH_2$; BIS), a cross-linking agent, in solution. The other components are buffer, an initiator (tetramethylethylenediamine, $(CH_3)_2NCH_2CH_2N(CH_3)_2$; TEMED), and the catalyst for the polymerisation (ammonium persulphate). Care is needed in the preparation of gels for electrophoresis of RNA since, for certain purposes, very dilute gels have to be used and these require ultra-pure acrylamide and BIS. Acrylamide is purified by recrystallisation from chloroform. A 7% (w/v) solution is filtered at 50°C and slowly cooled to −20°C. The crystals are collected by filtration in the cold. BIS is dissolved in the minimum volume of acetone at 50°C and recrystallised in the same way. AnalaR solvents are essential for these purifications.

There is also a problem in recovering the gels from the tubes. Polyacrylamide adheres to ordinary glass tubing and it is difficult to remove the gels without breaking them up. The tubing can be made of Perspex; alternatively, glass tubing can be cleaned thoroughly with chromic acid, carefully rinsed and dried, and then siliconised. The easiest way to siliconise glass tubing is to soak it for some time (approx. 1 h) in a dilute solution (about 1%) of hexamethyldisilazane or trimethylsilyl chloride in light petroleum (these reagents can be used many times over). The tubing is removed from the solution and baked in an oven at 110°C for several hours.

Gels are defined by the final concentration of polyacrylamide in them. To make them, two stock solutions are prepared: A, consisting of acrylamide and BIS; and B, consisting of tris, sodium acetate and EDTA (see *Table 1*). These solutions are mixed with water in varying proportions to give gels of different concentrations up to 5% (*Table 1*). Gels stronger than 5% are prepared in the same way except that acrylamide is omitted from solution A and the concentration of BIS is halved (*Table 1*).

The gel tubes (typically 7 to 10 cm long and 7 mm internal diameter) are stoppered at the bottom and clamped vertically. The mixture of solutions A and B with water is degassed under vacuum for approx. 0·5 min and 0·025 ml of TEMED and 0·25 ml of freshly-prepared 10% (w/v) ammonium persulphate are added to it. The mixture is pipetted into the tubes immediately since the gel sets quite quickly. After the gel has set the stopper in the bottom of the tube is removed. It is wise to replace the stopper with a little rubber O-ring that fits into the tube to stop the gel sliding out of it. For the 2% or 2·2% gels it is a good idea to have a disc of Millipore filter membrane under the gel since it is so dilute it can flow past the ring. The buffer for this type of electrophoresis is a 1:5 dilution of solution B. The electrophoresis is performed either in the cold or at room temperature. In the latter case it is possible to include a detergent in the buffer as a ribonuclease inhibitor (either SDS or TIPNS is suitable at about 0·2% w/v).

The gel is pre-run for 30 min at 8 V per cm; the current flow should be about 5 mA per tube. The current is switched off and the RNA (up to 0·1 mg) in electrophoresis buffer (up to 0·05 ml),

Table 1. PREPARATION OF POLYACRYLAMIDE GELS

Concentration of gel (%)	*Vol. of solution A*★ (ml)	*Vol. of solution B*† (ml)	*Vol. of water* (ml)
2·0	5·0	7·5	24·7
2·2	5·0	6·8	22·0
2·4	5·0	6·25	19·7
2·5	5·0	6·0	18·7
2·6	5·0	5·8	17·8
3·0	5·0	5·0	14·7
5·0	5·0	3·0	6·7
7·5	5·0	2·0	2·7

★ *For up to 5% gels*, solution A is 15% (w/v) acrylamide, 0·75% (w/v) BIS;
For > 5% gels, solution A is 0·375% (w/v) BIS.
† *For all gels*, solution B is 2·42% (w/v) tris, 0·82% (w/v) anhydrous sodium acetate, 0·185% (w/v) disodium EDTA, adjusted to pH 7·8 with acetic acid.

which contains enough sucrose (about 10% w/v) for the sample to form a tight band, is layered over the gel. The gel is run for up to 4 h at 8 V per cm. After electrophoresis, the rubber ring is removed from the bottom of the tube and the gel is gently slid out of it. It may need careful blowing to do this.

The most direct way of determining the distribution of the RNA is to scan the gel in a Chromoscan with ultraviolet optics. In the absence of such a device, the gel may be stained with 0·1% toluidine blue dissolved in 40% (v/v) aqueous 2-ethoxyethanol. The gel should be left in the stain for about 2 h and then destained in 40% (v/v) ethoxyethanol overnight. The stained bands can be recorded on a Chromoscan with visible optics. TIPNS is strongly absorbing in the u.v. so it can only be used in the electrophoresis buffer if the gels are to be stained.

For the assay of radioactivity in bands in the gel, or for recovery of material, the gels are frozen with solid CO_2 and sliced. For recovery of material, the slices should be 0·5 mm (or less) thick. Several procedures are available for recovery; in our experience, the simplest method is to allow the gel to soak in 0·1% (w/v) SDS in 0·5 M buffered salt solution for several hours at room temperature. The salt and SDS can be removed with a short column of Sephadex G-25. If the bands are stained, the toluidine blue is removed by extracting the slices with a little phenol.

It is possible to exemplify only some of the uses of gel electrophoresis for fractionating RNA. The positions to which zones migrate can be assumed to be a linear function of the logarithm of the molecular weight of the RNA, provided the nucleotide compositions of the species being compared are similar. Thus the molecular weights of many types of rRNA have been determined (Loening, 1968*a*) but the correlations are invalid in such cases as mammalian mitochondrial rRNA which is extremely rich in A and U (Dawid, 1970).

An example of the resolution of 23*S*, 16*S*, 5*S* and 4*S* RNA from *Esch. coli* is shown in *Figure 11*. For maximum resolution of high

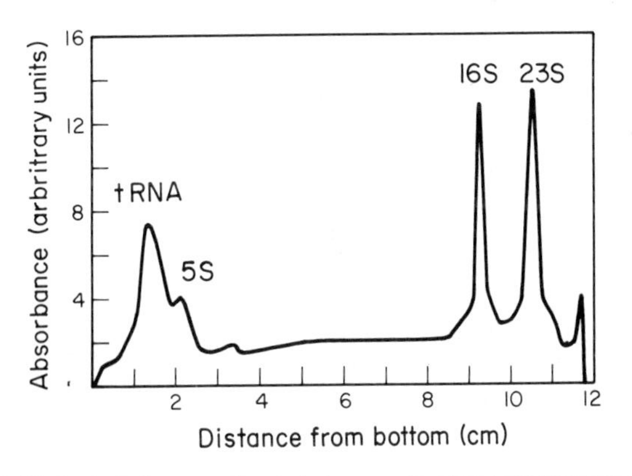

The gel was run for 2 h; for other details, see text. The stained gel was scanned in a Joyce–Loebel Chromoscan, using an orange filter (5–049); direction of migration, right to left

Figure 11. Fractionation of Esch. coli *RNA by electrophoresis on a 2·4% polyacrylamide gel (R. M. Mallon, unpublished data)*

molecular-weight RNA, such gels (or more dilute ones) run for a longer time are used. This technique is especially useful for the study of rRNA precursors in eukaryotic cells. These precursors consist of long molecules that contain the 18*S* and 28*S* sequences together with 'redundant' sequences which are excised from the molecule during 'maturation'. An example of the fractionation of these species is shown in *Figure 12*.

For maximum resolution of low molecular-weight RNA, stronger gels (4, 5 or 7·5%) are used. In addition to resolving 4*S* and 5*S* RNA, several other molecules have been characterised. In

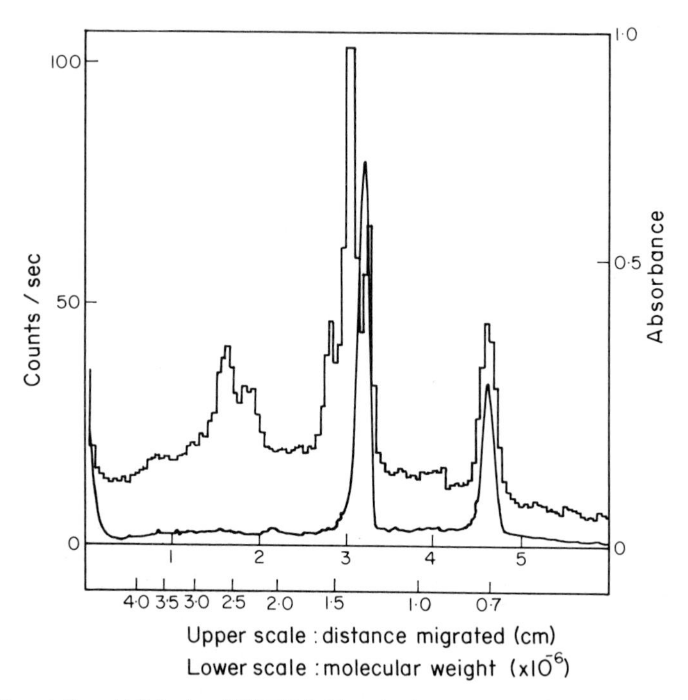

The yeast cells were labelled in vivo *with* [$^{32}PO_4^{3-}$] *for 7·5 min; the gel was run at 50V for 3 h. Direction of migration, left to right;* ———, *u.v. absorbance; histogram, radioactivity*

Figure 12. Fractionation of rRNA precursors by electrophoresis of RNA from the yeast Schizosaccharomyces pombe *on a 2·2% polyacrylamide gel (from Loening, 1970). This fractionation is particularly interesting as it contrasts the formation of an unlabelled aggregate (molecular weight, 2 × 10*^{6}*) with the highly-labelled rRNA precursors*

mammalian cells, these include several nuclear species and cytoplasmic 7*S* RNA which is associated with the 60*S* ribosomal subunit and is apparently specifically hydrogen-bonded to 28*S* rRNA. A survey of data relating to these species and electrophoretic patterns illustrating their isolation are to be found in Prestayko *et al.* (1970).

A final example of gel-electrophoretic analysis of RNA is the characterisation of the '9*S* RNA' obtained from reticulocyte polysomes by Gaskill and Kabat (1971). Evidence (outside the scope of

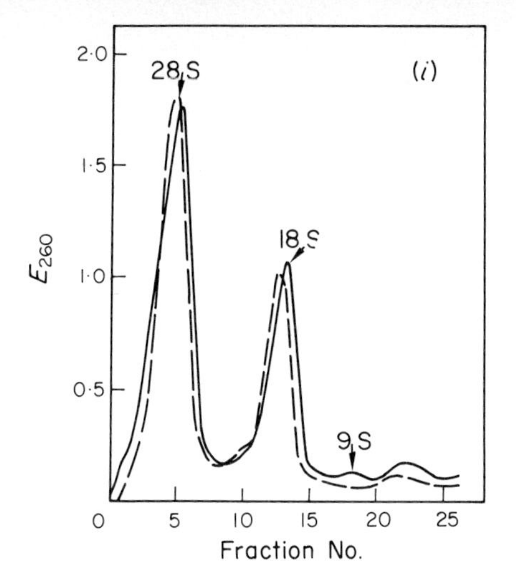

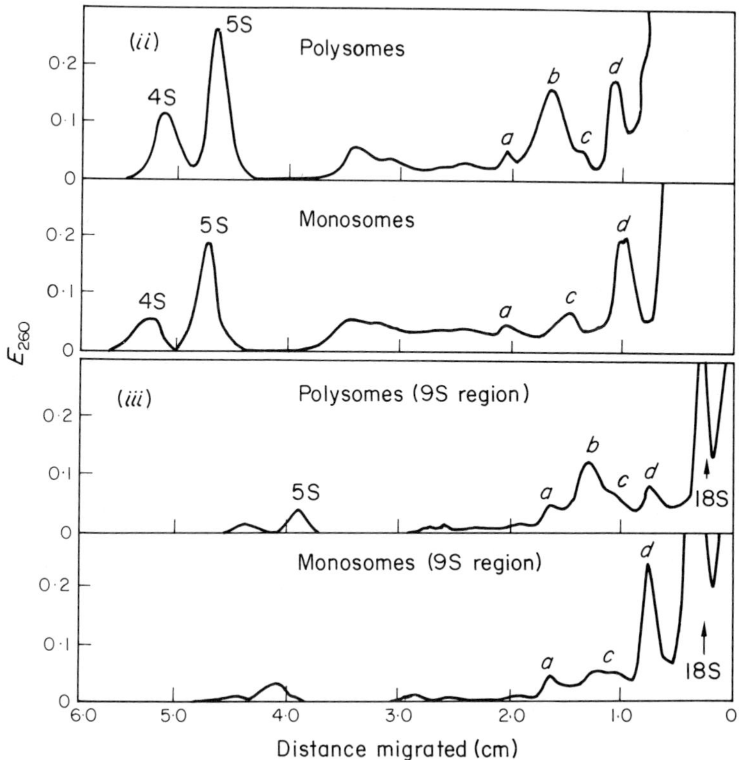

Ribonucleoprotein particles were prepared from rabbit reticulocytes and separated into ribosomes and polysomes; RNA was isolated from each separately.

(i) *Fractionation of RNA from polysomes (———) and ribosomes (. . . . . .) by sucrose density-gradient centrifugation. RNA from polysomes contains 9S material, that from ribosomes does not.*

(ii) *Fractionation of the same RNA preparations as in* (i) *on 4% polyacrylamide gels.*

(iii) *Fractionation of material from the 9S region of* (i) *on 4% polyacrylamide gels. Clearly component b in* (ii) *and* (iii) *is the 9S RNA molecule.*

Figure 13. Isolation of globin mRNA (from Gaskill and Kabat, 1971).

this review) points to the fact that this RNA is globin mRNA. From the electrophoretic mobility of the RNA, it appears to have a molecular weight of around 2×10^5, which is surprisingly high and implies that the mRNA contains more information than the sequence corresponding to globin alone. Examples of the fractionations obtained in this work are illustrated in *Figure 13*.

Acknowledgements. I am grateful to Dr I. H. Maxwell, Dr J. R. B. Hastings, Dr R. M. Malbon and Miss M. Brown for permission to quote unpublished results, and to authors, editors etc. concerned with *Figures 1, 2, 8, 12* and *13*. Work from the author's laboratory has been supported by the Science Research Council.

REFERENCES

BOCK, R. M. and CHERAYIL, J. D. (1967). In *Methods in Enzymology* (ed. Grossman, L. and Moldave, K.), vol. 12A, p. 638, New York and London (Academic Press)

DAWID, I. B. (1970). In *Control of Organelle Development; Symposium No. 24, Society for Experimental Biology* (ed. Miller, P. L.), p. 227, Cambridge (Cambridge University Press)

FEDORCSAK, I., NATARAJAN, A. T. and EHRENBERG, L. (1969). *Eur. J. Biochem.*, **10**, 450

GALIBERT, F., LARSEN, C. J., LELONG, J. C. and BOISSON, M. (1965). *Nature, London.*, **207**, 1039

GASKILL, P. and KABAT, D. (1971). *Proc. natn. Acad. Sci., U.S.A.*, **68**, 72

GILLAM, I., MILLWARD, S., BLEW, D., VON TIGERSTOM, M., WIMMER, E. and TENER, G. M. (1967). *Biochemistry*, **6**, 3043

HASTINGS, J. R. B., PARISH, J. H., KIRBY, K. S. and KLUCIS, K. (1965). *Nature, Lond.*, **208**, 645

HOLLEY, R. W. (1963). *Biochem. biophys. Res. Commun.*, **10**, 186

KIRBY, K. S., FOX-CARTER, E. and GUEST, M. (1967). *Biochem. J.*, **104**, 258

LOENING, U. E. (1968*a*). *J. molec. Biol.*, **38**, 355

LOENING, U. E. (1968*b*). In *Chromatographic and Electrophoretic Techniques* (ed. Smith, I.), vol. 2, p. 437, London (Heinemann)

LOENING, U. E. (1969). In *Organisation and control in procaryotic and eucaryotic cells; Symposium No. 20, Society for General Microbiology* (ed. Charles, H. P. and Knight, B. G. J. G.), p. 77, Cambridge (Cambridge University Press)

MANDELL, J. D. and HERSHEY, A. D. (1960). *Analyt. Biochem.*, **1**, 66

NOLL, H. (1967). *Nature, Lond.*, **215**, 360

NOLL, H. and STUTZ, E. (1968). In *Methods in Enzymology* (ed. Grossman, L. and Moldave, K.), vol. 12B, p. 129, New York and London (Academic Press)

OSAWA, S. and SIBATANI, A. (1967). In *Methods in Enzymology* (ed. Grossman, L. and Moldave, K.), vol. 12A, p. 678, New York and London (Academic Press)

PARISH, J. H. and KIRBY, K. S. (1966). *Biochim. Biophys. Acta*, **129**, 554

PARISH, J. H. and KIRBY, K. S. (1967). *Biochim. Biophys. Acta*, **142**, 273

PEARSON, R. L. and KELMERS, A. D. (1966). *J. biol. Chem.*, **241**, 767

PEARSON, R. L., WEISS, J. F. and KELMERS, A. D. (1971). *Biochim. Biophys. Acta*, **228**, 770

PRESTAYKO, A. W., TONATO, M. and BUSCH, H. (1970). *J. molec. Biol.*, **47**, 505

SEDAT, J. W. and SINSHEIMER, R. L. (1970). *Cold Spring Harb. Symp. Quant. Biol.*, **25**, 163

12

ISOPYCNIC CENTRIFUGATION OF DNA: METHODS AND APPLICATIONS

W. G. Flamm, M. L. Birnstiel and P. M. B. Walker

Equilibrium centrifugation in a density gradient is now commonly referred to as isopycnic banding or isopycnic centrifugation (Fisher *et al.*, 1964). The term is appropriate, since it implies that the material being centrifuged is driven to a point in the centrifuge cell or tube corresponding to its own solvated density, to a unique position in the density gradient where all effective centrifugal forces on it vanish. The density at this point is called the buoyant density which, in the case of DNA, depends upon chemical composition (Rolfe and Meselson, 1959; Sueoka *et al.*, 1959) and secondary structure (Schildkraut *et al.*, 1962; Vinograd *et al.*, 1963; Saunders and Campbell, 1965; Flamm *et al.*, 1967*a*). A linear relationship exists between the buoyant density of double-stranded DNA in solutions of CsCl and its guanine plus cytosine (GC) content. Schildkraut *et al.* (1962) have shown the mole fraction of guanine plus cytosine in native DNA to correspond to

$$\frac{\text{buoyant density} - 1{\cdot}660\ \text{g/cm}^3}{0{\cdot}098}$$

This equation, however, applies only to double-stranded DNA and only to DNAs containing the normal four bases, adenine, cytosine, guanine and thymine. Because single-stranded (Saunders and Campbell, 1965; Flamm *et al.*, 1967*a*) and denatured DNAs (Schildkraut *et al.*, 1962; Vinograd *et al.*, 1963; Corneo *et al.*, 1966) become more highly solvated in CsCl solution, they exhibit a higher buoyant density than double-stranded DNAs. Hence, buoyant-density data can provide useful information concerning the strandedness or helicity of DNA when its base composition is known. Thus, single strands can be identified and distinguished from double-stranded DNA, helix-coil transitions studied and renaturation reactions and products characterised.

The buoyant density of DNA can be altered experimentally in various ways. For example, incorporation of [^{15}N] or [^{2}H], or partial replacement of DNA-thymine with 5-bromouracil, during the biosynthesis of DNA causes a significant increase in its buoyant density (Meselson and Stahl, 1958; Simon, 1961; Chun and Littlefield, 1961, 1963; Dubnau *et al.*, 1965; Cutler and Evans, 1967*a*, *b*). Also, chelation with heavy metals, in particular mercuric (Davidson *et al.*, 1965; Nandi *et al.*, 1965) and silver ions, or intercalation of ethidium bromide (Radloff *et al.*, 1967), have been used to alter the buoyant density of isolated DNA.

Nuclear DNA from a higher organism has a range of buoyant densities because it has a GC content which varies by at least 10% and it may also have a satellite DNA (Kit, 1961, 1962). In the case of DNAs which are homogeneous in buoyant density (for example, DNA from simple organisms, certain satellite DNAs and possibly DNA from mitochondria and chloroplasts), the width of the band formed at equilibrium is inversely proportional to the solvated molecular weight of the DNA (Meselson *et al.*, 1957; Bond *et al.*, 1967). To distinguish between broad band widths due to heterogeneity in GC content and those due to low molecular weight, thermal denaturation studies are helpful (Flamm *et al.*, 1966*b*), since these are virtually independent of molecular weight (above 10^5 daltons) and recognise only variation in GC content.

Originally, Fisher *et al.* (1964) showed that fixed-angle rotors could be used for preparative isopycnic centrifugations, since reorientation of the density gradient is equivalent to the mechanical translation through 90 degrees with swing-out rotors. Fixed-angle rotors have a number of advantages over swing-out ones so far as isopycnic centrifugation is concerned. These include: (i) greater tube and rotor capacities; (ii) shorter gradient lengths and so shorter equilibrium times (see *Table 3*); (iii) better separation of DNA from RNA (*Figure 1*), since RNA sediments more quickly and pellets on the outer wall of the tube (*Figure 2*); and (iv) greater resolution of mixtures of DNA (Flamm *et al.*, 1966*a*).

This last advantage of fixed-angle density gradients arises from the relationship between the collected gradient volume and its corresponding density. In such gradients, the volume per unit radial distance is greatest at the top and decreases towards the bottom (*Figure 3*). Hence, the density–volume slope, that is, the

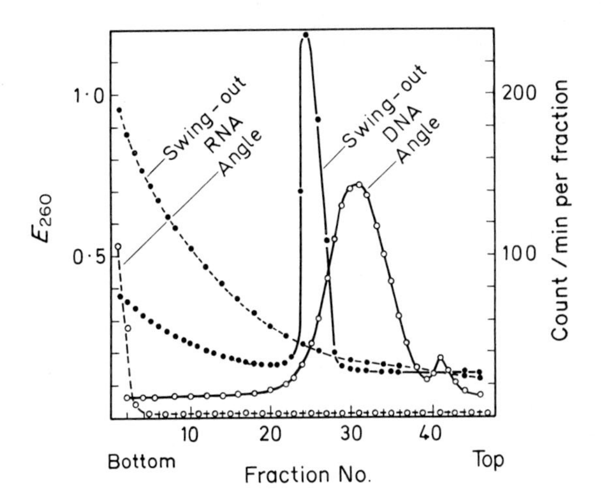

Each gradient contained 3 µg of E. coli *RNA (5000 counts/min/mg) and 50 µg of mouse-liver DNA; conditions of centrifugation specified in* Table 3

Figure 1. Absorbance (—) and radioactivity (– – –) profiles of a mixture of radioactive RNA and unlabelled DNA centrifuged to equilibrium in preparative CsCl density gradients in a fixed-angle (○) and a swing-out rotor (●)

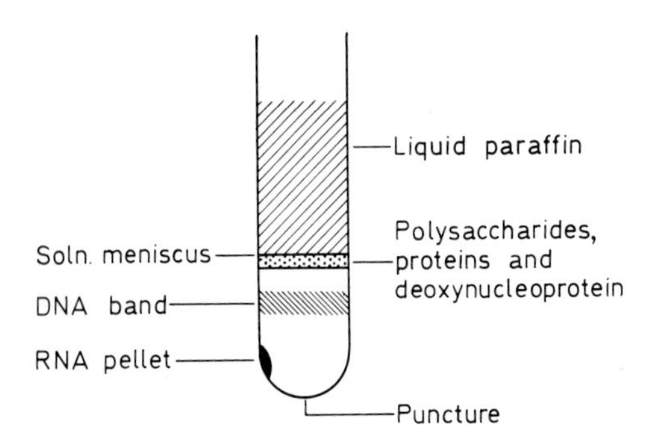

One mouse liver (approx. 1 g) was homogenised in 7·5 ml of 0·01 M *tris-HCl, pH 8·5 containing 1% (w/v) sodium dodecyl sulphate; enough solid CsCl was added to establish an initial density of 1·715 g/cm³; after preliminary low-speed centrifugation, mixture centrifuged to equilibrium (see Table 3)*

Figure 2. Diagram of a Spinco No. 30 tube following CsCl density-gradient centrifugation of a whole-cell homogenate

change in density with volume of collected solution, is least at the top of the gradient and greatest at the bottom, whereas in swing-out density gradients it is essentially constant (*Figure 4*). Fixed-angle density gradients have a far steeper density–volume slope at the bottom of the tube than at the top (0·03 and 0·008 g/cm^3 per ml, respectively, for CsCl density gradients in Spinco No. 40 tubes).

The reorientation has the effect of constricting bands near the bottom of the tube while expanding those near the top, as well as distances between bands (*Figure 3*). Band peaks (that is, points of greatest DNA concentration) are separated by greater volumes in

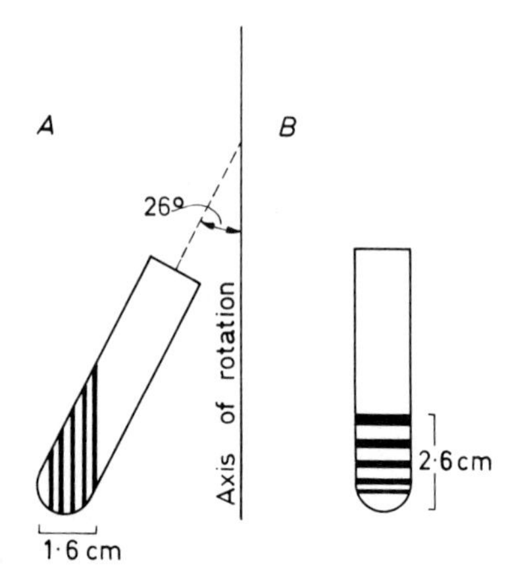

Figure 3. Scale diagram of a fixed-angle centrifuge tube in its at-speed position, A and in unloading position, B (from Flamm et al., *1966*a*). Note increase in band width upon changing tube positions from A to B*

fixed-angle than in corresponding swing-out density gradients. Hence, the former are less susceptible to disturbances which cause mixing, and larger fractions may be collected from them without sacrificing resolution. The practical virtue of these considerations is evident from *Figure 5* which shows the fractionation pattern for a mixture of three DNAs, each with a buoyant density differing by 0·1 g/cm^3 from its nearer neighbour, in both types of density gradients. A volume of 1·2 ml separates the peaks of *E. coli* DNA and mouse major-band DNA in fixed-angle (*Figure 5A*), but one of only 0·3 ml in swing-out density gradients (*Figure 5B*). Furthermore, only in the former are mouse satellite DNA and major-band DNA resolved, although not to the extent which would be predicted on the basis of the difference in their buoyant densities. This is due to the viscosity of the heavily-loaded zone of major-band DNA, which causes this band to drag along the tube walls as it is displaced downwards during unloading, resulting in tailing and consequent loss of resolution between it and the succeeding zone of satellite DNA. This effect also enhances the separation between

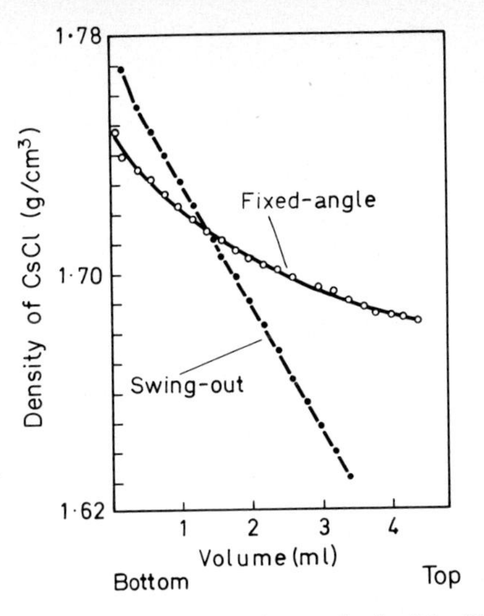

Fixed-angle gradient centrifuged in a Spinco No. 40 rotor, swing-out gradient in a Spinco SW 39L rotor; initial densities, 1·720 g/cm^3 for both gradients

Figure 4. Relationship between density and volume in the two types of rotor

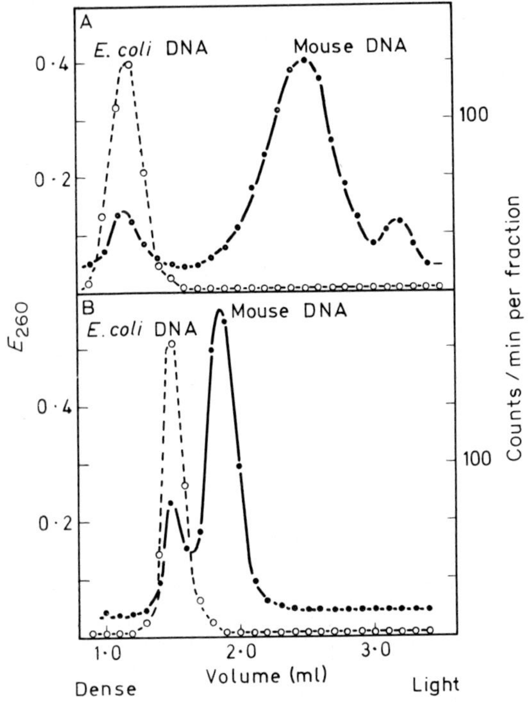

Each gradient had initial density of 1·710 g/cm^3; 3 μg of [^{14}C]DNA from E. coli *and 50 μg of unlabelled mouse-liver DNA; conditions of centrifugation as* Figure 1.

Figure 5. Absorbance (● —— ●) and radioactivity (○ — — ○) profiles of a mixture of [^{14}C]DNA from E. coli *and unlabelled DNA from mouse liver centrifuged to equilibrium in fixed-angle (A) and swing-out (B) CsCl density gradients*

the band of *E. coli* DNA, which is lightly loaded, and that of mouse major-band DNA.

Additional data regarding the relative resolving powers of the two types of density gradients, for a series of DNA concentrations, are summarised in *Table 1*. These data show that, for a given degree of resolution, a fixed-angle rotor tube has a capacity for DNA roughly ten times that of the corresponding swing-out type. This increased capacity is a consequence not only of the effect of re-orientation on the gradient, but also partly of the viscosity effect illustrated in *Figure 5*. The significance of the latter increases as the

Table 1. COMPARISON OF THE RESOLVING POWER AND CAPACITY OF FIXED-ANGLE AND SWING-OUT ROTORS (FROM FLAMM *et al.*, 1966*a*)

μg of DNA*/tube*		*Rotor type*	*Percentage purity of* DNA *peak*★		
E. coli	*mouse*		*E. coli*	*major*	*minor*
2·7	23	No. 40	100		91
	45		86	100	92
	90		68	100	77
	135		65	100	84
	225		79		65
	315		61		30
2·7	15	SW 39L	75	100	65
	45		34	100	33
	90		18	98	23
	135				16
	315		1	88	12

★ The percentage purity of labelled *E. coli* DNA from a preparative density-gradient centrifugation in a fixed-angle (Spinco No. 40) or a swing-out (Spinco SW 39L) rotor was determined from the change in its specific radioactivity before and after mixing and centrifuging; that of the major and minor mouse-liver DNA components by Spinco Model E analysis of pooled peaks; unfractionated mouse-liver DNA contains ~88% major and 12% minor component.

bands become more heavily loaded, but is always much greater in swing-out gradients due to their steeper density–volume slope.

To achieve a level of resolution with a swing-out density gradient comparable to that obtained with a fixed-angle one, it is necessary to use approximately one-tenth as much DNA per tube, and to collect about 4 times as many fractions. In the case of swing-out rotors such as the SW 39L or MSE 3 × 5, this involves using 4 to 5 μg of DNA per tube and collecting 200 fractions, each about 0·02 ml. Under these conditions even that fraction with the highest concentration of DNA (that is, the peak fraction) would contain less than 1 μg of DNA.

Although swing-out density gradients have been used to great advantage in the past (e.g. Rownd *et al.*, 1961; Giacomoni and

Spiegelman, 1962; Goodman and Rich, 1962; Yankofsky and Spiegelman, 1962, 1963; Chipchase and Birnstiel, 1963; Britten, 1965; Dubnau *et al.*, 1965; Wallace and Birnstiel, 1966), it is obvious that the use of fixed-angle rotors for isopycnic centrifugation has made it possible to isolate a particular DNA from a mixture of DNAs which differ only slightly in buoyant density.

Recently, it has been shown (Anet and Strayer, 1969*a*, *b*) that solutions of NaI have great potential for density-gradient centrifugation of DNA and have important advantages over CsCl solutions. Better resolution is obtained in gradients of NaI even in comparable centrifugal fields, apparently because shallower gradients are generated. Further, the dye ethidium bromide, which complexes with DNA to produce an intense colour, does not adversely affect the resolution of DNAs in NaI density gradients as it does in CsCl. Thus, DNA bands can be made visible to the naked eye, making drop-collection very convenient and easy. The cost of NaI is only about one-tenth that of CsCl which, in these austere times, is a real advantage as well. However, it does have one disadvantage. Iodide ion absorbs at 260 nm and solutions of NaI with a density of 1·55 g/cm^3 exhibit optical densities greater than 2. This would preclude the use of NaI in analytical centrifugation, which depends upon the u.v. absorption of DNA bands. This is particularly unfortunate since it will undoubtedly prolong the time required to learn as much about the behaviour of DNA in NaI as is known today of its characteristics in CsCl solutions. In any event, it is to be expected that the combination of NaI gradients with fixed-angle rotors will lead to new dimensions in studies of DNA.

EXPERIMENTAL PROCEDURES

PREPARATION OF CSCL DENSITY GRADIENTS

Unlike density-gradient sedimentation techniques where sucrose and similar substances have enjoyed such great popularity, isopycnic CsCl centrifugation does not require a preformed gradient. The requisite weight of CsCl crystals is simply added to a specified volume of solution containing the DNA (usually in 0·01 M tris-HCl, pH 8·4) to produce the initial density desired (*Table 2*). Alternatively, a useful relation for estimating the weight of CsCl contained in 100 g of solution at 25°C is

$$\text{Weight per cent} = 137{\cdot}48 - 138{\cdot}11(1/\rho^{25^\circ})$$

Table 2. WEIGHT OF CSCL TO BE ADDED TO EACH ML OF SOLUTION TO PRODUCE THE INITIAL DENSITY INDICATED

Initial density, g/cm³	CsCl, g	*Final** *volume of* CsCl *solution at 25°C,* ml
1·660	1·17	1·30
1·670	1·19	1·30
1·680	1·22	1·31
1·690	1·24	1·31
1·700	1·26	1·32
1·710	1·28	1·32
1·720	1·31	1·33
1·730	1·33	1·34
1·740	1·35	1·34
1·750	1·36	1·34
1·760	1·37	1·35

* Volume of solution containing DNA plus volume contributed by CsCl.

from which the weight of CsCl per weight of buffer or water can be calculated for any density corresponding to 30 to 60% (w/w) solutions (Vinograd and Hearst, 1962).

CHOICE OF INITIAL DENSITY

For preparative fixed-angle work, a starting density slightly higher than the most dense macromolecular component is used so that all bands will appear in the upper portion of the gradient (that is, the region of shallow density–volume slope). Usually, the density is made 0·010 to 0·020 density units higher than the buoyant density of the heaviest component. For analytical centrifugation, initial densities are set as close as possible to the buoyant density of the sample.

DETERMINING INITIAL DENSITY

Precise density information, accurate to $\pm 0{\cdot}001$ g/cm³, can be obtained with either a micropycnometer or a refractometer. When the latter is used, the linear relation between refractive index and initial density ($\rho_i^{25^\circ}$) is

$$\rho_i^{25^\circ} = 10{\cdot}8601 \times (\text{refractive index})^{25^\circ} - 13{\cdot}4974$$

which may be used for solutions of CsCl within the density range of 1·25 to 1·9 g/cm^3 (Vinograd and Hearst, 1962).

CORRECTING INITIAL DENSITY

Frequently the exact density is not arrived at on the first attempt and slight adjustments are necessary, particularly for fixed-angle gradients where the exact initial density is most crucial (within $\pm 0{\cdot}003$ g/cm^3). The following equations are guides to the amount of solid CsCl or buffer required to change the density of a given volume, V, by a specified increment, $\Delta\rho^{25^\circ}$

$$\text{g of CsCl} = \Delta\rho^{25^\circ} \times V \times 1{\cdot}32$$
$$\text{ml of buffer} = \Delta\rho^{25^\circ} \times V \times 1{\cdot}52$$

CENTRIFUGATION

The gradient is formed upon application of the centrifugal field. The time taken to establish an equilibrium density gradient for CsCl depends largely upon the length of the liquid column being centrifuged (Van Holde and Baldwin, 1958). The other time which concerns us is that needed for the DNA to reach equilibrium within the density gradient. This is also dependent upon column length (Meselson, 1957). In actual practice, we use the volumes (column lengths), speeds and times set down in *Table 3*. These conditions

Table 3. CONDITIONS USED TO ATTAIN ISOPYCNIC EQUILIBRIUM AT 25°C FOR A DNA OF MOLECULAR WEIGHT 10^7 DALTONS OR HIGHER

Rotor type	*Volume of soln. per tube,*★ ml	*Length of gradient,*† cm	*Speed,* rev/min $\times 10^{-3}$	*Time of run,* h
MSE 10 × 10‡	4·5	1·8	35	60
MSE 10 × 10‡	4·5	1·8	42	42
MSE 8 × 50‡	10·0	1·9	30	70
MSE analytical	0·65	1·2	45	20
Spinco No. 40§	4·5	1·6	33	60
Spinco SW 39L§‖	3·0	2·8	33	60
Spinco No. 30§	10·0	1·9	25	90

★ Except for the analytical, all tubes were overlayed with liquid paraffin, which helps to stabilise the gradient and prevents tube collapse.
† During centrifugation but uncorrected for compression.
‡ Polypropylene tubes
§ Nitrocellulose tubes
‖ Swing-out rotor

have evolved from practical experience over the past five years and do not necessarily represent optimal conditions rigorously derived from theoretical considerations. Obviously, the short gradient lengths possible in fixed-angle rotors are an advantage by facilitating more rapid attainment of equilibrium.

The time required to reach equilibrium can be further shortened by use of a technique known as 'gradient relaxation'. Essentially what one does is to 'overspeed' the rotor (Anet and Strayer, 1969*b*), that is, if equilibrium at 35 000 rev/min is desired and this would normally require 60 to 80 h of operation, one could instead centrifuge at a higher speed (e.g., 45 000 rev/min) then, after 18 h, readjust to the desired lower speed. In 24 h or so, equilibrium at the lower speed is attained and approximately 20 h of centrifugation time has been saved. The basic principle of the method is that equilibrium is more rapidly attained at higher rotational speeds; once equilibrium has been reached by 'overspeeding', a new equilibrium at a lower speed is quickly established by diffusion. Though our experience with the method has been limited to only a few operational conditions, the results obtained have been most satisfactory.

It should be remembered that the maximum speed at which any rotor may be operated is in part determined by the density of the solution being centrifuged. However, since fixed-angle gradients occupy only one-third of the tube's total volume (the remainder being occupied by liquid paraffin), no, or very little, correction is ordinarily needed for fixed-angle rotors.

GRADIENT CAPACITY

The maximum quantity of DNA per tube which can be processed depends upon the rotor and the conditions selected as well as on

Table 4. APPROXIMATE MAXIMUM QUANTITY OF DNA WHICH MAY BE USED WITH GOOD RESOLUTION (DENSITY DIFFERENCE = 0·005 g/cm³)

Rotor	*Speed**, rev/min × 10^{-3}	μg DNA/*tube*
SW 39L	33	<10
MSE 10 × 10	35	45
Spinco No. 40	33	45
MSE 8 × 50	30	150
Spinco No. 30	25	150

* Specified because at higher speeds the gradient becomes steeper, compressing the DNA band into a smaller volume and giving rise to viscosity disturbances during fractionation.

the nature of the experiment (*Table 4*). If two DNAs differing in buoyant density by 0·005 g/cm^3 are to be separated, only 50 to 150 μg of nucleic acid can be processed in one fixed-angle tube. If, on the other hand, the gradient is simply used to prepare or purify total cellular DNA and no fractionation is needed, the amount can be as much as 5 mg of DNA per tube (MSE 8 × 50 or Spinco No. 30 rotor).

GRADIENT RECOVERY AND FRACTIONATION

After centrifugation for the prescribed period at 25°C, the rotor is decelerated *without braking* and is then carefully removed from the centrifuge. To avoid disturbing the gradient, as much care as possible should be exercised in removing the rotor and extracting the tubes. For this reason, MSE polypropylene tubes are ideally suited in that they are sufficiently hard and inflexible to make removal from the rotor and other handling very easy. Following removal of the tube caps, the tubes are fitted to the simple device illustrated in *Figure 6*, which consists of two hypodermic syringes (20 ml and

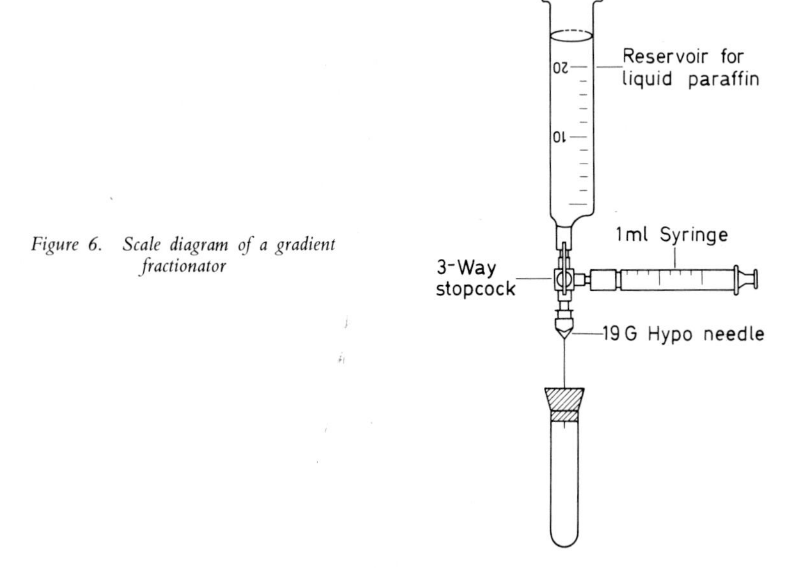

Figure 6. Scale diagram of a gradient fractionator

1 ml) attached to a 3-way stopcock (B-D Yale). The centrifuge tube containing one-third CsCl solution and two-thirds liquid paraffin is joined to the device with a rubber stopper and a 19G hypodermic needle. After a puncture is made in the bottom of the

centrifuge tube, CsCl solution is forced through the hole by displacement with liquid paraffin which enters the tube through the hypodermic needle from the 1 ml syringe. The volume of each fraction is precisely controlled and measured by that of liquid paraffin used to displace it. Alternatively, the tubes may be more simply unloaded by allowing the gradient to drip out under the force of gravity. Ordinarily, 45 fractions of 0·1 ml are collected which are then diluted to 0·2 ml with 0·1 M tris-HCl, pH 8·4. The extinction of these fractions at 260 nm may be conveniently determined in the Beckman DB spectrophotometer.

DETERMINING BUOYANT DENSITY (ANALYTICAL ONLY)

The buoyant density of a DNA sample ($\rho_s^{25°}$) may conveniently be determined from experiments where DNA of known density ($\rho_m^{25°}$) is included as a marker, according to the equation

$$\rho_s = \rho_m + \frac{\omega^2}{2\beta_o}(r_s^2 - r_m^2)$$

where r_m is the distance (cm) from the axis of rotation to the centre of the marker band; r_s (cm), that to the centre of the sample band; ω is 2π times the rotor speed (rev/sec); β_o varies with solution density: at densities of 1·65, 1·70, 1·75, 1·80 and 1·85 g/cm^3, $\beta_o \times 10^{-9}$ = 1·190, 1·190, 1·199, 1·215 and 1·236, respectively (Ifft *et al.*, 1961).

In the absence of a marker, the so-called 'isoconcentration distance' or root-mean-square position, r_e may be substituted for r_m, according to

$$r_e = [\tfrac{1}{2}(r_b^2 + r_a^2)]^{\frac{1}{2}}$$

where r_b is the distance (cm) from the centre of rotation to the bottom of the cell and r_a (cm), that to the meniscus. This is the point in the centrifuge cell where the density of CsCl equals that of the original solution (Vinograd and Hearst, 1962).

COMPUTING MOLE PER CENT GUANINE PLUS CYTOSINE (GC)

$$\text{GC} = \frac{\rho_s - 1{\cdot}660}{0{\cdot}098} \times 100$$

This equation is applicable to native, double-stranded DNA possessing only the four common bases, adenine, thymine, guanine and cytosine. The reproducibility of the measurement is better than 1% GC (Schildkraut *et al.*, 1962).

COMPUTING MOLECULAR WEIGHT FROM BAND WIDTH

For simple DNAs homogeneous in GC, the molecular weight, M, is related to the concentration distribution by

$$M = \frac{RT\rho_i}{(\mathrm{d}\rho/\mathrm{d}r)\sigma^2\omega^2 r_s}$$

where σ is the standard deviation of the Gaussian concentration distribution (one-half the band width (cm) at 6/10 peak height), R the gas constant, T the absolute temperature, and $(\mathrm{d}\rho/\mathrm{d}r)$ the density gradient (g/cm^3 per cm). The molecular weight of the sodium salt of DNA is 0·75 times that of the caesium salt (Meselson *et al.*, 1957).

ISOLATION AND PURIFICATION OF DNA BY ISOPYCNIC CENTRIFUGATION

Unfortunately, there does not seem to be a universally applicable procedure whereby DNA from any organism, tissue or organelle can be isolated and purified. There are, of course, many excellent methods (see review by Kirby, 1964), but no single one of them is ideally suited to all situations. The lack of general applicability is due to several factors, including (i) the different ways in which DNA is 'packaged' within an organism; (ii) the different types of proteins with which it may be associated or conjugated; (iii) variation in the nature of this association or binding; (iv) the proportion of DNA per unit dry weight of cellular material; (v) the relative amounts of the four main contaminants (protein, lipoprotein, polysaccharides and RNA); and (vi) variation in contaminating nucleases. Some or all of these factors may vary from organism to organism and, indeed, large differences may be encountered between the tissues of an organism, which are further accentuated if organelles (nuclei, nucleoli, mitochondria and chloroplasts) are studied.

The use of detergents for removing proteins and inhibiting nuclease activity (Marco and Butler, 1951; Kay *et al.*, 1952) is one feature most methods have in common. Phenol in combination with lipophilic salts, for example, dithiocarbamate (Kirby, 1962*a*),

4-aminosalicylate or 8-hydroxquinoline (Kirby, 1962*a*, *b*), is a popular alternative.

Since the procedures which liberate DNA from tissue also liberate RNA, the problem of separating these two nucleic acids is virtually universal, although its extent depends very much on the source of the nucleic acid. A general solution has been to add deoxyribonuclease-free ribonuclease (boiled ribonuclease) and then remove ribonucleotides by dialysis. The disadvantage of this method is that any deoxyribonuclease contamination within the nucleic acid preparation itself might become functional during incubation with ribonuclease. Other methods for eliminating RNA are also available (Mandell and Hershey, 1960; Colter *et al.*, 1962; Aldridge, 1960; Jones, 1963), but they also lack general applicability. One alternative is to centrifuge nucleic acid preparations in CsCl density gradients. The RNA is sedimented, since it has a buoyant density greater than the density of any CsCl solution, while DNA is usually banded in the upper half of the density gradient. The method has not enjoyed widespread use, mainly because long centrifugation times are needed (see *Table 3*) and, also, it was thought that swing-out rotors, which have a very limited total capacity, were required. Also, in these rotors, RNA is pelleted directly on the bottom of the tube where the puncture is usually made for collecting fractions. Elution of the RNA undoubtedly occurs as the gradient solution streams through the hole in the pellet and the bottom of the tube (see *Figure 1*).

However, since we know now that isopycnic centrifugations are not limited to swing-out rotors, the preparative potential of CsCl density gradients needs to be reconsidered. Obviously, tube and rotor capacities are greater and density-gradient lengths considerably shorter with fixed-angle rotors (*Table 3*). Hence, (i) more DNA can be processed per centrifugation; (ii) RNA sediments to the wall of the tube more rapidly than it does to the bottom; and (iii) RNA pellets form on the outer wall where they do not interfere with the ultimate dripping of the gradient (see *Figure 1*). Furthermore, because the density range of a fixed-angle density gradient is narrow (generally between 1·74 g/cm^3 at the bottom of the tube to 1·68 g/cm^3 at the top), polysaccharides, which are a common contaminant of DNA preparations (Ritossa and Spiegelman, 1965; Counts and Flamm, 1966) and certain deoxynucleoprotein contaminants (Counts and Flamm, 1966) are concentrated at the meniscus and easily eliminated. Another advantage of using isopycnic CsCl centrifugation as a final stage in the purification process is that strong salt disrupts most DNA–protein bonds, thereby reducing contamination of DNA by histones and residual proteins. In fact, DNA prepared by a modification of the Marmur (1961) procedure (Flamm *et al.*, 1966*a*) contained 2 to 3% of protein by weight (Bond *et al.*, 1967). However, after banding in a CsCl density gradient, contamination was reduced to less than 0·6% of protein by weight, and even this value grossly overestimated the extent of protein contamination (*Table 5*).

Because highly concentrated solutions of caesium salts serve to remove proteins from DNA and then render these insoluble so

Table 5. UPPER LIMIT OF CONTAMINATION* OF SATELLITE AND MAJOR-BAND DNA BY VARIOUS PROTEIN FRACTIONS (BOND *et al.*, 1967)

Preparation	*Radioactivity incorporated from*		*Percentage contamination of* DNA *by protein*			
			Leucine		*Lysine*	
	Leucine	*Lysine*				
	counts/min/mg		Major DNA	Satellite DNA	Major DNA	Satellite DNA
Major DNA	356	333				
Satellite DNA	286	360				
Protein						
total	101 000	61 200	0·35	0·29	0·54	0·59
soluble	90 000	72 000	0·39	0·32	0·46	0·50
ribosomal	119 000	55 000	0·30	0·24	0·60	0·65
nuclear	63 400	72 500	0·56	0·45	0·46	0·50
histones	63 700	52 000	0·56	0·45	0·64	0·69

* Eight-day-old mice (8–10 g) were labelled with either [^{3}H]d,l-leucine or [^{3}H]d,l-lysine. Each mouse received 100 μC per injection once a day for 5 days. On the last day, the animals were sacrificed and protein and DNA fractions were prepared (Flamm and Birnstiel, 1964*a*, *b*; Flamm *et al.*, 1966*a*) and specific radioactivities determined. Calculations of percentage contamination assume that all radioactivity found in the DNA fractions arises from protein.

that they quickly concentrate at the meniscus, it is possible to use CsCl during the initial stages of DNA preparation and purification. The method simply involves grinding the tissue in a 1% (w/v) solution of sodium dodecyl sulphate in a suitable buffer (tris, phosphate or versene, pH 8) and then adding the requisite weight of solid CsCl to establish the desired density, usually 1·70 to 1·76 g/cm^3 (see *Table 2*). At this point, DNA is released from the tissue, the solution becomes highly viscous and signs of protein precipitation are seen. By performing a relatively short (30 min) centrifugation at 10 000*g* most of the precipitated protein can be removed from the meniscus, and the solution is then ready for isopycnic centrifugation according to the conditions outlined in *Table 3*. The results of one such centrifugation are shown in *Figure 2*. Polysaccharides and glycogen with a density of 1·67 g/cm^3 (Counts and Flamm, 1966; Anderson *et al.*, 1966) are banded at the meniscus along with protein (density, 1·2 to 1·4 g/cm^3)

(Anderson *et al.*, 1966) and deoxyribonucleoprotein contaminants (density 1·60 to 1·64 g/cm^3). DNA is found in the upper two-thirds of the gradient while RNA is compressed into a pellet on the outer wall of the tube (*Figure 2*).

In addition to the simplicity of the method, the high salt concentration has the desirable effect of inhibiting nucleases which are then separated from the position of the DNA band by centrifugation. For this reason, we have applied the method to those tissues where endogenous nuclease activity is high (mouse skin, intestinal epithelium, lung and myeloma tissue), or in situations where maintenance of the highest possible molecular weight was necessary (mitochondrial, nucleolar and, occasionally, L-cell DNA). In these instances, the buoyant densities of the resultant DNA were such as to suggest that deproteinisation was complete.

We are not suggesting that the use of caesium salt density gradients necessarily solves all of the many problems involved in isolating and purifying DNA from the varied sources available. However, the advantages and simplicity of the method make it an attractive alternative to those in current use.

FRACTIONATION OF DNA AND ISOLATION OF SATELLITE BANDS

It is often difficult to define exactly what is meant by the term 'fractionation' as applied to DNA. For instance, in viruses and bacteria where the entire genome consists of a single DNA molecule, the DNA must be broken into fragments before it can be fractionated. Although there is certainly no difficulty in doing this, the question arises as to what is then really fractionated? If, for example, these random DNA fragments are simply separated on the basis of molecular size, each fraction is likely to contain precisely the same genetic information. Whether or not this should be called fractionation depends upon the point of view, but certainly from a biological vantage point we would want each fraction to be unique in terms of its genetic information.

The problem confronting us is two-fold. First, every nucleotide sequence of a given length behaves physico-chemically very much like any other of comparable length and, second, there are thousands of these sequences within a bacterial genome, each one of which is slightly different from the others. To separate any one from the entire group of sequences present is obviously an enormous task.

FRACTIONATION IN TERMS OF ORDER OF DNA SYNTHESIS

Many micro-organisms are known to start DNA replication at one fixed point on the chromosome. Dubnau *et al.* (1965), O'Sullivan and Sueoka (1967) and Cutler and Evans (1967*a*, *b*) have all made use of this property to isolate specific parts of the genome by introducing a heavy isotope into the DNA of a synchronously-growing

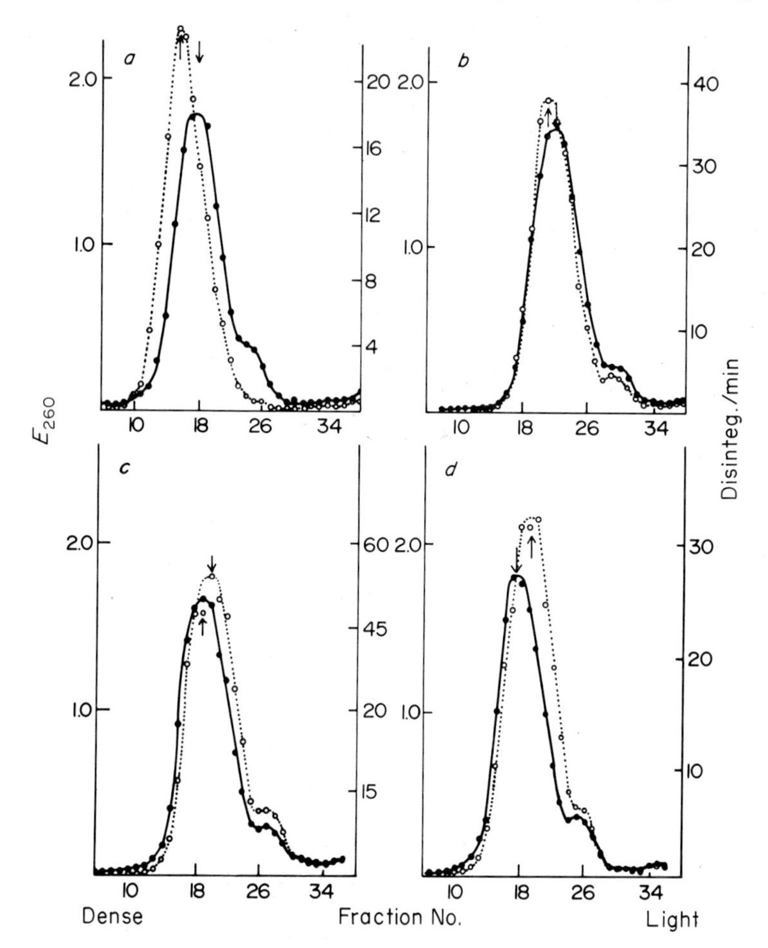

Four 150 ml suspension cultures (5 × 10⁵ cells/ml), which had been blocked in metaphase, were incubated for 1 h with 100 μCi of [³H]thymidine (a), 2 h; (b), 5 h; (c), 7 h; and (d), 9 h after reversal of the block. DNA was extracted by the sodium dodecyl sulphate–CsCl method and centrifuged to equilibrium in CsCl gradients as described in the text

Figure 7. Absorbance (●——●) and radioactivity (○.... ○) profiles of DNA from mouse lymphoma cells grown in tissue culture (from Flamm et al., *1971)*

culture of bacteria. In higher organisms there are many replication points per chromosome; however, the method is still applicable (Taylor, 1966).

Recently, it has been shown that in mouse lymphoma cells the nucleotide sequences of DNA which replicate in early S-phase are enriched in GC content (Flamm *et al.*, 1971; Tobia *et al.*, 1970; Bostock and Prescott, 1971). In fact, as indicated by *Figure* 7, the GC content of newly-replicated sequences gradually decrease as synchronised cells progress through S-phase. Further, it is clear that the entire mouse satellite fraction is replicated only near the end of S-phase. These observations, coupled with cytogenetic studies, assert that specific sequences (or genes) are only replicated at certain specified times in S-phase. Thus, even with the complicated mammalian genome, it becomes possible to isolate specific genetic regions by the use of a density label and appropriately synchronised cells.

FRACTIONATION BY BASE COMPOSITION

DNA from higher organisms is more heterogeneous in base composition than bacterial DNA and will therefore form a broader,

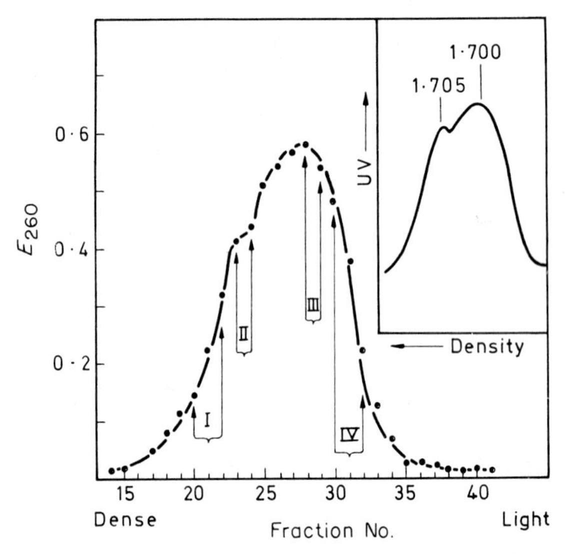

Conditions of centrifugation specified in Table 3. *Arrows indicate fractions pooled for subsequent analysis (see* Figures 9, 14, Table 6). *Insert: microdensitomer tracing of u.v. photograph of the same DNA (2 μg), centrifuged to equilibrium in an MSE analytical ultracentrifuge (45 000 rev/min for 18 h at 25°C); initial densities: preparative gradient, 1·715, analytical gradient, 1·705 g/cm³*

Figure 8. Absorbance profile of a typical preparative CsCl density-gradient centrifugation (MSE 10 × 10 rotor) of 50 μg of DNA from guinea-pig liver

and occasionally skewed, band in a CsCl density gradient (*Figure 8*). If various parts of this band are selected and rerun in an analytical centrifuge, each fraction has a unique buoyant density (*Figure 9*) and, as expected, the denser fractions possess a higher GC content and T_m (mid-point of a thermally-induced helix-coil transition)

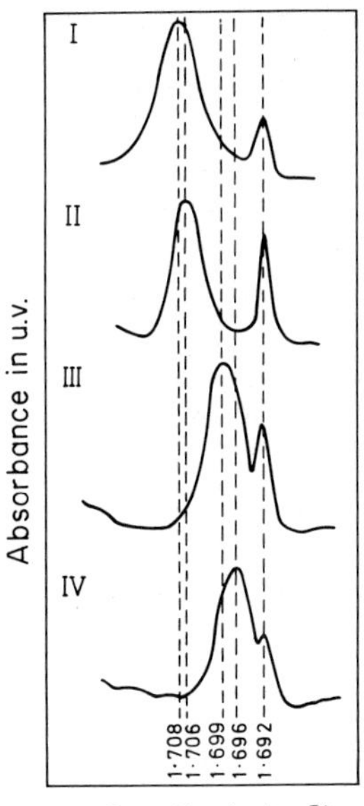

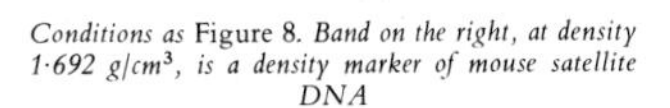
Conditions as Figure 8. *Band on the right, at density 1·692 g/cm³, is a density marker of mouse satellite DNA*

*Figure 9. Microdensitometer tracings of u.v. photographs of pooled guinea-pig DNA fractions (*Figure 8*) centrifuged to equilibrium in an MSE analytical ultracentrifuge*

Table 6. BUOYANT DENSITY, T_m AND BASE COMPOSITION OF GUINEA-PIG DNA (FLAMM, MCCALLUM AND WALKER, UNPUBLISHED)

Fraction★	*Density*, g/cm³	T_m†, °C	G+C, mole per cent
I	1·708	78·2	46·0
II	1·706	77·5	43·0
III	1·699	74·3	38·5
IV	1·696	73·5	34·0
Total DNA	1·700	74·9	39·5

★ Fractions correspond to those of *Figure 8*.
† In 0·06 M sodium phosphate, pH 6·8.

than the lighter (*Table 6*). Each fraction is, therefore, a large population of sequences with different average properties, and the method gives a separation comparable to those of thermal chromatography (Bolton and McCarthy, 1962; Miyazawa and Thomas, 1965; Walker and McLaren, 1965).

FRACTIONATION BASED ON MULTIPLE DNA SEQUENCES

Good examples of multiple DNA sequences are found in mitochondria and chloroplasts. There are many of these organelles in each cell and the molecular weight of the DNA from them is relatively small, which limits the possible amount of heterogeneity. If the GC content of this DNA is also different from that of the remainder of the DNA, it forms a sharp separate band in CsCl density gradients which allows the separation of highly purified DNA from mitochondria (Rabinowitz *et al.*, 1965) and chloroplasts (Ray and Hanawalt, 1965; Edelman *et al.*, 1965). The rapid rate at which these cytoplasmic DNAs renature after thermal denaturation is consistent with their presumed homogeneity, size and relatively small number of base pairs per mitochondria or chloroplast (Borst and Rittenberg, 1966; Wells and Birnstiel, 1967). Thus, DNA can be isolated from these organelles because of its small size, which confers homogeneity in base composition and,

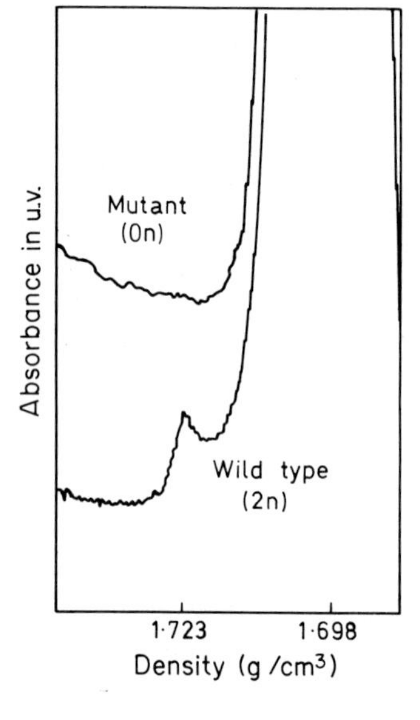

CsCl density gradients (initial density 1·720 g/cm³) overloaded with either 87 μg of mutant DNA or 69 μg of wild-type DNA to reveal the minor satellite component present in wild-type; tracings displaced vertically to facilitate comparison; both gradients centrifuged in Beckman Model E analytical ultracentrifuge for 26–48 h at 44 770 rev/min

Figure 10. Microdensitometer tracings of u.v. photographs of DNA from mutant and wild type Xenopus laevis *(Birnstiel* et al., *1966)*

in turn, allows the possibility of the composition being different from the bulk of the nuclear DNA.

A more complicated situation occurs in nuclear DNA which also has multiple copies within a single genome. There is also considerable evidence that the degree of repetition of different sequences varies considerably. Perhaps the most understandable example of multiple copies is that of ribosomal RNA cistrons, that is, the stretches of DNA that code for ribosomal RNA. We can easily understand why there is a need to have multiple copies of these cistrons since so much of the cell's RNA is ribosomal and thus produced by these cistrons.

In amphibian cells, ribosomal cistrons have the same high GC content (60 to 70%) as the ribosomal RNA, despite the fact that the average GC content of animal DNA is low, usually around 40%. This difference in base composition is reflected by a correspondingly large difference in buoyant densities (0·022 g/cm^3), thus making possible their isolation by CsCl density-gradient centrifugation. These stretches of DNA are well resolved from the total mass of nuclear DNA, as shown in *Figure 10*, in which the unfractionated DNA of two different genotypes of *Xenopus laevis* are compared in the analytical ultracentrifuge. The mutant, which is known to lack ribosomal RNA cistrons, is used here as a control to illustrate that the satellite band at 1·723 g/cm^3 is only present in the genotype which possesses these cistrons. The satellite has been further identified as representing the ribosomal RNA cistrons by RNA–DNA hybridisation studies.

For these experiments the satellite material was first enriched by an isopycnic CsCl centrifugation in either a Spinco No. 30 or an MSE 8 × 50 rotor. The fractions containing satellite DNA were pooled, diluted 4-fold with water and pelleted at 105 000*g* for 18 h. The pellets were redissolved in buffer and further purified by recycling through a second density gradient. Analytical centrifugation of the purified satellite DNA shows a single homogeneous component of density 1·723 g/cm^3 (*Figure 11*), proving that even minor DNA components representing no more than 0·3% of the total DNA can be isolated in a high state of purity. Small as the value of 0·3% seems, it nevertheless represents a total of 1200 individual cistrons or coding units for ribosomal RNA per *Xenopus laevis* genome.

As was demonstrated by Walker and McLaren (1965), one-tenth of the total mass of nuclear DNA in mouse cells undergoes very rapid renaturation following thermal denaturation, even at low concentrations of DNA. In fact, Waring and Britten (1966) have shown that this fraction of mouse DNA renatures some fifteen times faster than simian virus 40 DNA, which only has 6000 base pairs, thus implying that the rapidly-renaturing fraction consists of a single nucleotide sequence of only 300 to 400 base pairs. Since it represents at least 10% of the total DNA in mouse cells, there must be about one million copies of this short sequence per mouse genome.

The possibility that this fraction might be identified with either viral or cytoplasmic DNA, or with the DNA sequences coding for ribosomal RNA, was discounted on the basis of the unusually large proportion of rapidly-renaturing DNA. On the same basis, however, several workers (Britten, 1965; Flamm *et al.*, 1966*b*) suggested that it might be identified with the satellite band previously seen in

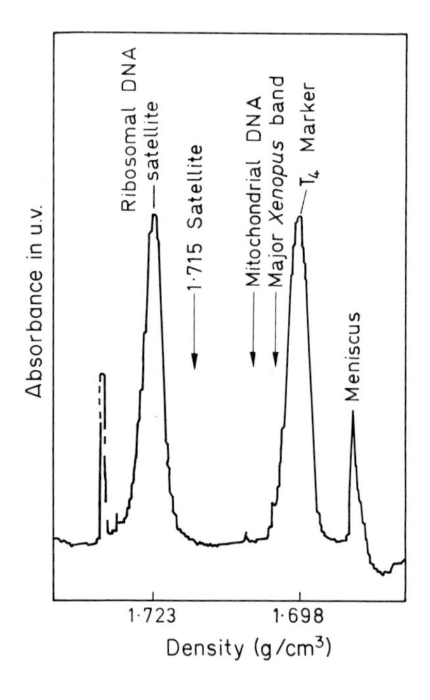

Equivalent amounts of Xenopus laevis *satellite DNA and T4 DNA (approx. 1 μg) centrifuged to equilibrium in CsCl density gradient; initial density, 1·720 g/cm^3, under conditions for* Figure 10

Figure 11. Microdensitometer tracing of a u.v. photograph of purified ribosomal DNA satellite (from Birnstiel, 1967)

equilibrium CsCl density gradients (Kit, 1961) and shown to represent approximately 10% of the total DNA. Britten (1965) provided additional support for this view by showing that, after renaturation, this fraction possessed a buoyant density very similar to that of native mouse satellite DNA (1·691 g/cm^3). The issue was finally settled when native mouse satellite DNA was isolated in a highly-purified form by isopycnic CsCl centrifugation (*Figure 12*) and shown to possess the renaturation characteristics and kinetics of the rapidly-renaturing fraction (Bond *et al.*, 1967; Flamm *et al.*, 1967*a*). Furthermore, it has been shown that the major-band DNA, comprising the other 90% and banding at a higher buoyant density, renatures at a rate a thousand to a million times slower.

As shown in *Figure 12*, it is necessary to recycle a previously-enriched satellite DNA preparation in order to free it completely of contamination by the major-band DNA. Using such highly-purified material, we have been able to show beyond doubt that the satellite fraction is composed of unique nucleotide sequences. For instance, we know that the total DNAs of some other rodents (*Peromyscus*, *Apodemus*, guinea pig and rat) do not possess this sequence, even though considerable sequence homology exists between mouse major-band DNA and these other DNAs (Flamm *et al.*, 1967*b*; Flamm *et al.*, 1969*a*, *b*).

Despite the fact there now exists an impressive mass of information concerning the metabolism, molecular weight, chemical composition, genomal distribution, sequence repetition and specificity

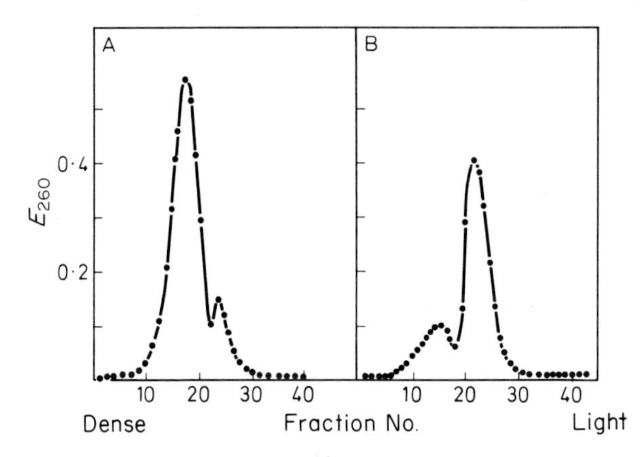

Both density gradients, of initial density 1·710 g/cm^3, centrifuged in an MSE 10 × 10 rotor (see Table 3)

Figure 12. Absorbance profiles of whole mouse DNA (A) and a recycled satellite fraction (B) obtained from the first centrifugation

of mouse satellite DNA, its biological function is still obscure. We can say, however, that it is not the only example of a rapidly-renaturing nuclear DNA within the animal kingdom, since other rodents, as well as non-rodent mammals, have similar DNA fractions though with different nucleotide sequences (Polli *et al.*, 1965, 1966; Flamm *et al.*, 1969*a*, *b*). In fact, our experience with guinea-pig DNA proves that a highly-repetitive fraction can easily possess a totally different base composition with a correspondingly varying buoyant density.

As can be seen from *Figure 13*, the highly-repetitive sequences are again associated with the satellite band but, unlike mouse, guinea-pig satellite DNA is more dense than the majority of the DNA and not nearly so well separated from it. Hence, repeated isopycnic centrifugations of satellite-enriched preparations of guinea-pig DNA can never yield a completely purified fraction, unless its ability to renature rapidly is exploited. This is done by thermally denaturing the fractions at low concentrations of DNA (less than 10 μg/ml) and then incubating them under conditions (*Figure 14*) which lead to selective renaturation of only the repetitive sequences. The renatured material can then be isolated by CsCl density-gradient centrifugation. The method depends on the fact that buoyant densities are increased by denaturation and decreased by renaturation. Hence the contaminant (major-band DNA),

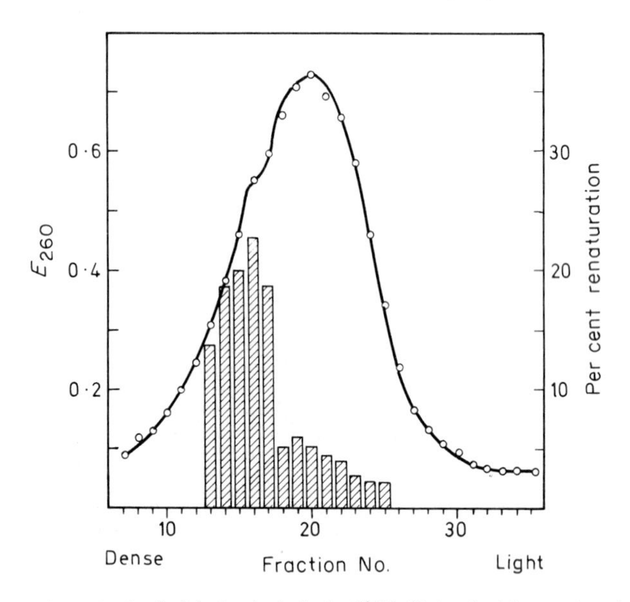

Bars show percentage renaturation of each fraction when incubated at 60°C for 10 min at 1 μg/ml concentration as described by McCallum and Walker (1967); for conditions of centrifugation in MSE 10 × 10 rotor see Table 3

Figure 13. Absorbance profiles of guinea-pig DNA (Flamm, McCallum and Walker, unpublished)

which remains denatured, has a higher buoyant density than the satellite DNA, which has fully renatured. This is shown by the bimodal distribution of fractions I and II (*Figure 8*), as seen in *Figure 14*, where the light peak corresponds to the rapidly-renaturing DNA. Fractions III and IV (*Figure 8*) lack the satellite DNA and consequently show no evidence of renaturation, only a typical unimodal distribution (*Figure 14*, III and IV).

The purification of renatured satellite DNA as outlined above can be greatly simplified by application of hydroxyapatite chromatography (McCallum and Walker, 1967) in which the ability of this material to discriminate between denatured and renatured forms of DNA is used. This procedure also provides a valuable means of determining the rate of renaturation even at very low concentrations of DNA (Flamm *et al.*, 1967*a*). In fact, it has been possible to compare accurately the renaturation rates of mouse and guinea-pig satellite DNAs which, despite their differences in sequence, density and base composition, are kinetically indistinguishable.

Corneo and his co-workers (Corneo *et al.*, 1968) have developed a procedure whereby guinea-pig satellite can be isolated in a highly-purified state without denaturing the DNA. This procedure involves the use of Ag_2SO_4 in Cs_2SO_4 density gradients.

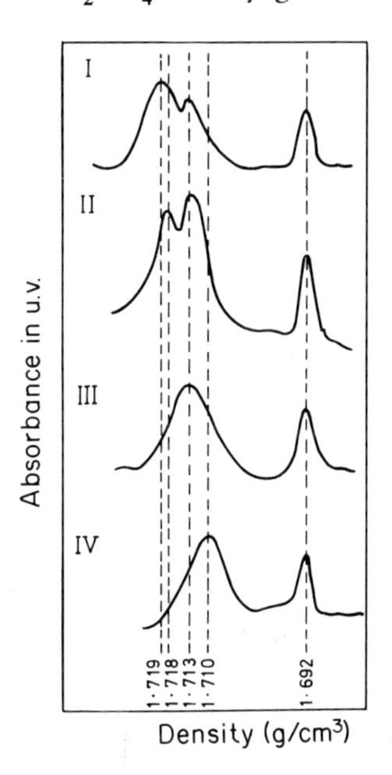

Each fraction heat-denatured, then incubated in 0·12 M sodium phosphate, pH 6·8, at 60°C at DNA concentration of 5–10 µg/ml for 10 min; initial densities 1·705–1·720 g/cm³; band on right, at density 1·692 g/cm³, is a density marker of mouse satellite DNA

Figure 14. Microdensitometer tracings of u.v. photographs of the pooled guinea-pig fractions (and conditions) of Figure 8, *centrifuged to equilibrium in an MSE analytical ultracentrifuge*

It is still not clear what the deciding factors are in determining how much Ag^+ will combine with specific sequences. However, the data available make it clear that more is involved than just base ratios. Possibly stretches of repeated purines or repeated pyrimidines, or some other feature of the nucleotide sequence, is critical but it is still too early to say. In any event, there is no question that the $Ag^+ - Cs_2SO_4$ procedure is an exceedingly powerful technique and well worth further exploration. There are, in fact, indications from D. D. Brown's laboratory that both 4*S* and 5*S* RNA cistrons can be isolated by this approach.

Guinea-pig satellite DNA, like mouse satellite, is built of a short nucleotide sequence of 50 to 100 base pairs, repeated perhaps three or four million times in each cell (Flamm *et al.*, 1969*b*). Again, the highly-repetitive sequences appear to be unique, since they fail to interact with DNA from other sources, although guinea-pig major-band DNA will do so (Flamm *et al.*, 1969*a*, *b*).

We now recognise that certain organisms may possess a rapidly-renaturing fraction without actually exhibiting a satellite band like those found in mouse and guinea pig. Rat DNA is one such example, where apparently the highly-repetitive sequences have a buoyant density at, or very near, the peak of the major-band DNA. A satellite DNA does appear, however, after this DNA is subjected to the kind of denaturation–renaturation treatment applied to the guinea-pig fractions (*Figure 14*). Clearly, therefore, rapidly-renaturing fractions (that is, those with repetitive sequences) may be heavier than, lighter than, or have a density similar to, the heterogeneous sequences which comprise the vast majority of nuclear DNA.

SEPARATION AND ISOLATION OF COMPLEMENTARY STRANDS OF DNA

One of the most recent and exciting applications of isopycnic density-gradient centrifugation concerns the separation and isolation of complementary strands of DNA from alkaline solutions of CsCl. The potential usefulness of alkaline CsCl density gradients for separating single strands of DNA was suggested some time ago by Vinograd *et al.* (1963), who recognised that only the N–H bonds of DNA-thymine and guanine are titrated by alkali at pH 12·5 and, therefore, only thymine and guanine are responsible for the increases in buoyant density observed in alkaline CsCl. This is because, for each thymine or guanine residue titrated by OH^-, an additional Cs^+ ion is added to single-stranded (denatured) DNA. Because of the high effective density of Cs^+, the buoyant density of the polymer is increased in accordance with the extent of titration or, when fully titrated, with the thymine plus guanine (TG) content of the DNA strand. This has been elegantly substantiated by Wells and Blair (1967), who have shown that poly d(AC), which is not titrated by alkali, has the same buoyant density in either neutral or alkaline solutions, whereas poly d(TG) is 0·048 density units heavier after full titration. Therefore, if one strand of a double-stranded DNA possesses a higher TG content than the complementary strand, the strands should be separable by isopycnic centrifugation in alkaline CsCl.

One such separation is illustrated by the fractionation profile in *Figure 15*, which shows that mouse satellite DNA forms two well-resolved bands in a preparative alkaline CsCl density gradient.

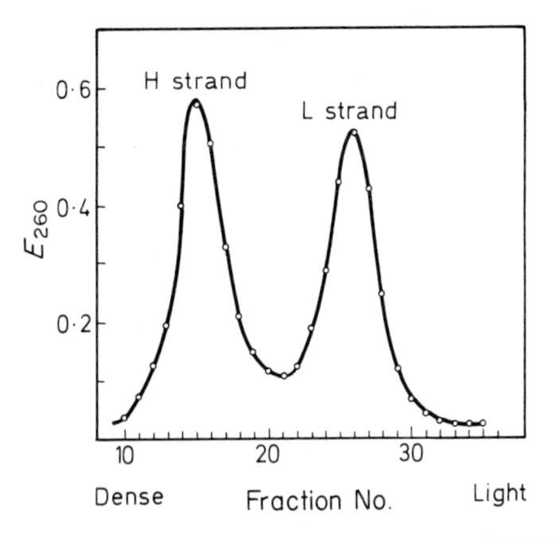

To 60 μg of purified mouse satellite DNA in 3·2 ml of 0·01 M tris-HCl, pH 8·5, were added 0·1 ml of 1% (w/v) sodium dodecyl sulphate, 0·1 ml of 1·0 N NaOH, 4·615 g of CsCl. Initial density of 1·755–1·770 g/cm³ and pH of 12·3–12·6 were obtained; density gradient centrifuged in MSE 10 × 10 rotor for 20–24 h at 42 000 rev/min at 25°C

Figure 15. Absorbance profile of preparative alkaline CsCl density-gradient centrifugation of mouse satellite DNA

Table 7. BASE COMPOSITION OF MOUSE SATELLITE DNA (FLAMM *et al.*, 1967*a*)

Bases	*Duplex*	*H-strand* mole per cent	*L-strand*
Guanine	16·4	12·5	19·9
Adenine	32·2	20·7	44·2
Cytosine	17·8	22·1	14·3
Thymine	33·6	44·7	21·6
G+C	34·2	34·6	34·2
T+G	50·0	57·2	41·5
Purine/pyrimidine ratio	0·95	0·50	1·79

Nucleotide analysis of the isolated heavy (H) and light (L) strands confirms the prediction that the heavier strand would have a higher proportion of thymine plus guanine (57 to 58%) than the complementary light strand (41 to 42%; *Table* 7). A similar kind of experiment was previously reported by Siegel and Hayashi (1967), who succeeded in isolating the complementary strands of the replicative form of ϕX-174 DNA by alkaline density-gradient centrifugation. Again, the heavy strand contained more thymine plus guanine.

This kind of distributional bias between complementary strands of DNA has also been seen in other bacteriophages, for instance, SP-8, α, TP-84 (Marmur and Cordes, 1963; Marmur and Greenspan, 1963; Marmur *et al.*, 1963; Saunders and Campbell, 1965). We would not, however, expect the DNA of higher plants and animals, or even of bacteria, to possess a demonstrable interstrand bias of this type, since these DNAs should be sufficiently heterogeneous (in terms of nucleotide sequences) to average out interstrand differences. On the other hand, certain DNA fractions have been isolated from higher plants and animals which are highly homogeneous, and these, along with DNA from viruses and bacteriophages, are likely to show an interstrand bias and are good candidates for possible strand separation and isolation of complementary strands. As shown by *Figures 15* and *16*, mouse satellite

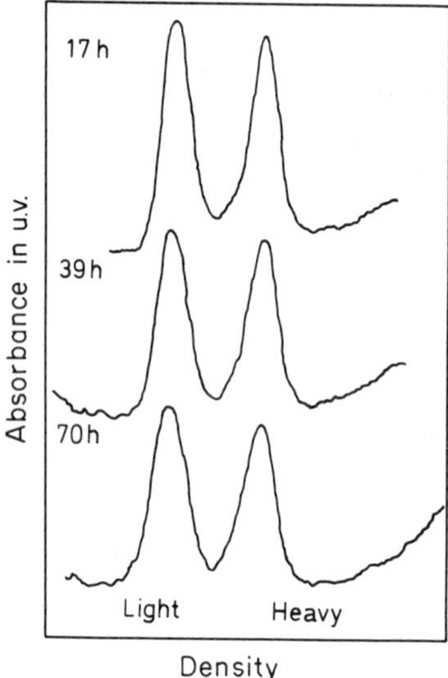

A single gradient, initial density 1·755 g/cm³, containing 3 µg of satellite DNA photographed after 17, 39 and 70 h of centrifugation; conditions as Figure 8

Figure 16. Microdensitometer tracings of u.v. photographs of alkaline CsCl density-gradient centrifugation of mouse satellite DNA

DNA, which is known to be remarkably homogeneous, gives rise to a bimodal distribution in alkaline CsCl density gradients, and, as evidenced by *Figure 17*, guinea-pig satellite DNA gives rise to five bands. Four of these bands (the two lightest and the two heaviest) have been shown (Flamm *et al.*, 1969*b*) to be derived from two separate, but highly repetitive, nucleotide sequences.

These have been referred to as the guinea-pig α and β satellite DNAs. The two extreme bands in *Figure 17* are derived from the α satellite while the two relatively sharp bands immediately adjacent to, and appearing on either side of, the broad band have been referred to as the β satellite (Flamm *et al.*, 1969*b*). It should also be pointed out that Corneo and his co-workers have succeeded in

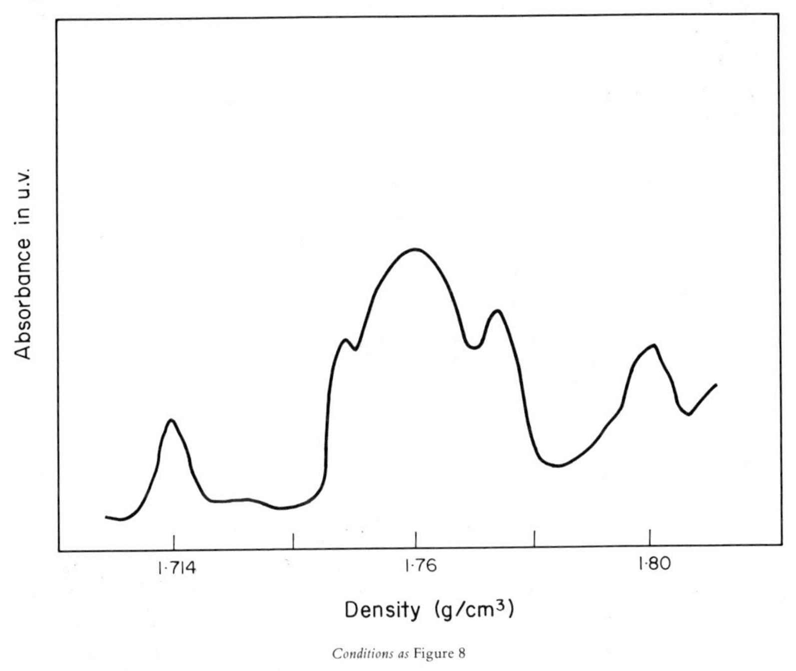

Conditions as Figure 8

Figure 17. Microdensitometer tracing of satellite-enriched guinea-pig DNA centrifuged to equilibrium in an alkaline gradient of CsCl

isolating both types in the duplex form using the $Ag_2SO_4-Cs_2SO_4$ procedure mentioned earlier (Corneo *et al.*, 1970). The rather broad band appearing in the middle of the tracing in *Figure 17* represents main-band contamination and a third satellite.

As well as satellite DNAs, mitochondrial DNA from various species exhibits a similar pattern in alkaline density gradients (J. Vinograd and H. E. Bond, personal communications). Hence, certain select DNAs, either from the cytoplasm or nucleus of higher organisms, will undergo strand separation and form a separate band for each of the two complementary strands.

The stability of DNA in alkaline solutions is, of course, crucial since narrow band widths and good resolution depend upon the maintenance of high molecular weight during centrifugation. Mouse satellite DNA is extremely resistant to alkaline hydrolysis, and only a slight broadening of the heavy and light strands is evident after 70 h of centrifugation (*Figure 16*). This result is in good agreement with those of Vinograd *et al.* (1963), who observed no more than one to two hydrolytic cleavages of T4 DNA in 64 h and of Siegel and Hayashi (1967), who report 0·5 hits every 40 h in circular ϕX-174 DNA during alkaline density-gradient centrifugation. However, this rather high degree of resistance to alkaline hydrolysis in concentrated solutions of CsCl may not apply to all DNAs. Vinograd (personal communication) has experienced some difficulties with mitochondrial DNAs breaking down shortly after displaying a bimodal distribution.

For alkaline CsCl density-gradient centrifugation, centrifuge tubes of polypropylene are best because of their greater resistance to high pH. It is, however, necessary to add a detergent to these gradients in order to prevent absorption of single-stranded DNA onto the walls of the centrifuge tube. We have, in fact, observed that the heavy strand of mouse satellite DNA is absorbed to an appreciably greater extent than the light strand, but that addition of sodium dodecyl sulphate (200 μg/ml of CsCl gradient) successfully combats this effect. In the presence of this detergent we find recoveries to be quantitative with equivalent amounts of DNA in each of the two modes.

During the course of many alkaline CsCl centrifugation studies we discovered that the choice of CsCl (that is, both grade and manufacturer) is critical. In fact, Harshaw's optical grade of CsCl appears obligatory if the procedure outlined in the legend to *Figure 15* is to be followed. Other grades of CsCl apparently contain heavy-metal ion contaminants, which have the effect of increasing the density of alkaline DNA (pH 12·5) to a point where it will no longer band. This difficulty has been overcome by including a chelating agent such as EDTA in the CsCl gradient. Accordingly, 1·7 ml of 0·02 M EDTA solution, pH 9·5, is added to 1·7 ml of 0·01 M tris-HCl, pH 8·6, containing 500 μg of sodium dodecyl sulphate and 200 to 500 μg DNA; 0·1 ml of 1 M NaOH is then added (pH 12·4) and the solution brought to a density of 1·755 g/cm^3 with AnalaR CsCl (B.D.H. Ltd.).

There is now sufficient evidence from various studies on isolated single strands of DNA (Marmur and Cordes, 1963; Saunders and Campbell, 1965; Siegel and Hayashi, 1967; Flamm *et al.*, 1967) to indicate that the exact buoyant density of single-stranded DNA in neutral CsCl is related to two factors, (i) the mole per cent guanine plus cytosine and, (ii) the ratio of TG to AC. Since the individual complementary strands of a DNA duplex must have the same GC content, the difference between their respective buoyant densities depends solely on the ratio of TG to AC. Hence, heavy strands contain a greater proportion of thymine plus guanine, the corresponding light ones contain a higher proportion of adenine plus cytosine. In fact, present data show a linear relationship between the thymine plus guanine content and the increase in buoyant density observed when a single strand dissociates from the native double-stranded DNA.

REFERENCES

ALDRIDGE, W. G. (1960). *Nature, Lond.*, **187**, 323

ANDERSON, N. G., HARRIS, W. W., BARBER, A. A., RANKIN, C. T. and CANDLER, E. L. (1966). *Natn. Cancer Inst. Monog.*, **21**, 253

ANET, R. and STRAYER, D. R. (1969). *Biochem. biophys. Res. Comm.*, (*a*) **34**, 328; (*b*) **37**, 52

BIRNSTIEL, M. L. (1967). *Ciba Symp. Cell Differentiation*, p. 178

BIRNSTIEL, M. L., WALLACE, H., SIRLIN, J. and FISCHBERG, M. (1966). *Natn. Cancer Inst. Monog.*, **23**, 431

BOND, H. E., FLAMM, W. G., BURR, H. E. and BOND, S. (1967). *J. mol. Biol.*, **27**, 289

BOSTOCK, C. and PRESCOTT, D. (1971). *Expl Cell Res.*, **64**, 267

BRITTEN, R. J. (1965). *Yearb. Carnegie Instn*, p. 313

CHUN, E. H. L. and LITTLEFIELD, J. W. (1961). *J. mol. Biol.*, **3**, 668

CHUN, E. H. L. and LITTLEFIELD, J. W. (1963). *J. mol. Biol.*, **7**, 245

CHIPCHASE, M. and BIRNSTIEL, M. (1963). *Proc. natn. Acad. Sci., U.S.A.*, **50**, 1101

COLTER, J. S., BROWN, R. A. and ELLEM, K. A. O. (1962). *Biochim. biophys. Acta*, **55**, 31

CORNEO, G., GINELLI, E. and POLLI, E. (1970). *Biochemistry*, **9**, 1565

CORNEO, G., GINELLI, E., SOAVE, C. and BERNARDI, G. (1968). *Biochemistry*, **7**, 4373

CORNEO, G., MOORE, C., SANADI, D. R., GROSSMAN, L. I. and MARMUR, J. (1966). *Science, N.Y.*, **151**, 687

COUNTS, W. B. and FLAMM, W. G. (1966). *Biochim. biophys. Acta*, **114**, 628

CUTLER, R. G. and EVANS, J. E. (1967). *J. mol. Biol.*, **26**, (*a*) 81; (*b*) 91

DUBNAU, D., SMITH, I. and MARMUR, J. (1965). *Proc. natn. Acad. Sci., U.S.A.*, **54**, 724

EDELMAN, M., SCHIFF, J. A. and EPSTEIN, H. T. (1965). *J. mol. Biol.*, **11**, 769

ERIKSON, R. L. and SZYBALSKI, W. (1964). *Virology*, **22**, 111

FISHER, W. D., CLINE, G. B. and ANDERSON, N. G. (1964). *Analyt. Biochem.*, **9**, 477

FLAMM, W. G., BERNHEIM, N. J. and BRUBAKER, P. E. (1971). *Expl Cell Res.*, **64**, 97

FLAMM, W. G. and BIRNSTIEL, M. L. (1964*a*). In *The Nucleohistones* (ed. by Bonner, J. and Ts'o, P. O. P.), p. 230, San Francisco (Holden-Day Inc.)

FLAMM, W. G. and BIRNSTIEL, M. L. (1964*b*). *Expl Cell Res.*, **33**, 616

FLAMM, W. G., BOND, H. E. and BURR, H. E. (1966*a*). *Biochim. biophys. Acta*, **129**, 310

FLAMM, W. G., BOND, H. E., BURR, H. E. and BOND, S. (1966*b*). *Biochim. biophys. Acta*, **123**, 652

FLAMM, W. G., MCCALLUM, M. and WALKER, P. M. B. (1967*a*). *Proc. natn. Acad. Sci., U.S.A.*, **57**, 1729

FLAMM, W. G., MCCALLUM, M. and WALKER, P. M. B. (1967*b*). *Biochem. J.*, **104**, 38P

FLAMM, W. G., WALKER, P. M. B. and MCCALLUM, M. (1969*a*). *J. mol. Biol.*, **40**, 423

FLAMM, W. G., WALKER, P. M. B. and MCCALLUM, M. (1969*b*), *J. mol. Biol.*, **42**, 441

GIACOMONI, D. and SPIEGELMAN, S. (1962). *Science, N.Y.*, **138**, 1328

GOODMAN, H. M. and RICH, A. (1962). *Proc. natn. Acad. Sci., U.S.A.*, **48**, 2101

IFFT, J. B., VOET, D. H. and VINOGRAD, J. (1961). *J. physical Chem.*, **65**, 1138
JONES, A. S. (1963). *Nature, Lond.*, **199**, 280
KAY, E. R. M., SIMMONS, N. S. and DOUNCE, A. L., *J. am. chem. Soc.*, **74**, 1724
KIRBY, K. S. (1962*a*). *Biochim. biophys. Acta*, **55**, 382
KIRBY, K. S. (1962*b*). *Biochim. biophys. Acta*, **55**, 545
KIRBY, K. S. (1964). *Progr. Nucleic Acid Res. and mol. Biol.*, **3**, 1
KIT, S. (1961). *J. mol. Biol.*, **3**, 711
KIT, S. (1962). *Nature, Lond.*, **193**, 274
LOWRY, O. H., ROSEBURGH, J., FARR, A. L. and RANDALL, R. J. (1951). *J. biol. Chem.*, **193**, 265
MANDELL, J. D. and HERSHEY, A. D. (1960). *Analyt. Biochem.*, **1**, 66
MARKO, A. M. and BUTLER, G. C. (1951). *J. biol. Chem.*, **190**, 165
MARMUR, J. (1961). *J. mol. Biol.*, **3**, 208
MARMUR, J. and CORDES, S. (1963). In *Informational Macromolecules* (ed. by Vogel, H. J., Bryston, V. and Lampen, J.), p. 79, New York (Academic Press)
MARMUR, J. and GREENSPAN, C. M. (1963). *Science, N.Y.*, **142**, 387
MARMUR, J., GREENSPAN, C. M., PALECEK, E., KAHAN, F., LEIRNE, J. and MANDEL, M. (1963). *Cold Spring Harbor. Symp. Quant. Biol.*, **28**, 191
MCCALLUM, M. and WALKER, P. M. B. (1967). *Biochem. J.*, **105**, 163
MESELSON, M. (1957). *Thesis*, California Institute of Technology, Pasadena, California, U.S.A.
MESELSON, M. and STAHL, F. W. (1958). *Proc. natn. Acad. Sci., U.S.A.*, **44**, 671
MESELSON, M., STAHL, F. W. and VINOGRAD, J. (1957). *Proc. natn. Acad. Sci., U.S.A.*, **43**, 581
NANDI, U. S., WANG, J. C. and DAVIDSON, N. (1965). *Biochemistry*, **4**, 1687
O'SULLIVAN, A. and SUEOKA, N. (1967). *J. mol. Biol.*, **27**, 349
POLLI, E., CORNEO, G., GINELLI, E. and BIANCHI, P. (1965). *Biochim. biophys. Acta*, **103**, 672
POLLI, F., GINELLI, E., BIANCHI, P. and CORNEO, G. (1966). *J. mol. Biol.*, **17**, 305
RABINOWITZ, M., SINCLAIR, J., DESALLE, L., HASELKORN, R. and SWIFT, H. H. (1965). *Proc. natn. Acad. Sci., U.S.A.*, **53**, 1126
RADLOFF, R., BAUER, W. and VINOGRAD, J. (1967). *Proc. natn. Acad. Sci., U.S.A.*, **57**, 1514
RAY, D. S. and HANAWALT, P. C. (1965). *J. mol. Biol.*, **11**· 760
RITOSSA, F. M. and SPIEGELMAN, S. (1965). *Proc. natn. Acad. Sci., U.S.A.*, **53**, 737
ROLFE, R. and MESELSON, M. (1959). *Proc. natn. Acad. Sci., U.S.A.*, **45**, 1039
ROWND, R., LANYI, J. and DOTY, P. (1961). *Biochim. biophys. Acta*, **53**, 225
SAUNDERS, G. F. and CAMPBELL, L. L. (1965). *Biochemistry*, **4**, 2836
SCHILDRAUT, C. L., MARMUR, J. and DOTY, P. (1962). *J. mol. Biol.*, **4**, 430
SIEGEL, J. E. D. and HAYASHI, M. (1967). *J. mol. Biol.*, **27**, 443
SIMON, E. H. (1961). *J. mol. Biol.*, **3**, 101
SUEOKA, N. (1959). *Proc. natn. Acad. Sci., U.S.A.*, **45**, 1480
SUEOKA, N., MARMUR, J. and DOTY, P. (1959). *Nature, Lond.*, **183**, 1429
TOBIA, A., SCHILDKRAUT, C. L. and MAIO, J. J. (1970). *J. mol. Biol.*, **54**, 499
VAN HOLDE, K. E. and BALDWIN, R. L. (1958). *J. physical Chem.*, **62**, 734
VINOGRAD, J. and HEARST, J. E. (1962). *Fortschritte der Chemie Organischer Naturstoffe*, **20**, 373
VINOGRAD, J., MORRIS, J., DAVIDSON, N. and DOVE, W. (1963). *Proc. natn. Acad. Sci., U.S.A.*, **49**, 12
WALKER, P. M. B. and MCLAREN, A. (1965). *Nature, Lond.*, **208**, 1175
WALLACE, H. and BIRNSTIEL, M. L. (1966). *Biochim. biophys. Acta*, **114**, 296
WARING, M. and BRITTEN, R. J. (1966). *Science, N.Y.*, **154**, 799
WELLS, R. and BIRNSTIEL, M. L. (1967). *Biochem. J.*, **105**, 53P
WELLS, R. D. and BLAIR, J. E. (1967). *J. mol. Biol.*, **27**, 273
YANKOFSKY, S. A. and SPIEGELMAN, S. (1962). *Proc. natn. Acad. Sci., U.S.A.*, **48**, 146
YANKOFSKY, S. A. and SPIEGELMAN, S. (1963). *Proc. natn. Acad. Sci., U.S.A.*, **49**, 538

INDEX